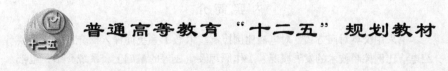

普通高等教育"十二五"规划教材

电液比例控制技术

Electro-hydraulic Proportional Control Technology

宋锦春　编著

北　京

冶金工业出版社

2022

内 容 简 介

本书采用汉英对照方式，对电液比例控制技术做了系统而深入的介绍。在简要介绍电液比例控制技术的发展概况、工作原理及组成等的基础上，系统介绍了电液比例控制系统的重要元件、控制方法和基本回路等理论知识，如比例电磁铁、电液比例控制阀、电液比例容积控制及电液比例控制基本回路等，其中许多内容涉及电液比例控制领域的新产品、新技术；最后介绍了电液比例控制系统的设计与分析方法，以及编者多年来在电液比例控制领域积累的科研工程应用实例。

本书可作为高等学校机械工程类专业双语教材，也可供相关领域的科技工作者和工程技术人员参考。

图书在版编目（CIP）数据

电液比例控制技术：汉英对照/宋锦春编著. —北京：冶金工业出版社，2014.8（2022.7重印）

普通高等教育"十二五"规划教材

ISBN 978-7-5024-6644-2

Ⅰ.①电… Ⅱ.①宋… Ⅲ.①电液伺服系统—比例控制—双语教学—高等学校—教材—汉、英 Ⅳ.①TH137.5

中国版本图书馆 CIP 数据核字（2014）第 170589 号

电液比例控制技术

出版发行	冶金工业出版社		电　　话	(010)64027926
地　　址	北京市东城区嵩祝院北巷39号		邮　　编	100009
网　　址	www.mip1953.com		电子信箱	service@mip1953.com

责任编辑　宋　良　任咏玉　美术编辑　吕欣童　版式设计　孙跃红
责任校对　李　娜　责任印制　禹　蕊

北京虎彩文化传播有限公司印刷

2014年8月第1版，2022年7月第3次印刷

710mm×1000mm 1/16；23印张；445千字；350页

定价 48.00元

投稿电话　(010)64027932　投稿信箱　tougao@cnmip.com.cn
营销中心电话　(010)64044283
冶金工业出版社天猫旗舰店　yjgycbs.tmall.com

（本书如有印装质量问题，本社营销中心负责退换）

前　言

电液比例控制技术是继液压伺服技术之后发展起来的一门新兴学科，是液压技术的重要组成部分。随着工业自动化技术的不断发展，电液比例控制技术越来越广泛地应用到各种机械自动化装备中。

编者总结三十年来的专业教学和科研工作中大量的工程实际应用经验，结合自身留学经历，完成了本双语教材的编写工作。书中汉英相关内容相互对照，利于学生积累这门新兴科学技术的专业英文词汇；结合编者多年教学经验，对相关理论知识结构进行合理编排，利于读者对所学内容的理解与掌握；将计算机仿真技术应用于比例控制系统的分析，并给予了较为详细的论述，使读者学会电液比例系统的设计与分析方法；注重结合工程应用，书中给出的各项实例基本上都是编者科研工作的成功范例，有利于读者解决工程实际问题。

本书采取汉英对照的方式，以电液比例控制技术为主线，注重阐述电液比例系统的典型元件、控制方法与重要回路，并且多处选用编者多年工程项目的应用实例，实现了汉语讲解与英语表述、理论知识与实际应用的有效结合。书中液压职能符号统一采用国家最新标准，所涉及的液压元件尽可能选用最新产品，如最新出现的高性能电液比例阀等，注重结合技术发展，关注学术前沿。读者可以通过本书对电液比例控制系统的原理与组成、设计与分析及该技术目前发展水平与应用情况有较全面的了解。

本书由黑龙江科技大学侯清泉审稿，他提出了非常宝贵的意见和建议；博士研究生倪克、张凯、蔡衍、袁聪、李松、任广安等做了图表公式整理、中英文校对、格式排版等工作，在此表示衷心的谢意！

限于编者水平，书中难免存在不足之处，诚望读者斧正。

作　者

2014年4月

目 录

第1章 概论

Chapter 1 Introduction ········ 1

1.1 电液比例技术的发展概况（Overview of the development of electro-hydraulic proportional technology） ········ 1

1.2 电液比例控制的概念（Concept of electro-hydraulic proportional control） ········ 5

1.3 电液比例控制系统的工作原理及组成（Operating principle of electro-hydraulic proportional control system and its constitution） ········ 7

   1.3.1 工作原理（Operating principle） ········ 7

   1.3.2 电液比例系统的组成（Composition of the electro-hydraulic proportional system） ········ 8

1.4 电液比例控制系统的分类（Classification of electro-hydraulic proportional control system） ········ 11

第2章 比例电磁铁

Chapter 2 The Proportional Electromagnet ········ 14

2.1 电-机械转换元件的作用及形式（Function and form of electro-mechanical conversion component） ········ 14

   2.1.1 电-机械转换元件的形式（Form of the electro-mechanical conversion component） ········ 14

   2.1.2 电-机械转换元件的要求（Demands of electric machinery conversion component） ········ 15

   2.1.3 比例电磁铁概述（Summary of the proportional electromagnet） ········ 15

2.2 电磁铁的结构与工作原理（Structure and working principle of electromagnet） ········ 16

   2.2.1 普通螺线管型电磁铁（Ordinary solenoid electromagnet） ········ 16

   2.2.2 单向比例电磁铁（One-way proportional electromagnet） ········ 17

2.3 比例电磁铁的特性（Characteristics of proportional electromagnet）······ 18
   2.3.1 电磁铁的吸力特性（Attracting characteristic of proportional electromagnet）·············· 18
   2.3.2 电磁铁的负载特性（Load characteristic of electromagnet）········· 22
2.4 比例电磁铁的分类与应用（Classification and application of the proportional electromagnet）·············· 24
   2.4.1 力控制型比例电磁铁（Force control type）············ 25
   2.4.2 行程控制型比例电磁铁（Stroke control type）············ 27
   2.4.3 位置调节型比例电磁铁（Position control type）············ 28
2.5 比例电磁铁的初步设计（Preliminary design of proportional electromagnet）·············· 31

第3章 电液比例控制阀

Chapter 3 Electro-hydraulic Proportional Control Valve ········ 34

3.1 概述（Summary）·············· 34
3.2 电液比例压力控制阀（Electro-hydraulic proportional pressure control valve）·············· 37
   3.2.1 比例溢流阀（Proportional relief valve）············ 38
   3.2.2 比例减压阀（Proportional pressure reducing valve）············ 47
3.3 电液比例方向控制阀（Electro-hydraulic proportional directional control valve）·············· 58
   3.3.1 直动型比例方向阀（Direct operated proportional directional valve）·············· 58
   3.3.2 先导型比例方向阀（Pilot operated proportional directional valve）·············· 59
   3.3.3 中位机能与阀芯结构（Structure and the center configuration of the valve spool）·············· 64
3.4 电液比例流量控制阀（Electro-hydraulic proportional flow control valve）·············· 66
   3.4.1 利用方向阀作流量控制（Directional valve used for flow control）·············· 67
   3.4.2 节流阀（Throttle valve）············ 68
   3.4.3 比例调速阀（Proportional flow control valve）············ 71
   3.4.4 定差溢流型比例调速阀（Pressure compensated flow bypass control valve）·············· 72

3.5 压力补偿器（Pressure compensator） ………………………… 73
3.5.1 进口压力补偿器（Inlet pressure compensator） ……………… 74
3.5.2 出口压力补偿器（Outlet pressure compensator） …………… 76
3.6 电液比例复合阀（Electro-hydraulic proportional composite valve） …… 79

第 4 章 电液比例容积控制
Chapter 4 Electro-hydraulic Proportional Volume Control ………… 83

4.1 容积泵的基本控制方法（The basic control method of capacity pump） ……………………………………………………………… 83
4.1.1 流量适应控制（Flow adaptive control） …………………… 84
4.1.2 压力适应控制（Pressure adaptive control） ………………… 97
4.1.3 功率适应控制（Power adaptive control） …………………… 100
4.1.4 恒功率控制（Constant power control） ……………………… 104
4.2 比例排量变量泵和变量马达（Proportional displacement variable pump and variable motor） ……………………………………… 108
4.2.1 位移直接反馈式比例排量调节（Shift direct feedback proportional displacement adjustment） ……………………… 111
4.2.2 位移-力反馈式比例排量调节变量泵（Shift-force feedback proportional displacement regulation variable capacity pump） …… 114
4.2.3 位移-电反馈型比例排量调节（Shift-electronic feedback proportional displacement regulation） ……………………… 116
4.2.4 一种新型比例排量控制式 PV 泵（A new proportional displacement control PV pump） ………………………………………………… 117
4.3 电液比例压力调节型变量泵（Electro-hydraulic proportional pressure regulation variable pump） ……………………………………… 118
4.3.1 工作原理及结构（Working principle and structure） ………… 118
4.3.2 先导式比例压力调节变量叶片泵的特性分析（Characteristic analysis of pilot proportional pressure regulation variable vane pump） ……… 120
4.4 电液比例流量调节型变量泵（Electro-hydraulic proportional flow regulation variable pump） ………………………………………… 125
4.4.1 稳流量调节控制原理（Steady flow regulation control principle） … 127
4.4.2 泵的特性分析（Analysis of pump characteristics） …………… 128
4.4.3 带流量适应的比例流量调节型变量泵（Proportional flow regulation variable pump with flow adaption） ……………… 130

4.5 电液比例压力和流量调节型变量泵（Electro-hydraulic proportional pressure and flow regulation variable pump） …………………… 132

    4.5.1 压力补偿型比例压力和流量调节（Pressure compensation proportional pressure and flow regulation） …………………… 132

    4.5.2 电反馈型比例压力和流量调节（Electrical feedback proportional pressure and flow regulation） ………………………………… 135

4.6 二次静压调节技术（Secondary static pressure regulation technology） … 138

    4.6.1 二次调节静液传动的概述（Outline of secondary regulation hydrostatic transmission） ……………………………………… 139

    4.6.2 二次调节技术的发展（The development of secondary regulation technology） ……………………………………………… 143

    4.6.3 二次调节静液传动的工作原理（Operating principle of secondary regulation hydrostatic transmission） …………………… 145

    4.6.4 二次调节系统的转速控制（Rotating speed control of secondary regulation system） ………………………………………… 147

    4.6.5 二次调节系统的转矩控制（Torque control of secondary regulation system） ………………………………………………… 149

    4.6.6 二次调节系统的功率控制（Power control of secondary regulation system） ………………………………………………… 151

    4.6.7 二次调节静液传动系统的特点（Characteristics of secondary regulation hydrostatic transmission system） ………………… 155

    4.6.8 二次调节技术的主要应用（Main applications of secondary regulation technology） ……………………………………… 156

第5章 电液比例控制基本回路

Chapter 5 Basic Electro-Hydraulic Proportional Control Circuits …… 165

5.1 电液比例压力控制回路（Electro-hydraulic proportional pressure control circuit） ………………………………………………… 165

    5.1.1 比例溢流调压回路（Proportional relief pressure-regulating circuit） ………………………………………………………… 166

    5.1.2 比例减压回路（Proportional pressure reducing circuit） ……… 169

    5.1.3 比例容积式调压回路（Proportional volumetric pressure regulating circuit） …………………………………………… 171

    5.1.4 比例调压回路的应用（Applications of proportional pressure regulating circuit） …………………………………………… 173

5.2 电液比例速度控制回路（Electro-hydraulic proportional speed control circuit） ………………………………………………… 174
 5.2.1 比例节流调速回路（Proportional throttle regulating circuit） …… 174
 5.2.2 比例容积调速回路（Proportional volumetric speed control circuit） …………………………………………………………… 176
 5.2.3 比例容积节流式流量调节回路（Proportional volumetric throttling flow regulating circuit） ……………………………………… 177
 5.2.4 电液比例速度控制回路的应用（Applications of electro-hydraulic proportional speed control circuit） ……………………………… 179
5.3 电液比例方向及速度控制回路（Electro-hydraulic proportional direction and speed control circuit） ……………………………………… 183
 5.3.1 对称执行器比例方向控制回路（Proportional directional control circuit of the symmetric actuator） …………………………… 185
 5.3.2 非对称执行器的比例方向控制回路（Asymmetric actuator's proportional directional control circuit） …………………………… 188
 5.3.3 比例差动控制回路（Proportional differential control circuit） …… 190
 5.3.4 其他比例方向阀控制的实用回路（Other practical control circuits using proportional directional valve） ……………………… 193
 5.3.5 比例方向速度控制回路的应用（Applications of proportional direction speed control circuit） …………………………………… 198
5.4 比例复合回路（Proportional compound circuit） ……………………… 205
 5.4.1 比例压力-流量复合阀调压调速回路（Pressure and speed control circuit with proportional pressure-flow compound valve） … 205
 5.4.2 比例压力/流量调节型变量泵回路（Proportional pressure/flow regulating variable pump circuit） ………………………………… 206
5.5 应用于比例节流的压力补偿回路（Pressure compensating circuit using proportional throttle） ……………………………………… 207
 5.5.1 进口节流压力补偿（Inlet throttle pressure compensation） ……… 208
 5.5.2 出口节流压力补偿（Outlet throttle pressure compensation） …… 211

第6章 比例放大器

Chapter 6 Proportional Amplifier ……………………………………… 212

6.1 概述（Summary） ……………………………………………………… 212
 6.1.1 比例放大器的基本技术要求（Fundamental requirements of the proportional amplifier） ……………………………………… 212

6.1.2 比例放大器的分类（Classification of the proportional amplifier） … 214
6.1.3 比例放大器的主要电路（Main circuits of the proportional amplifier） … 216
6.1.4 运算放大器简介（Introductions of operational amplifier） … 219
6.2 电源电路（Power circuit） … 220
 6.2.1 整流电路（Rectification circuit） … 222
 6.2.2 滤波电路（Filter circuit） … 222
 6.2.3 稳压电路（Voltage stabilizing circuit） … 224
 6.2.4 串联型稳压电路（Serial regulating circuit） … 224
 6.2.5 集成稳压器（Integrated voltage regulator） … 224
6.3 控制信号发生电路（Control signal generation circuit） … 226
 6.3.1 周期信号发生器（Periodic signal generator） … 227
 6.3.2 非周期信号发生器（Nonperiodic signal generator） … 229
6.4 比例放大器电路（Proportional amplifier circuit） … 231
 6.4.1 比例运算放大电路（Proportional operation amplification circuit） … 232
 6.4.2 调节器电路（Regulator circuit） … 236
 6.4.3 功率放大级（Power amplifier stage） … 243
6.5 比例放大器的使用及调整（Usage and adjustment of the proportional amplifier） … 246

第7章 电液比例控制系统的分析与设计

Chapter 7 Analysis and Design of Electro-hydraulic Proportional Control System … 253

7.1 电液比例控制系统的设计内容与步骤（Contents and steps of designing an electro-hydraulic proportional control system） … 253
7.2 电液比例控制系统的方案拟订（Programming of electro-hydraulic proportional control system） … 254
 7.2.1 确定比例控制系统的控制方式（Determine control mode of the proportional control system） … 255
 7.2.2 确定比例控制系统的控制系统类型（Determine the type of proportional control system） … 255
 7.2.3 确定比例控制系统的控制信号类型（Determine the type of control signals in proportional control system） … 256
 7.2.4 确定比例控制系统的控制元素类型（Determine the type of control elements in proportional control system） … 257

7.2.5 确定比例控制系统的检测元件类型（Determine the type of detection components in proportional control system） ·············· 257
7.3 电液比例控制系统的静态分析（Static analysis of electro-hydraulic proportional control system） ·············· 258
   7.3.1 比例控制系统供油压力的选择（Selection of oil pressure of the proportional control system） ·············· 259
   7.3.2 确定液压执行器规格尺寸及阀规格（Determine the size of hydraulic actuators and the specification of valve） ·············· 260
   7.3.3 液压缸-负载系统固有频率的估算（Estimating the nature frequency of the hydraulic cylinder – load system） ·············· 264
7.4 电液比例控制系统的动态分析（Dynamic analysis of electro-hydraulic proportional control system） ·············· 266
   7.4.1 闭环控制系统开环增益的组成（Composition of open loop gain of closed loop control system） ·············· 267
   7.4.2 闭环控制系统开环增益的影响（Influence of open-loop gain of closed-loop control system） ·············· 268
   7.4.3 闭环控制系统开环增益的分配（Distribution of the open loop gain of closed-loop control system） ·············· 269
   7.4.4 闭环控制系统的动态数学模型（Dynamic mathematical model of closed-loop control system） ·············· 272
7.5 电液比例控制系统的静态性能分析（Static performance analysis of electro-hydraulic proportional control system） ·············· 277
   7.5.1 输入误差（Input error） ·············· 277
   7.5.2 指令输入引起的稳态误差（Steady-state error caused by command input） ·············· 279
   7.5.3 扰动输入引起的稳态误差（Steady state error caused by interference input） ·············· 280
   7.5.4 元件误差（Components error） ·············· 280
7.6 电液比例控制系统的动态性能分析（The dynamic performance analysis of electro-hydraulic proportional control system） ·············· 282
   7.6.1 MATLAB仿真工具软件介绍（Introduction of MATLAB simulation tool） ·············· 283
   7.6.2 闭环位置控制系统仿真实例（Example for closed-loop position control system simulation） ·············· 284
7.7 电液比例控制系统的液压能源选择（Choice of hydraulic energy of

electro-hydraulic proportional control system) ·················· 295
 7.7.1 比例系统常用液压油源（Common hydraulic oil source of proportional system）·················· 296
 7.7.2 液压油源与负载的匹配（Matching of hydraulic energy and load）·················· 296
7.8 液压伺服与比例控制系统设计举例（Instance of hydraulic servo and proportional control system design）·················· 297

第8章 电液比例控制技术的工程应用

Chapter 8 Engineering Application of Electro-hydraulic Proportional Control Technology ·················· 301

8.1 电液比例控制技术在钢管水压试验机上的应用（Application of electro-hydraulic proportional control technology in pipe hydrostatic tester）·················· 301

8.2 飞机拦阻器电液比例控制系统（Electro-hydraulic proportional control system of aircraft arresting system）·················· 305
 8.2.1 改进前的液压系统（Original hydraulic system）·················· 306
 8.2.2 纯机液拦阻器的不足（Disadvantages of pure mechanical-hydraulic aircraft arresting system）·················· 306
 8.2.3 改进后的液压系统（Improved hydraulic system）·················· 307

8.3 电液比例控制技术在 CVT 中的应用（Application of the electro-hydraulic proportional control technology in CVT）·················· 309

8.4 管拧机浮动抱钳夹紧装置电液比例系统（Electronic-hydraulic proportional control system of floating clamping unit of pipe-wrenching machine）·················· 312
 8.4.1 工作原理（Operating principle）·················· 312
 8.4.2 液压站（Hydraulic station）·················· 313

8.5 带钢对中装置电液比例控制系统（Electro-hydraulic proportional control system of strip steel alignment device）·················· 315

8.6 电渣炉电极进给电液比例控制系统（Electro-hydraulic proportional control system of electrode feeding system for electroslag remelting furnace）·················· 319
 8.6.1 电渣重熔原理（Electroslag remelting principle）·················· 319
 8.6.2 电渣炉主要功能机构（Main mechanisms of electroslag remelting furnace）·················· 320

8.6.3　电极进给装置电液比例控制系统（Electro-hydraulic proportional control system for electrode feeding device） ……………………… 322

8.7　矫直机比例系统设计（Design of Proportional control system of straightening machine） ……………………………………………… 324

8.8　比例技术在压铸机液压系统中的应用（Application of proportional technology in die casting machine hydraulic system） ……………… 326

8.9　风力发电机变桨距比例控制系统（Wind driven generator variable propeller pitch proportional control system） …………………… 330

8.10　异型断面盾构设备电液比例控制系统（Electro-hydraulic proportional control system for special-section tunnel boring device） ………… 333

　　8.10.1　异型断面盾构技术背景（Background of special-section tunnel boring technology） ……………………………………………… 333

　　8.10.2　异型断面盾构设备主要功能机构（Main mechanisms of special-section tunnel boring device） ………………………………… 335

　　8.10.3　几何关系推导（Derivation of geometrical relationship） ………… 336

　　8.10.4　异型断面盾构设备电液比例系统（Electro-hydraulic proportional system of special-section tunneling device） ……………………… 337

附录　英语专业词汇 ……………………………………………………………… 340

参考文献 …………………………………………………………………………… 348

第1章 概 论
Chapter 1　Introduction

1.1　电液比例技术的发展概况（Overview of the development of electro-hydraulic proportional technology）

传统的液压控制方式是开关型控制。这是迄今为止用得最多的一种液压控制方式。它通过电磁驱动或手动驱动来实现对液压流体的通、断和方向控制，从而实现被控对象的机械化和自动化。但是这种方式无法实现对液流流量、压力连续、按比例地控制，同时控制的响应比较低、精度差，换向时冲击比较大，因此在许多场合下不宜采用。第二次世界大战期间，对于以飞机、火炮等军事装备为对象的控制系统，要求快速响应、高精度等高性能指标，促进了电液伺服控制技术的迅速发展。它可根据输入信号（如电流）的大小连续、按比例地改变液流的流量、压力和方向。克服开关型控制的缺点，实现高性能的控制要求。

The traditional way of hydraulic control is the on-off control, which has been the most widely used control method so far. In this on-off control method, there are two ways to control the hydraulic fluid: the electromagnetic drive way and the manual drive way. The hydraulic fluid can be controlled to get through or off, thus the controlled object can achieve mechanization and automation. However, in this on-off control method, it is impossible that the flow and the pressure of the fluid can be controlled continuously and proportionally. Besides, in this on-off control method, the response speed is slow, and the control accuracy is low and the impact brought by hydraulic fluid direction changing is great. For these reasons, the on-off control method cannot be used in many occasions. During World War II, the control systems of aircrafts, artilleries and other military equipments were required to meet high-performance targets such as fast response high precision, and it caused rapid development of electro-hydraulic servo control. In electro-hydraulic servo control system, the flow, pressure and direction of the fluid can be controlled continuously and proportionally, according to the strength of the input signal (such as the current). This electro-hydraulic servo control system has overcame the shortcomings of the on-off control method, and it has achieved the requirements of high performance control.

20世纪60年代，电液伺服控制日趋成熟，开始向民用工业推广。同时，液压伺服系统暴露出了它致命的弱点：元件的制造精度要求过高，对油液污染十分敏感，因此系统制造和维护的成本较高，并且其能量损失（阀压降）较大。因为一般工业控制系统对精度要求不太高，也不需要响应太快速，维护简单，成本低廉，系统对油液污染度要求和普通液压系统差不多，于是人们就考虑如何发展一种廉价的伺服控制，这便促进了电液比例控制技术的研究和发展。

In the 1960s, the technology of electro-hydraulic servo control became mature day by day, and it was rapidly promoted toward the civilian industry. But at the same time, hydraulic servo system exposed its biggest weakness: high requirement of precision for components and considerable sensitiveness to oil pollution, which caused high cost of manufacture and maintenance, and great loss of energy (valve pressure drop). However, in most industrial control systems, the requirement of precision is not required so strictly, nor is the speed of respond required so quickly. These control systems are easy to maintain and the cost is low, what's more, they are not sensitive to oil pollution, which are similar to ordinary hydraulic systems. So people turned to develop a low-cost servo control system, promoting the research and development of electro-hydraulic proportional control technology.

比例控制技术的发展大致可以划分为三个阶段：

The development of the proportional control technology can be divided into three stages:

第一阶段：1967年，瑞士Beoringer公司率先生产出KL型比例复合阀，标志着液压比例技术的诞生。到1970年代初，日本油研公司研制出压力和流量两种比例阀并获得了专利。这段时间，主要是以比例型电-机械变换器，例如比例电磁铁、伺服电机、动圈式力矩马达等，取代普通液压阀中的手动调节装置和普通电磁铁，实现电液比例控制，而阀内的结构和设计准则几乎没有什么变化。从性能上说，其频宽1~5Hz，滞环4%~7%，多数只用于开环控制。

The first stage: In 1967, the Swiss Beoringer company firstly produced the KL proportional composite valve, which was marked as the birth of hydraulic proportional technology. In the early 1970s, the Japanese company, YUKEN, produced two kinds of proportional valve (pressure and flow), and applied for patents. In this period of time, manual adjustment device and ordinary electromagnet were mainly replaced by proportional electro-mechanical converter such as proportional solenoid, servo motor, dynamic torque motor etc. to realize electro-hydraulic proportional control, with the

1.1 电液比例技术的发展概况
(Overview of the development of electro-hydraulic proportional technology)

structures and design standards almost keeping the same. In terms of performance, the bandwidth was about 1~5Hz and the hysteresis was about 4%~7%, most of which could only be used for open-loop control.

第二阶段：从1975年到1980年，比例技术进入其发展的第二阶段，比例器件普遍采用了各种内反馈回路，同时研制了耐高压的比例电磁铁，与之配套的比例放大器也日趋成熟。从性能上说，比例阀的频宽已达5~15Hz，滞环缩小到3%左右，不仅用于开环控制，也广泛用于各种闭环控制系统中。

The second stage: From 1975 to 1980, the development of proportional control technology entered its second stage. Various internal feedback loops had been commonly used in proportional devices. Meanwhile, high pressure solenoid was developed and the development of amplifiers became mature day by day. In terms of performance, the bandwidth of proportional valves had reached to 5~15Hz and hysteresis had been narrowed to 3%. The proportional technology had begun to be not only used for open-loop control systems, but also widely used in a variety of closed-loop control systems.

第三阶段：20世纪80年代以来，比例技术进入了飞速发展阶段，并取得了长足的进步，具体体现在：

The third stage: Since the 1980s, the proportional technology had entered the stage of rapid development, and made great progress. For example:

（1）设计原理进一步完善，通过液压、机械以及电气的各种反馈手段，使比例阀的性能（如滞环、频宽等）同工业伺服阀接近，只是因制造成本所限，尚存在一定的中位死区。

Design principles had been improved, and through various hydraulic, mechanical and electrical feedback methods, the performances of proportional valve such as hysteresis, bandwidth etc. were closer and closer to industrial servo valves, though there still existed some median dead zone because of the limit of manufacturing costs.

（2）比例技术同插装技术结合，开发出二通、三通比例插装阀。

Proportional technology was combined with the cartridge technology, developing two-way and three-way proportional cartridge valves.

（3）研制出各种将比例阀、传感器、电子放大器和数字显示装置集成在一块的机电一体化器件。

The electro-mechanical devices had developed, in which proportional valve, sensors, electronic amplifier and digital display were combined.

（4）将比例阀同液压泵、液压马达等组合在一起，构成节能的比例容积器件。

The proportional valves were combined with the hydraulic pumps and the hydraulic motors, and that constitute energy-saving proportional volume devices.

尤其是近年来，比例控制技术被迅速、广泛地应用于各种工业控制，各种比例器件在许多国家形成了系列化、标准化的产品，例如德国已有30%的普通阀市场被比例阀取代。除了模拟式的电液比例元件外，人们也注重于开发出各种数字式的比例元件。数字式液压元件也是今后比例技术发展的一个重要分支。

Especially, in recent years, the proportional control technology has been rapidly and widely used in various industrial control systems. And the proportional devices have been developed in serial and standard in many countries. For example, over 30% of the market of the normal valves have been replaced by proportional valves in Germany. And people have paid more attention to the development of various digital proportional components besides the analogous electro-hydraulic proportional components. Digital hydraulic components will become one of most important branches of proportional control technology in the future.

电液比例技术还出现了一些整体闭环控制即全程电反馈的电液比例元件。其中有各种比例阀、比例容积控制、恒功率控制、恒流量控制、恒压力控制动力源等。

Electro-hydraulic proportional components with overall closed-loop control which means the electrical feedback in the whole course such as various proportional valves, proportional volume control, constant power control, constant flow control, constant pressure control, etc. also appear in the field of electro-hydraulic proportional technology.

此外，以德国 Bosch 公司为代表推出的高性能闭环比例阀，采用了高响应直流比例电磁铁和相应的放大器，含内置位置检测电子装置和反馈闭环，采用零开口四边滑阀，其输出稳态特性中无中位死区，滞环仅 0.3%，频宽达 200Hz，其性能已与伺服阀无异。

In addition, as a representative, the German Bosch company has introduced high-performance closed-loop proportional valves, which used high response DC proportional solenoids, corresponding amplifiers, built-in position detection electronic devices and feedback loops, with zero-opened four-sided slide valve. There is no dead zone in its steady-state characteristics and the hysteresis is just 0.3% and the bandwidth is 200Hz, which has the same performance as servo valves.

现在，一些比例阀把传感器、测量放大器、控制放大器和阀复合在一起，使得结构更紧凑，性能进一步提高。

Nowadays, some proportional valves are electro-mechanical components which combine the sensors, measuring amplifiers and control amplifiers with valves. This makes the structure more compact and improves the performance.

1.2 电液比例控制的概念（Concept of electro-hydraulic proportional control）

由于比例阀具有上述众多优点，因此它获得了远比伺服阀更为广泛的工业应用。可以预料到的是，比例元件将作为普通的液压元件而大量应用，并与传统的液压阀分享工业市场。

As the proportional valves have many advantages that have been mentioned above, they achieve more extensive industrial application than servo valves. It is expected that proportional components will be widely applied as common components, and it will share the industrial markets with the traditional hydraulic valves.

从广义上讲，凡是输出量，如压力、流量、位移、速度、加速度等，能随输入信号连续地按比例地变化的控制系统，都称为比例控制系统。从这个意义上说，伺服控制也是一种比例控制。

Generally speaking, proportional control system is the one that output signal such as pressure, flow, displacement, velocity and acceleration, etc. can change continuously and proportionally with the input signal. In this sense, servo control is also a kind of proportional control.

但是通常所说的比例控制系统，是特指介于开关控制和伺服控制之间的一种新型控制系统。与开关控制系统比较，它能实现连续、比例控制，并且控制精度高、反应速度快；与伺服控制系统比较，由于比例阀是在普通工业用阀的基础上改造而成的，因此加工精度不高，制造成本低廉，抗污染性能好，几乎同开关型控制差不多。虽然在控制精度、反应速度等控制性能比伺服阀和伺服系统差，但能满足大多数工业控制的要求，并且阀内压降小，因此能节省能耗，降低发热量。

However, normally, proportional control system is considered as a new kind of control system between the on-off control and servo control. Compared with the on-off control system, it can achieve the continuous and proportional control, and it has high control accuracy and quick response. Compared with the servo control system, the proportional control system is almost similar with the on-off control. Since proportional valves are transformed from common industrial valves, so they have low cost and good anti-pollution performance, but their machining accuracy are not so good. Though proportional valves and systems have lower control performance such as control precision, response speed, etc. than servo valves and servo systems, they still can meet the requirements of most industrial control systems. Besides, the valve pressure drop is low, so it saves energy and reduces heat.

比例控制主要用于开环系统，伺服控制主要用于闭环系统。伺服控制装置总是带有内反馈，任何检测到的误差都会引起系统状态的改变，而这种改变正是强迫这个误差为零。误差为零时，伺服系统处于平衡状态，直到新的误差被检测出来。

The proportional control system is mainly used in open-loop control system, and the servo control system is mainly used in closed-loop system. Servo control devices always have internal feedback. Any error detected can cause the change of system state, and this change is to force the error to zero. When the error is zero, servo system is in balance until a new error is detected.

在伺服控制系统中，平衡状态控制信号（误差）理论上为零，而比例控制系统却永远不会为零。

The control signal (error) under balanced state is theoretically zero in the servo control system, but it can never be zero in the proportional control system.

比例系统中的主控元件可以有无限种状态，分别对应受控对象的无限种运动；而开关控制中，控制元件只有两种状态。在工程实际应用中，由于多数被控对象仅需要有限的几种状态，因而开关控制也有可取之处。开关元件通常简单可靠，不存在系统的不稳定性。

In the proportional system, the main control elements may have infinite states, in corresponding to infinite states of controlled elements. However, in on-off control system, the control elements only have two states. In actual engineering application, as most controlled elements only need several states, on-off control does have its merits, with simple and reliable on-off elements, while without instability in the system.

在模拟信号比例控制中，如果需要用计算机来控制，则必须具有 A/D、D/A 接口元件与计算机连接。近年来已经开发出了一些比例元件（电液数模转换器），其输出量与脉冲数、脉宽或脉冲频率成比例。其优点是抗污染能力强，滞后时间短，重复性好，能与数字计算机直接连接，是电液比例技术中的一个新领域。

In the analog signal proportional control system, if the computer is needed, A/D, D/A interface elements must be connected with the computer. In recent years, some proportional elements such as electro-hydraulic analog-to-digital converter have been invented, whose outputs are proportional to pulse number, pulse width or pulse frequency. With the advantages in contamination resistance, shorter delay time, excellent repeatability and connection to the computer directly, it is a new field in electro-hydraulic proportional technology.

1.3 电液比例控制系统的工作原理及组成（Operating principle of electro-hydraulic proportional control system and its constitution）

1.3.1 工作原理（Operating principle）

电液比例控制可以分为开环控制和闭环控制。当通过电液比例阀进行开环控制时（如图1-1所示），输入电压 U_1 经电子放大器放大后，驱动比例电磁铁，使之产生一个与驱动电流 I 成比例的力 F_D，去推动液压控制阀阀芯。液压控制阀输出一个大功率的液压信号（压力 p 和流量 q），使执行元件拖动负载以所期望的速度 v 运动。改变输入信号 U_1 的大小，便可改变负载的运动速度 v。

Electro-hydraulic control can be divided into open-loop control and closed-loop control. As Fig. 1-1 shows, in open-loop control, the input voltage U_1 is amplified by the amplifier, and drives the proportional electromagnet, to create a force F_D which is proportional to the driving electricity I, and pushes the spool of the hydraulic control valve. Then the hydraulic control valve outputs a high power hydraulic signal (pressure p and flow q) to drive the load with an expected velocity v. We can change the velocity v of the load through changing the input signal U_1.

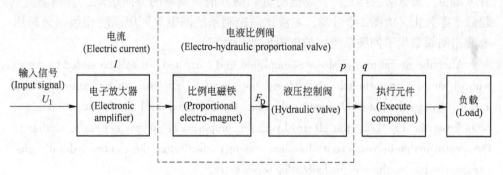

图 1-1　电液比例开环控制系统方框图

(Fig. 1-1　Block diagram of electro-hydraulic open-loop control system)

若需提高控制性能，可以采用闭环控制，见图1-2。这时，可在开环控制的基础上增加一个测量反馈元件，不断测量系统的输出量 v，并将它转换成一个与之成比例的电压 U_2，反馈到系统的输入端，同输入信号 U_1 比较，形成偏差 e。

If we want to improve the control performance, we can use closed-loop control system, as shown in Fig. 1-2. Then, based on the open-loop control system, a measuring and feedback element will be added, which measures the output v and transforms it into voltage U_2 proportionally. The voltage U_2 returns to the input and compares with the input signal U_1, forming deviation signal e.

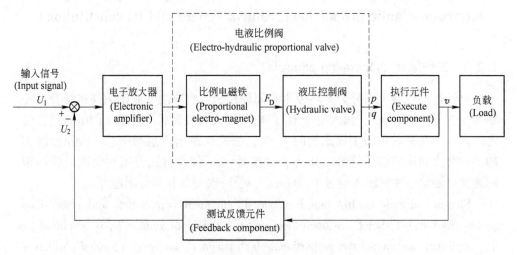

图 1-2 电液比例闭环控制系统方框图

(Fig. 1-2 Block diagram of electro-hydraulic closed loop control system)

此偏差信号 e 经放大、校正后，加到电液比例阀上，放大成大功率的液压信号 p 和 q，去驱动执行元件，以拖动负载朝着消除偏差的方向运动，直到偏差 e 趋近于零为止。由图 1-2 可知，电液比例控制系统同电液伺服系统相似，只是用电液比例阀取代了伺服系统中的电液伺服阀而已。

After the deviation signal e was amplified and corrected, it will be added to the electro-hydraulic proportional valve. Then, the signal is converted to hydraulic signal p and q, driving the power element to eliminate the deviation until the deviation e tends to be 0. From the Fig. 1-2, the electro-hydraulic proportional control system is similar to the electro-hydraulic servo control system, merely substituting the electro-hydraulic proportional valve for the electro-hydraulic servo valve.

1.3.2 电液比例系统的组成 (Composition of the electro-hydraulic proportional system)

电液比例控制系统，尽管其结构各异，功能也不相同，但都可归纳为由功能相同的基本单元组成的系统，如图 1-3 所示。

In the electro-hydraulic proportional control system, though differing in structures and functions, it can be categorized to the system comprised by basic units of the same function, as shown in Fig. 1-3.

1.3 电液比例控制系统的工作原理及组成
(Operating principle of electro-hydraulic proportional control system and its constitution)

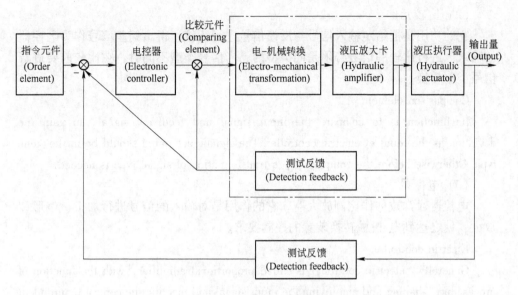

图 1-3 电液比例控制系统
(Fig. 1-3 Electro-hydraulic proportional control system)

图 1-3 中虚线所示为可能实现的检测与反馈。包含了外反馈回路控制系统的才称为闭环控制系统，不包含外反馈的称为开环系统。如果存在比例阀本身的内反馈，也可以构成实际的局部小闭环控制。但一般不称其为闭环系统。

In the Fig. 1-3, the dashed line represents possible measuring and feedback. Those systems including external feedback are called closed-loop control system, otherwise open-loop control system. If the proportional valve has internal feedback itself, it can also constitute local small closed-loop control. But in general, it can not be called closed-loop control system.

组成电液比例控制系统的基本元件有（Basic elements of electro-hydraulic proportional control system）：

（1）指令元件

它是产生与输入控制信号的元件，也称编程器或输入电路。在有反馈信号存在的情况下，它给出与反馈信号有相同形式和量级的控制信号。它也可以是信号发生装置或程序控制器。

Order element

It is the generating and inputting element of the given control signal, also known as the programming unit or input circuit. With the feedback signal, it will give the control signal with the same type and order as that feedback signal. It can be either signal generator or programming controller.

第 1 章 概　　论
Chapter 1 Introduction

(2) 比较元件

它的功用是把给定输入信号与反馈信号进行比较，得出偏差信号作为电控器的输入。进行比较的信号必须是同类型的，当信号类型不同时，在比较前要进行信号类型转换。

Comparison element

Its function is to compare the input signal and feedback signal, to gain the deviation as the input of electric controller. The signals compared should be in the same type. Otherwise before the comparison, a transformation of signal type is needed.

(3) 电控器

电控器通常被称作比例放大器。它的作用是对输入的信号进行加工、整形和放大，使之达到电-机械转换装置的控制要求。

Electric controller

Generally, electric controller is called proportional amplifier, with the function of processing, shaping and amplifying the input signal and meeting the control requirement of electro-mechanical conversion device.

(4) 比例阀

它作为整个系统的功率放大部分，可分为电-机械转换器、液压放大元件两部分，还可能带有阀内的检测反馈元件。

Proportional valve

As the part of power amplification of the whole system, it can be divided into two categories: electro-mechanical transformer and hydraulic amplifier (some even with a measuring and feedback element inside).

(5) 液压执行器

通常指液压缸或液压马达，它是系统的输出装置，用于驱动负载。

Hydraulic actuator

Generally, known as the hydraulic cylinder or the hydraulic motor, it is the system output unit, used for driving the load.

(6) 检测反馈元件

对于闭环控制，需要加入检测反馈元件，用于检测被控量或中间变量的实际值，得出系统的反馈信号。从框图中可见，检测元件还有内、外环之分。内环检测元件通常包含在比例阀内，用于改善比例阀的动、静态特性；外环检测元件直接检测输出量，用于提高整个系统的性能和控制精度。

Measuring and feedback element

In closed-loop control system, a measuring and feedback element is needed, which will measure the controlled object or the actual value of intermediate variable and get the feedback signal of system. From the block diagram, the measuring elements are classified into external and internal loop. For internal loop measuring elements, they are inside the proportional valves, to improve the dynamic and static characteristics of the valves. For external measuring elements, they are used for measuring the output directly, to improve performance and control accuracy of the whole system.

1.4 电液比例控制系统的分类（Classification of electro-hydraulic proportional control system）

电液比例控制系统可以从不同的角度按很多方式来进行分类。

Electro-hydraulic proportional control system can be classified from different viewpoints according to a number of ways.

电液比例控制系统的分类 (Classification of electro-hydraulic proportional control system)
- 被控量是否有反馈 (Feedback (YES or NO))
  - 开环比例控制 (Open-loop proportional control)
  - 闭环比例控制 (Closed-loop proportional control)
- 信号类型 (Signal type)
  - 模拟式控制 (Analogue control)
  - 数字式控制：脉宽调制、脉码调制、脉数调制 (Digital control: Pulse width, code and number modulation)
- 比例元件类型 (Type of proportional element)
  - 比例节流控制：适用功率较小的系统 (Proportional throttling control system: small power)
  - 比例容积控制：适用功率较大的场合 (Proportional volume control system: large power)

目前，最常用的分类方式是按被控对象（量或参数）来进行分类。则电液比例控制系统可以分为：

At present, the most commonly used classification method is based on the controlled object (quantity or parameter). Then, electro-hydraulic control system can be divided as follows:

比例流量控制系统（Proportional flow control system）;

比例压力控制系统（Proportional pressure control system）;

比例流量压力控制系统（Proportional flow & pressure control system）;

比例速度控制系统（Proportional velocity control system）;

比例位置控制系统（Proportional position control system）;

比例力控制系统（Proportional force control system）;

比例同步控制系统（Proportional synchronization control system）。

第 1 章 概　　论
Chapter 1　Introduction

电液比例阀是介于开关型液压阀与伺服阀之间的一种液压元件。与电液伺服阀相比，其优点是价廉、抗污染能力强。除了在控制精度及响应快速性方面不如伺服阀外，其他方面的性能和控制水平与伺服阀相当，其动、静态性能足以满足大多数工业应用的要求（表 1-1）。与传统的液压控制阀比较，虽然价格较贵，但因其可实现模拟量控制而得到推广。

The electro-hydraulic proportional valve is a type of hydraulic element between on-off valve and servo valve. Compared with the electro-hydraulic servo valve, its advantages are lower in price and its contamination resistance is better. Except control accurately and response rapidly, the electro-hydraulic proportional control valve can match with electro-hydraulic servo control valve in other performance and control levels. Its dynamic and static performance can fulfill most requirements in industrial application (Table 1-1). Electro-hydraulic proportional valves are widely used because they can control analog quantity, although they are more expensive, compared with traditional hydraulic control valves.

电液比例控制的主要优点为（Main advantages of electro-hydraulic proportional control）：

- 操作方便，容易实现电气远程控制（Easy to operate, easy to realize remote electrical control）。
- 自动化程度高，容易实现编程控制（Highly automation, easy to realize programmable control）。
- 工作平稳，控制精度较高（Working smoothly, high control accuracy）。
- 结构简单，使用元件较少，对污染不敏感（Simple structure, using less elements and no sensitive to pollution）。
- 系统的节能效果好（Good energy conservation）。

主要缺点为（Main disadvantages）：

- 成本较高（High cost）。
- 技术较复杂（Difficulty in technology）。

表 1-1　电液比例阀与开关阀、伺服阀的特性比较

(Table 1-1　Comparison among on-off valve, proportional valve and servo valve)

	开关阀 (On-off valve)	比例阀 (Proportional valve)	伺服阀 (Servo valve)
介质过滤精度/μm (Filter rating of medium)	25	25	5
阀内压力损失/MPa (Pressure loss in valve)	<0.5	0.5~2	7

1.4 电液比例控制系统的分类
(Classification of electro-hydraulic proportional control system)

续表 1-1

	开关阀 (On-off valve)	比例阀 (Proportional valve)	伺服阀 (Servo valve)
控制功率/W (Control power)	15~40	10~25	0.05~5
频宽/Hz (Bandwidth)	<10	10~70	20~200
滞环/% (Hysterics)	—	3	0.1~0.5
重复精度/% (Repeatability precision)	—	0.5~1	0.5~1
中位死区 (Mid-position dead zone)	有 (Yes)	有 (Yes)	无 (No)
温度漂移(20~60℃)/% (Temperature drift)	—	5~8	2~3
价格比 (Price ratio)	1	3~5	10

第 2 章 比例电磁铁
Chapter 2　The Proportional Electromagnet

电磁铁是将输入的电信号转变为力和位移的器件。大功率的液压阀利用电磁铁改变滑阀的位置，从而改变液体通过阀的路径。比例电磁铁是新型电磁铁，它的衔铁位置取决于电信号的强度，可以停在两个极限位置之间的任一位置上。比例电磁铁使阀较容易地发展为输出流量或输出压力可调节的模拟阀。在许多应用方面，比例阀与电控相结合，是替代传统伺服阀的廉价元件。

The electromagnet is the component which converts electric signal into force and displacement signal. A high-power hydraulic valve changes the flowing path of the liquid in it by changing the position of its sliding spool controlled by the electromagnet. The new kind of electromagnet is a proportional one in which the position of its armature depends on the strength of the electrical signal. It can be at any position between the two limit positions. Valves can be easily developed to be simulation valves whose output flow and output pressure are adjustable by using proportional electromagnets. In many applications, proportional valves combined with electric control system are cheaper alternative to the traditional servo valves.

2.1　电-机械转换元件的作用及形式（Function and form of electro-mechanical conversion component）

2.1.1　电-机械转换元件的形式（Form of the electro-mechanical conversion component）

目前，生产上应用的电-机械转换元件大多采用电磁式设计，并且利用电磁力与弹簧力相互平衡的原理，实现电-机械的比例转换。最常见的有直流伺服电机、步进电机、力矩马达、动圈式力马达以及动铁式力马达。后者更一般地称为比例电磁铁。从目前的使用情况来看，应用最为广泛的还是比例电磁铁，它目前已经成为最主要的电-机械转换元件。

At present, the electric machinery conversion component applied in production

2.1 电-机械转换元件的作用及形式
(Function and form of electro-mechanical conversion component)

mostly adopts electromagnetic design and utilizes equilibrium electric machinery principle of electromagnetic force and spring force to achieve proportional conversion. Dc servo motors, stepping motors, torque motors, moving-coil force motors and moving-iron force motors are most widely used. The latter are more generally called proportional electromagnets. The most widely used electric machinery components are the proportional electromagnets in current situation, which have already become the main electric machinery conversion one.

2.1.2 电-机械转换元件的要求 (Demands of electric machinery conversion component)

在电液比例技术中，对作为阀的驱动装置的电-机械转换元件的基本要求有以下几点：

The basic demands of the conversion component as valve driving device are as follows in electro-hydraulic proportional technology：

(1) 具有水平吸力特性，即输出的机械力与电信号大小成比例，与衔铁的位移无关。

It is in possession of aclinic attraction characteristic, its output force is proportional to strength of electrical signal and has nothing to do with the displacement of the keeper.

(2) 有足够的输出力和行程，结构紧凑、体积小。

There must be enough output force and stroke with compact structure and small size at the same time.

(3) 线性好，死区小，灵敏度高，滞环小。

It must possess the quality of fine linearity, little dead zone, high sensitivity and tiny hysteresis.

(4) 动态性能好，响应速度快。

It must possess the quality of good dynamic performance and fast response.

(5) 长期工作中温升不会过大，并在允许温升下仍能工作。

The temperature rise must not be too high during long working time, and it should work normally at the state of permitted temperature rise.

(6) 能承受液压系统高压，抗干扰性好。

It can bear the high pressure of hydraulic system and the resistance to disturbance must be strong.

2.1.3 比例电磁铁概述 (Summary of the proportional electromagnet)

比例控制的核心是比例阀。比例阀的输入单元是电-机械转换器，它将输入

第 2 章 比例电磁铁
Chapter 2　The Proportional Electromagnet

信号转换成机械量。

The key component of proportional control technology is the proportional valve. Its input unit is the electric machinery conversion device which converts the input electrical signal into mechanical one.

比例电磁铁根据法拉第电磁感应原理设计,能使其产生的机械量(力或力矩和位移)与输入电信号(电流)的大小成比例,再连续地控制液压阀阀芯的位置,实现连续地控制液压系统的压力、方向和流量。

The proportional electromagnet which is designed according to Faraday electromagnetic induction principle can make mechanical signal (force, torque and displacement) proportional to input electrical signal (current), and it can control the position of spool successively. The above characteristics will make sure that the pressure, direction and flow of hydraulic system change continuously.

比例电磁铁同电液伺服系统中伺服阀的力矩马达或力马达相似,是一种将电信号转换成机械力和位移的电-机械转换器。比例电磁铁是电子技术与液压技术的连接环节,是一种直流行程式电磁铁,可产生一个与输入量(电流)成比例的输出量:力和位移。其性能的好坏,对电液比例阀的特性有着举足轻重的作用。

The proportional electromagnet which is similar to the torque motor or force motor of the servo valve in electro-hydraulic servo system is a kind of converters converting electrical signal to force and displacement (signal). The proportional electromagnet is a link between electronic technology and hydraulic technology. The proportional electromagnet is a kind of DC stroke electromagnets. It generates an output signal (force and displacement) which is proportional to the input signal (current). Its performance has an extremely important effect on the characteristics of the electro-hydraulic proportional valve.

2.2　电磁铁的结构与工作原理
(Structure and working principle of electromagnet)

2.2.1　普通螺线管型电磁铁 (Ordinary solenoid electromagnet)

普通甲壳型螺线管电磁铁如图 2-1 所示,由外壳 2、激磁线圈 3、挡板 4、衔铁 6 组成。当线圈通有直流电流时,线圈便在铁芯中产生磁场,并形成闭合的磁力线路。电磁铁存在两个间隙,一个是工作间隙 5,另一个是非工作间隙 1。在电磁铁吸合过程中形成两个变化的磁通,即主磁通 Φ 和变化的磁通 Φ_L。衔铁 6 所受到的吸力主要由两部分组成。主磁通产生的力称为端面力,而漏磁通产生的

2.2 电磁铁的结构与工作原理
(Structure and working principle of electromagnet)

力称为螺管力。对于图示结构,两个力的方向是一致的。这两个力的合力就构成了总的电磁力。

The ordinary shell type solenoid electromagnet shown in Fig. 2-1 is made up of a shell 2, a field coil 3, a baffle 4 and a armature 6. When flowed through by DC, the coil generates magnetic field in the core which forms a close magnetic circuit. There are two gaps in the electromagnet. One is working gap 5 and the other is non-working gap 1. Two changing fluxes are created. One is called the main magnetic flux Φ and the other is called the changing magnetic flux Φ_L. The suction imposing on the armature 6 is composed of two parts. One is the face force created by the main magnetic flux and the other is solenoid force created by the changing magnetic flux. The directions of the two forces are the same. The resultant force of the two forces is the total electromagnetic force.

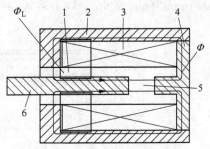

图 2-1 普通螺线管型电磁铁
(Fig. 2-1 The ordinary solenoid electromagnet)
1—非工作间隙(Non-working gap);
2—外壳(Shell); 3—激磁线圈(Field coil);
4—挡板(Baffle); 5—工作间隙(Working gap);
6—衔铁(Armature)

2.2.2 单向比例电磁铁 (One-way proportional electromagnet)

直流比例电磁铁是电液比例控制器件中应用最广泛的电-机械转换器。它也是装甲式螺线管电磁铁,由于可动部分是衔铁,所以是一种动铁式马达,具有结构简单、价格低、功率/重量比大、能够输出较大的力和位移等优点。按比例电磁铁输出位移的形式,除最常见的单向直动式外,还有双向直动式。回转式比例电磁铁应用较少。

The DC proportional electromagnet is the most widely used electricity-machine conversion component among electro-hydraulic proportional control devices. It is armored solenoid electromagnet and moving-iron motor because its moving part is the armature. It can export greater force and displacement with simple structure, low price and large power/weight ratio. The proportional electromagnets are divided into one-way direct operated electromagnet and two-way direct operated electromagnet according to its form of output displacement. The rotary proportional electromagnet is seldom used.

线圈通电后形成的磁路,经壳体、导向套锥端到轭铁而产生斜面吸力;另一路是直接由衔铁端面到轭铁的输出力(如图2-2所示)。

After the coil is energized, the following magnetic circuit generates bevel suction by going through the shell, cone point of the guide sleeve and then the yoke iron; the other output force of the yoke iron is generated by the armature face (in Fig. 2-2).

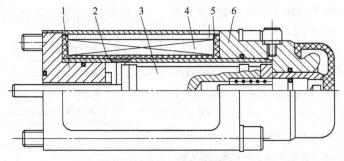

图 2-2 单向比例电磁铁

(Fig. 2-2 The one-way proportional electromagnet)

1—轭铁(Yoke iron); 2—导向套锥端(Cone end of guide sleeve); 3—衔铁(Armature);
4—线圈(Coil); 5—导向套(Guide sleeve); 6—壳体(Shell)

2.3 比例电磁铁的特性(Characteristics of proportional electromagnet)

2.3.1 电磁铁的吸力特性(Attracting characteristic of proportional electromagnet)

电磁铁(直线力马达)是依靠电磁系统产生的电磁吸力,使衔铁对外做功的一种电动装置。其基本特性可表示为衔铁在运动中所受到的电磁力 F_m 与它的行程(位移) x 之间的关系,即 $F_m=f(x)$。这个关系称为吸力特性。对比例电磁铁,要求它具有水平的吸力特性。

The electromagnet (linear force motor) is a kind of electric device which makes the armature do work by the electromagnetic force generated in the electromagnetic system. Its basic characteristics can be expressed as the relationship between electromagnetic force F_m working on the moving armature and the displacement x of the armature, namely $F_m=f(x)$. The relationship is called the attraction characteristic. The proportional electromagnet needs to possess the aclinic attraction characteristic.

A 比例电磁铁基本结构(Basic structure of proportional electromagnet)

由图 2-3 可知,比例电磁铁主要由推杆、衔铁、轴承环、隔磁环、导套和限位环组成。导套前后两段由导磁材料制成,中间用一段非导磁材料(隔磁环)焊接。衔铁前端有推杆,用于输出力或者位移;后端装有由弹簧和调节螺钉组成的调零机构,可在一定范围内对比例电磁铁特性曲线进行调整。

2.3 比例电磁铁的特性
(Characteristics of proportional electromagnet)

The basic structure of the proportional electromagnet is shown in Fig. 2-3. The proportional electromagnet is made up of push rod, armature, bearing collar, septal magnetic collar, guide sleeve, limit collar. The front and the back of the guide sleeve whose middle part is welded with non conduction material (septal magnetic collar) is made of conduction material. The push rod is in front of the armature exports force or displacement; The adjusting zero mechanism is composed of the spring and adjusting screw so that the armature can govern the characteristic curve of the proportional electromagnet in a certain range.

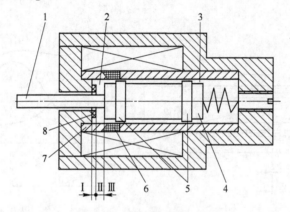

图 2-3 比例电磁铁基本结构

(Fig. 2-3 Basic structure of proportional electromagnet)

Ⅰ—吸合区 (Attracting zone); Ⅱ—工作行程区 (Working zone); Ⅲ—空行程区 (Idle zone);
1—推杆 (Push rod); 2—工作间隙 (Working gap); 3—非工作间隙 (Non-working gap);
4—衔铁 (Armature); 5—轴承环 (Bearing collar); 6—隔磁环 (Septal magnetic collar);
7—导套 (Guide sleeve); 8—限位环 (Limit collar)

B 比例电磁铁工作原理 (Working principle of the proportional electromagnet)

当给比例电磁铁控制线圈通入一定电流时，在线圈电流控制磁势的作用下，形成两条磁路（如图 2-4a 所示）。一条磁路 Φ_1 由前端盖盆形极靴底部，沿轴向工作气隙，进入衔铁，穿过导套后段和导磁外壳回到前端盖极靴。另一磁路 Φ_2 经盆形极靴锥形周边（导套前段），径向穿过工作气隙进入衔铁，而后与 Φ_1 相汇合。由于电磁作用，磁通 Φ_1 产生了通常的端面力 F_{M1}，磁通 Φ_2 则产生了一定数量的附加轴向力 F_{M2}（如图 2-4b 所示）。两力的综合，就得到了整个比例电磁铁的输出力 F_M。在工作区域内，电磁力 F_M 相对于衔铁位移基本呈水平力特性关系。

The working principle of the proportional electromagnet: When the control coil of the proportional electromagnet is powered/energized, two magnetic circuits are formed by the magnetic potential of the coil current (as shown in Fig. 2-4a). One magnetic circuit Φ_1 returns to the front end cover pole shoe via the bottom of front end cover basin shape pole shoe, working gap, armature, back of guide sleeve and conduction shell. The other one is combined with Φ_1 via conical surrounding of basin shape pole shoe (front of guide sleeve), working gap and armature. flux Φ_1 generates the face force F_{M1} because of electromagnetic while flux Φ_2 generates a certain amount of additional axial force F_{M2} (as shown in Fig. 2-4b). The combination of the two is the output force F_M of the proportional electromagnet. Electromagnetic force F_M is mainly constant force relative to the displacement of armature in working stroke zone.

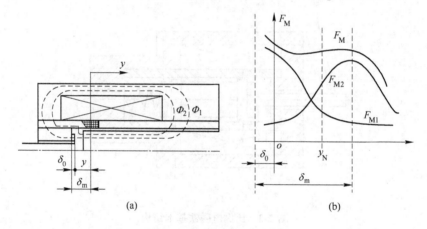

图 2-4 比例电磁铁基本原理

(Fig. 2-4 Basic principle of the proportional electromagnet)

(a) Φ_1, Φ_2 的磁路示意图 (Magnetic circuit of Φ_1 and Φ_2);

(b) 位移-力特性 (Characteristic diagram of displacement and forces)

图 2-5 所示为普通电磁铁与比例电磁铁的静态吸力特性。所谓静态吸力特性,就是在稳态过程中得到的吸力特性。与它相对应的是动态特性。

The static attractive characteristics of the ordinary and proportional electromagnet is shown in Fig. 2-5. The static attractive characteristic is obtained in a steady process. The corresponding one is dynamic characteristic.

如图 2-5 所示,比例电磁铁在整个工作行程内,力-位移特性并不全是水平特性。它可以分为三个区段。在工作气隙接近于零的区段,输出力急剧上升,称为吸合区Ⅰ。这一行程区段不能正常工作,因此在结构上用加限位片的方法将其排除,使衔铁不能移动到该区段内。当工作气隙过大,电磁铁输出力明显下降的区段,称为空行程区Ⅲ。除吸合区Ⅰ和空行程区Ⅲ外的区段,称为工作区Ⅱ。比例电磁铁必须具有水平吸力特性,即在工作区内,其输出力的大小只与电流有关,与衔铁位移无关。

2.3 比例电磁铁的特性
(Characteristics of proportional electromagnet)

As shown in Fig. 2-5, force-displacement characteristic is not totally aclinic in the entire working stroke of the proportional electromagnet. It can be divided into three parts. The zone I is called attraction zone where the output force increases rapidly and the working gap is close to zero. The proportional electromagnet cannot work normally in this zone. The armature cannot move to this zone by using a limit piece. The zone is called idle stroke zone where the output force decreases rapidly and the working gap is too large. The left zone II is called working stroke zone. The proportional electromagnet needs to possess aclinic attraction characteristic which means that the output force is only relative to the current and has nothing to do with the displacement of the armature in the working stroke zone.

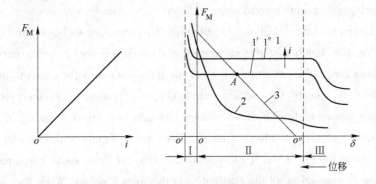

图 2-5 比例电磁铁的静态吸力-位移特性
(Fig. 2-5 Static suction-displacement characteristic of the proportional electromagnet)
1—比例电磁铁 (Proportional electromagnet); 2—普通电磁铁 (Ordinary electromagnet);
3—弹簧负载曲线 (Spring load curve)

C 比例电磁铁的工作过程 (Working process of the proportional electromagnet)

以弹簧负载为例，如图2-5所示。衔铁始终受到电磁力与弹簧力的作用，电磁力吸引衔铁，所以其方向向左，弹簧力方向向右。衔铁初始位置在 o″点，观察曲线 1′，此时弹簧还没有被压缩，所以电磁力大于弹簧力，衔铁向左运动；随着衔铁的运动，弹簧逐渐被压缩，弹簧力逐渐变大，直到弹簧力等于电磁力，衔铁将不再移动，最后停止在 A 点。

The working process of the proportional electromagnet is shown in Fig. 2-5 and takes spring load as an example. The armature is affected by electromagnetic force whose direction is left and spring force whose direction is right. Observing curve 1′, the initial position of Armature is at o″ when the spring is not compressed. The spring force is larger than electromagnetic force at present, therefore the armature will be moved left. The spring will be gradually compressed during the moving process of the Armature. The spring force will become larger and larger until it is equal to the electromagnetic force. The armature will be stopped at A at last.

若电磁铁的吸力不显水平特性，弹簧曲线与电磁力曲线簇只有有限的几个交点，这意味着不能进行有效的位移控制。在工作范围内，不与弹簧曲线相交的各电磁力曲线中，对应的电流在弹簧曲线以下，不会引起衔铁位移；在弹簧曲线以上时，若输出这样的电流，电磁力将超过弹簧力，将衔铁一直拉到极限位置为止。相反，若电磁铁具有水平特性，那么在同样的弹簧曲线下，将与电磁力曲线簇产生许多交点。在这些交点上，弹簧力与电磁力相等，就是说，逐渐加大输入电流时，衔铁能连续地停留在各个位置上。

If the proportional electromagnet does not possess the aclinic attraction characteristic, the spring curve and electromagnetic force curve family will intersect at limited spots. Displacement control will not be effective. The armature will not move when the current curve is a member of the current curves which are under spring curve in the whole working zone; the armature will be at the limit position if the current curve of the exact current is above the spring curve. On the contrary, if the proportional electromagnet possesses the aclinic attraction characteristic and the spring is the same one, the spring curve and electromagnetic force curve family will intersect at many spots. The spring force is equal to electromagnetic force at these spots. The armature will be continuously stopped at all the positions that the spots represent when the current becomes larger and larger.

直动式比例电磁铁的吸力特性有以下特点：在其工作行程范围内，具有基本水平的位移-力特性，因此一定的控制电流对应一定的输出力，即输出力与输入电流成比例。

The static attractive of the direct operated proportional electromagnet is provided with the following characteristics: the proportional electromagnet should possess aclinic displacement-force characteristic with qualified accuracy in the working zone so that a certain current corresponds to a certain output force which means output force is proportional to the output current.

2.3.2 电磁铁的负载特性 (Load characteristic of electromagnet)

电磁铁在运动过程中，必然要克服机械负载和阻力而做功。对于普通电磁铁，一般都要求电磁吸力大于负载反力；而对于比例电磁铁，则要求衔铁处于电磁吸力与负载反力相平衡的状态，只有这样，电磁铁才能正常工作。为使电磁铁可靠地工作，应使吸力特性与负载特性有良好的配合。常见的负载反力的特性如图2-6所示。

2.3 比例电磁铁的特性
(Characteristics of proportional electromagnet)

During moving, the electromagnet has to overcome mechanical load and resistance and do work. It is demanded that electromagnetic attraction force is larger than load reaction for the ordinary electromagnet. The proportional electromagnet works normally because the armature keeps the electromagnetic attraction force and the load reaction balance. Attraction characteristic should be matched with load characteristic to make the electromagnet reliable. The curve of common load reaction characteristic is shown in Fig. 2-6.

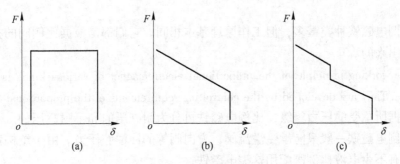

图 2-6 典型负载反力特性

(Fig. 2-6 Curve of typical load reaction characteristic)
(a) 恒力负载特性 (Constant force load characteristic); (b) 弹簧负载特性 (Spring load characteristic);
(c) 多级弹簧负载特性 (Multistage spring load characteristic)

对于吸合型电磁铁,在吸合过程中,电磁吸力特性曲线应在负载反力曲线的上方;而在释放运动中,负载反力必须大于电磁产生的剩磁力。

The curve of electromagnetic attraction characteristic should be above the curve of load reaction for the suction electromagnet during suction process; the load reaction must be larger than the remaining electromagnetic force during release process.

比例电磁铁在工作过程中,电磁力总是与负载力相平衡,吸力特性曲线有很多条,而负载多为弹性负载。它工作时吸力特性与负载特性的配合情况如图 2-5 所示。负载弹簧的特性曲线与多条吸力特性曲线相交。对应不同的输入电流,电磁力的曲线水平上下平移,而它与弹簧特性曲线的相交点便是对应电流下的工作点。由图中可以看出,当电流改变时,工作点也改变。比例电磁铁正是利用这一特性来实现电-机械信号的比例转换。

The electromagnetic force should always be equal to the load force during the working process of the electromagnet. There are many attraction characteristic curves. The load is normally elastic load. Its attraction characteristic curves matched with load characteristic curves is shown in Fig. 2-5. The curve of load spring characteristic intersects with many attraction characteristic curves. Electromagnetic force curve will be moving up and down as the current changing. The intersection of spring characteristic curve and load characteristic curve is the working point of corresponding current. It can be seen that the working point can be changed when the current is changed. The proportional electromagnet make electricity-mechanical signal transfer by the way of this characteristics.

2.4 比例电磁铁的分类与应用 (Classification and application of the proportional electromagnet)

比例电磁铁种类繁多，但工作原理基本相同。它们都是根据比例阀的控制需要开发出来的。

The working principle of the proportional electromagnet of various kinds is mostly the same. They are developed by the controlling requirements of the proportional valves.

根据所承受的压力等级，比例电磁铁可分为耐高压的和不耐高压的。不耐高压的比例电磁铁一般只能承受溢流阀、方向阀等的回油腔压力。由于结构比较简单，仍有不少电液比例阀配用这类电磁铁。

The proportional electromagnet can be divided into two types of high-pressure resistant and high-pressure nonresistant, according to its pressure resistance level. The non enduring high pressure type can normally endure the pressure of the return oil cavity of relief valve and directional valve. This type is still used because of its simple structure.

根据所输出的运动参数，比例电磁铁可分为直线运动的和旋转运动的（角位移式）。直线运动的比例电磁铁应用最为广泛。

The proportional electromagnet can be divided into linear motion type and rotary motion type according to its output motion parameters. The linear motion type is the most widely used one.

根据参数的调节方式和它们与所驱动阀芯的连接形式，比例电磁铁可分为力控制型、行程控制型和位置调节型三种基本类型。本书按照这三种情况分类叙述其各自的特点和应用。

The proportional electromagnet can be divided into force control type, stroke control type and position control type according to its parameters' accommodation mode and connection form of the driving valve spool. The characteristics and application of the three types are shown respectively in this book.

此外，还有双向极化式耐高压比例电磁铁、插装阀式比例电磁铁、防爆比例电磁铁、内装集成比例放大器的比例电磁铁等等。有的比例电磁铁在其尾部装有手动应急装置，当控制线路发生故障导致比例电磁铁断电时，能手动控制比例阀。

In addition, there are bidirectional polarized enduring high pressure type,

2.4 比例电磁铁的分类与应用
(Classification and application of the proportional electromagnet)

cartridge valve type, explosion-proof type and inner integration type and so on. Manual emergency device is fixed at the tail of some proportional electromagnet. The proportional valve can be controlled by hand when it is powered off for the control circuit's breaking down.

2.4.1 力控制型比例电磁铁 (Force control type)

力控制型比例电磁铁的基本特性是力-行程特性。在力控制型比例电磁铁中，衔铁行程没有明显变化时，改变电流 I，就可以调节其输出的电磁力。由于在电子放大器中设置有电流反馈环节，在电流设定值恒定不变而磁阻变化时，可使磁通量不变，进而使电磁力保持不变。

The basic characteristic of the force control type is force-stroke characteristic. The stroke of the keeper is unconspicuously changed. The output electromagnetic force can be adjusted by changing current I. Because of the current feedback link set in electronic amplifier, the magnetic flux can remain unchanged when current set value is constant and reluctance is changing, thus keeping electromagnetic force a constant.

在控制电流不变时，电磁力在其工作行程内保持不变。如图 2-7 所示，这类电磁铁的有效工作行程约为 1.5mm。

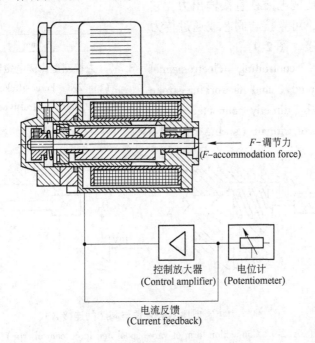

图 2-7 力控制型比例电磁铁原理图
(Fig. 2-7 Principle diagram of force control type)

第 2 章 比例电磁铁
Chapter 2 The Proportional Electromagnet

When control current is constant, electromagnetic force remain unchanged in working stroke zone. As shown in Fig. 2-7, effective working stroke of force control type is about 1.5mm.

由于行程较小,力控制型电磁铁的结构很紧凑。正由于其行程小,可用做比例压力阀或比例方向阀的先导级,将电磁力转化为液压力。

The structure of force control type is compact because of its short stroke. Because of its short stroke, it can be used as pilot stage of proportional pressure valve and proportional direction valve which convert electromagnetic force into hydraulic force.

这种比例电磁铁是一种可调节型直流电磁铁,在其衔铁腔中,充满了工作油液。

This kind of proportional electromagnet is a kind of adjustable DC electromagnet. The chamber of the keeper is filled with working oil.

(1) 力控制型比例电磁铁原理图和力-行程曲线(图 2-8)(Principle diagram and force-stroke curve of force control type (Fig. 2-8));(2) 力控制型比例电磁铁与阀芯的连接方式(Connection form of valve spool and force control type)。

力控制型比例电磁铁直接输出力,它的工作行程短,可直接与阀芯或通过传力弹簧与阀芯连接(图 2-9)。

The force controlling electromagnet exports force directly, and its working stroke is short. It can be directly connected to the valve spool with or without a spring (Fig. 2-9).

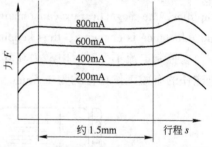

图 2-8 力控制型比例电磁铁力-行程曲线

(Fig. 2-8 Force-stroke curve of force control type)

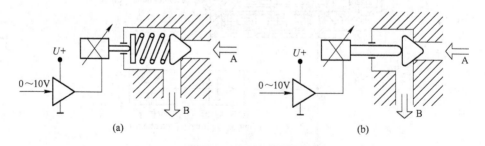

图 2-9 力控制型比例电磁铁与阀芯的连接方式

(Fig. 2-9 Connection form of valve spool and force control type)
(a) 比例电磁铁通过传力弹簧作用在阀芯上(Connected by transfer spring);
(b) 比例电磁铁直接作用在阀芯上(Connected directly)

2.4.2 行程控制型比例电磁铁 (Stroke control type)

在行程控制型比例电磁铁中，衔铁的位置由一个闭环调节回路进行调节。只要电磁铁在其允许的工作区域内工作，其衔铁位置就保持不变，而与所受反力大小无关。使用行程调节型比例电磁铁，能够直接推动诸如比例方向阀、流量阀及压力阀的阀芯，并将其控制在任意位置上。电磁铁的行程，因其规格而异，一般在 3~5mm 之间。

The position of the armature is adjusted by a close loop in the stroke control type. As long as the electromagnet is working in its permitted zone, the position of its armature keeps unchanged and has nothing to do with the reaction. By using the stroke control type, the valve spool of proportional flow valve, proportional direction valve and proportional pressure valve can be pushed and stopped at any position. The stroke of electromagnet is different because of its specification. It is normally between 3mm and 5mm.

如前所述，行程控制型比例电磁铁主要用来控制四通比例方向阀。配上电反馈环节后，电磁铁的滞环及重复误差均较小。此外，作用在阀芯上的液动力也受到抑制（与各种可能产生的干扰力相比，电磁力较小）。

As what mentioned before, the stroke control type is mainly used to control four-way proportional direction valve. Its hysteresis and repetitive error are all quite small with the using of electric feedback. In addition, the flow force exerting on valve spool can be restrained (the electromagnetic force is quite small compared with other possible disturbing forces).

在先导阀中，受控压力作用在一个较大的控制面积上，因此供使用的调节力要大得多，而干扰力的影响并不大。因此，先导式比例阀也可不带电反馈机构。

Due to the fact that the controlled pressure exerts on a larger area in the pilot valve, the available adjusting force is much larger than that in the valves mentioned before, but the percentage of the effect of interference force is not so large. Thus, the pilot proportional valve can also work without electric feedback mechanism.

(1) 行程控制型比例电磁铁工作原理图（图 2-10） (Principle diagram of stroke control type (Fig. 2-10))。

(2) 行程控制型比例电磁铁结构示意图（图 2-11） (Schematic diagram of stroke control type (Fig. 2-11))。

行程控制型比例电磁铁是在力控制型比例电磁铁的基础上，将弹簧布置在阀芯的另一端得到的。

Based on the force control type, the stroke control type is obtained by the spring equipped on the other side of the valve spool.

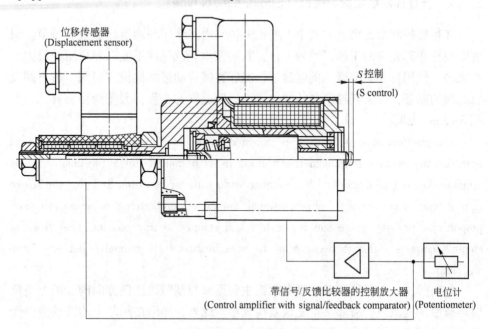

图 2-10 行程控制型比例电磁铁工作原理图
(Fig. 2-10 Principle diagram of stroke control type)

2.4.3 位置调节型比例电磁铁 (Position control type)

比例电磁铁衔铁的位置通过位移传感器检测，与比例放大器一起构成位置反馈系统，就形成了位置调节型比例电磁铁。只要电磁铁运行在允许的工作区域内，其衔铁就保持与输入电信号相对应的位置不变，而与所受反力无关，即它的负载刚度很大。这类位置调节型比例电磁铁多用于控制精度要求较高的比例阀上。在结构上，除了衔铁的一端接上位移传感器（位移传感器的动杆与衔铁固定联接）外，其余与力控制型、行程控制型比例电磁铁基本相同。

The position of proportional electromagnet is detected by displacement sensor and it constitutes a feedback system with proportional amplifier. These form the position control type. As long as the electromagnet is working in its permitted working range, the position of its armature keeps unchanged and has nothing to do with the reacting force. Its load stiffness is large. The position control type is often used on the proportional valve whose control precision is demanded higher. The structure of this type is the same as the other types except that its armature is linked with displacement sensor (the dynamic rod of displacement sensor is fixed with armature).

2.4 比例电磁铁的分类与应用
(Classification and application of the proportional electromagnet)

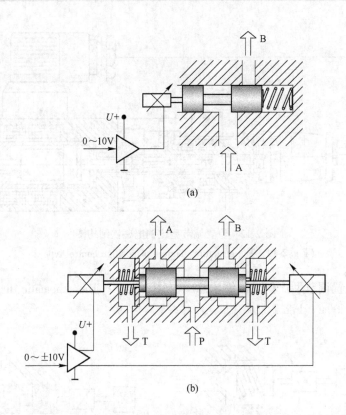

图 2-11 行程控制型比例电磁铁结构示意图
(Fig. 2-11 The schematic diagram of stroke control type)
(a) 单个使用的行程控制型比例电磁铁 (Single using);
(b) 成对使用的行程控制型比例电磁铁 (Pairs using)

位置调节型比例电磁铁用在比例方向阀和比例流量阀上，可控制阀口开度；用在比例压力阀上，可获得精确的输出力。这种比例电磁铁具有很高的定位精度，负载刚度大，抗干扰能力强。由于这类比例电磁铁是一个位置反馈系统，故要与配套的比例放大器一起使用。

The valve port opening can be controlled in proportional flow valve and proportional direction valve of this type. The precise output force can be obtained in the proportional direction valve. The positioning accuracy of this type is high, the load stiffness of this type is large and the anti-interference ability is strong. The type need using with proportional amplifier because it is a position feedback system.

(1) 位置调节型比例电磁铁结构图 (图 2-12) (The structure diagram of position control type (Fig. 2-12))。

第 2 章 比例电磁铁
Chapter 2 The Proportional Electromagnet

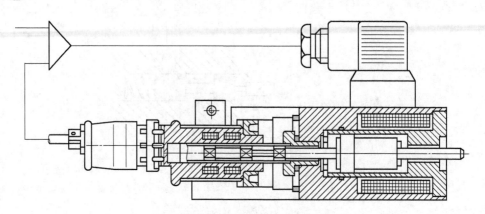

图 2-12　位置调节型比例电磁铁结构图

(Fig. 2-12 The structure diagram of position control type)

（2）位置调节型比例电磁铁示意图（图 2-13）（The schematic diagram of position control type (Fig. 2-13)）。

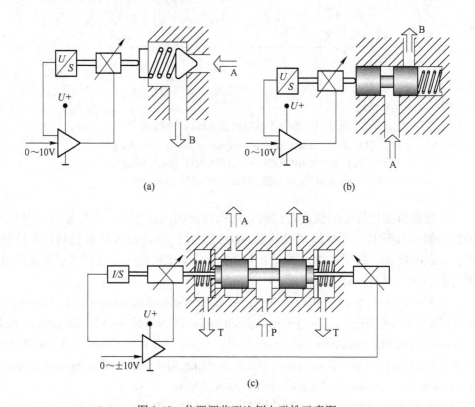

图 2-13　位置调节型比例电磁铁示意图

(Fig. 2-13 The schematic diagram of position control type)

2.5 比例电磁铁的初步设计 (Preliminary design of proportional electromagnet)

比例电磁铁的初步设计涉及的基本方程主要有四个，即电磁吸力方程、磁势方程、电压方程和发热方程。这些方程反映了结构尺寸和物理参数之间的基本关系。此外还有表征电磁铁尺寸参数的合理取值范围的关系式。下面分别讨论。

The basic equations concerning the preliminary design of proportional electromagnet are as follows. They are electromagnetic suction equation, magnetic potential equation, voltage equation and heating equation. These equations reflect the basic relationship between structure size and physical parameters. In addition, there is still the relational expression indicating reasonable numerical value range of electromagnet size parameters, which will be talked about separately as follows.

A 电磁铁的吸力方程 (The electromagnetic suction equation)

作为初步计算，计算基础为麦克斯韦公式，采用等效气隙磁导法较为简便。

As preliminary calculation, the computing base is Maxwell formular and equivalent gas gap magnetic conductivity method is adopted to be simple.

$$F_m = \frac{B_0^2 S_0}{2\mu_0} \tag{2-1}$$

式中 B_0——等效气隙处的磁感应强度 (The magnetic induction intensity at equivalent gas gap);

S_0——等效气隙端面积 (The area at equivalent gas gap end);

μ_0——空气磁导率 (Air permeability)。

B 磁势方程 (The magnetic potential equation)

磁势方程反映了电磁铁正常工作时所需要的激磁势值。利用磁势方程可求出线圈所需要的激磁安匝数：

The magnetic potential equation shows the needed shock magnetic potential valve during the normal working of electromagnet. Exciting ampere turns can be calculated by using the equation：

$$IN = B_0 S_0 (\sum R_g + \sum R_p) \tag{2-2}$$

式中 R_p, R_g——气隙磁阻和导磁体磁阻 (The reluctance of gas gap and magnetizer)。

C 电压方程 (The voltage equation)

电流通过线圈便产生磁势，并引起发热。为了确定线圈的参数（圈数、线径、线圈电流等），使线圈能够在额定电压下产生足够的磁势（IN），这时需要用到线圈电压方程：

When the current goes through coil, the magnetic potential is generated and lead to heating. The voltage equation is needed in order to determine coil parameters (laps, diameter and current), and make the coil under rated voltage and create enough magnetic potential:

$$U=IR_x=IN\frac{\rho_x \pi D_x}{g_x}=\frac{1}{2}IN\frac{\rho_x \pi (D_w+D_n)}{g_x} \quad (2\text{-}3)$$

式中　R_x——线圈总电阻（The resistance of coil）；

　　　I——线圈电流（The coil current）；

　　　D_x——线圈平均直径（The mean diameter of coil）；

　　　D_w——线圈外径（The external diameter of coil）；

　　　D_n——线圈内径（The inner diameter of coil）；

　　　N——线圈圈数（The coil laps）；

　　　ρ_x——线圈导线在相应温度下电阻率（The resistivity of wire under corresponding temperature）。

D　发热方程（The heating equation）

发热方程是为了校核电磁铁线圈在长期工作条件下，温升是否超过其允许值。比例电磁铁通常按长期工作计算温升：

The heating equation is used to check whether temperature rising exceed its permitted value when electromagnet coil is under long term operation. The proportional electromagnet is often calculated on the basis of its long term operation condition:

$$\theta=\frac{\rho_s}{2\times 10^4 \mu_s f_t b_s}\left(\frac{IN}{l_s}\right)^2 \quad (2\text{-}4)$$

式中　μ_s——散热系数（Heat release coefficient）；

　　　f_t——线圈填充系数（Coil filled coefficient）；

　　b_s, l_s——线圈厚度及长度（The thickness and length of coil）；

　　　ρ_s——工作温度下的电阻率，查表可得（The resistivity under working temperature, look-up table）。

E　比例电磁铁的设计步骤（The design steps of proportional electromagnet）

设计的已知条件通常是：最大电磁吸力 F_m，初始气隙 δ_0，推杆直径 d_0，线圈电压 U_0 及允许温升 $[\theta]$。

The designed known conditions are often: maximum electromagnetic attraction F_m initial gas gap δ_0, rod diameter d_0, coil voltage U_0 and permitted temperature rising $[\theta]$.

2.5 比例电磁铁的初步设计
(Preliminary design of proportional electromagnet)

综上所述，设计步骤如下：

To sum up, the design steps are as follows:

(1) 求出结构因素 K_φ，查出电磁感应强度 B_0。

Find out structural factors K_φ and electromagnetic intensity B_0.

(2) 求出衔铁的外径 d_1 和盆底极靴（导套）的外径 d_2。半径间隙 δ 由实验确定，初步设计取 $\delta = 0.02$。

Find out the outer diameter d_1 of keeper and the outer diameter d_2 of basin bottom pole shoe (guide sleeve). The radius gap δ is determined by experirment. Preliminarily adopt $\delta = 0.02$.

(3) 计算所需安匝数 IN。线圈最大工作电流可取 800mA，也可取到 1.2A，由此确定线圈匝数。

Calculate the needed ampere turns IN. The maximum working current can be 800mA or 1.2A to determine the number of turns of coil.

(4) 利用发热方程式初步计算线圈长度 l_{so}，取 $b_s = l_s/5$，得出线圈厚度 b_s。考虑到导套的外径及绝缘材料厚度，初定线圈的内、外直径 D_n 和 D_w。

Calculate loop length l_{so} by using the heating equation. Coil thickness b_s is obtained when $b_s = l_s/5$. The inner and outer diameter of coil (D_n and D_w) is preliminarily determined considering the outer of guide sleeve and the thickness of insulating material.

(5) 利用电压方程式计算导线截面积 g_x，计算出线圈能否容纳必要的匝数。

Calculate sectional area g_x of wires by using voltage equation and find out whether coil can hold necessary number of turns.

(6) 确定比例电磁铁的其余结构尺寸。

Confirm the other structure dimensions of proportional electromagnet.

(7) 进行必要的验算（例如温升等）工作。

Carry on necessary checking (for example temperature rising).

第 3 章 电液比例控制阀
Chapter 3 Electro-hydraulic Proportional Control Valve

3.1 概述（Summary）

电-液比例控制阀能与电子控制装置组合在一起，从而十分方便地对各种输入、输出信号进行运算和处理，实现复杂的控制功能。同时它又具有抗污染、低成本以及响应较快的优点，在液压控制工程中获得越来越广泛的应用。

Combined with electronic control units, electro-hydraulic proportional control valves can be very convenient to process various the input and output signals and to realize complex control operation. Furthermore they have the advantage of anti-pollution, low cost and rapid response, therefore the electro-hydraulic proportional control valves are more and more widely used in the hydraulic control engineering.

最常见的分类方法是按其控制功能来分类，可以分为比例压力控制阀、比例流量控制阀、比例方向阀和比例复合阀。前两者为单参数控制阀，后两者为多参数控制阀。

According to specific control object, they are usually divided into proportional pressure control valves, proportional flow control valves, proportional direction valves and proportional composite valves. The former two are single parameter control valves, while latter are multi-parameter control valves.

按压力放大级的级数来分，又可以分为直动式和先导式。直动式是由电-机械转换元件直接推动液压功率级，由于转换元件的限制，它的控制流量都在 15 L/min 以下。先导控制式比例阀由直动式比例阀与能输出较大功率的主阀级构成，流量可达到 500L/min，插装式比例阀流量更可以达到 1600L/min。

The valves can be categorized into direct- or pilot-operated control type based on the number of pressure amplification stages. The hydraulic power stage of the direct type is driven by the electro-mechanical conversion component directly. The flow of the type is usually below 15 L/min due to the limitation of the conversion components. A pilot operated proportional valve is made up of a direct proportional valve and a main valve

3.1 概述 (Summary)

stage that can produce higher power, the flow of which can reach 500 L/min, and the cartridge type can even reach as large as 1600 L/min.

按比例控制阀内含的级间反馈参数或反馈物理量的形式可以分为带反馈或不带反馈型。

The proportional valves can be categorized into with and without feedback types with regard to the forms of the feedback parameters or the feedback physical quantities that contained in the valves.

反馈型又可以分为流量反馈、位移反馈和力反馈。

The feedback signal can also be flow, displacement and force.

比例阀按其主阀芯的形式来分，又可以分为滑阀式和插装式。

According to the type of the main spool, proportional valves can also be divided into spool or cartridge type.

图 3-1 为一个闭环比例系统框图，点划线方框内为电液比例阀的组成部分。由图可以看出比例阀在系统中所处的地位以及与电控器、液压执行器之间的关系。

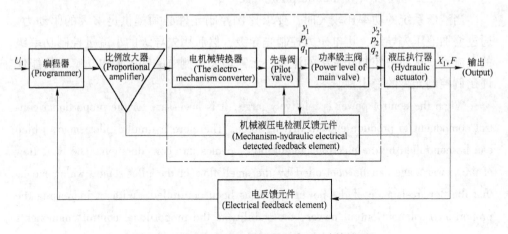

图 3-1 闭环的电液比例控制系统及比例阀框图

(Fig. 3-1 Closed-loop of electro-hydraulic proportional control system and proportional valve block diagram)

The Fig. 3-1 shows the block diagram of a closed loop proportional system. The dotted line frame indicates the components of the electro-hydraulic proportional valve. The diagram also tells the roll that the proportional valve plays in the system, its relation with the electric controller and the hydraulic actuator.

从电液比例阀的原理框图中可以看出，它主要由以下几部分组成。

From the electro-hydraulic proportional valve block diagram, the components that make up the valve includes.

A 电-机械转换元件(Electro-mechanical conversion component)

它的作用是把经过放大后的输入电流信号成比例地转换成机械量,从而控制流阻。普遍采用电磁式设计。最常见的电-机械转换元件是比例电磁铁,还有直流伺服电机、步进电机、力矩马达及动圈式马达等。比例电磁铁是电子技术和比例液压技术的连接环节。比例电磁铁产生与输入量(电流)成正比的输出量:力和位移。

Such component converts the amplified input signal into mechanical quantity proportionally to control the flow resistance. Electromagnetic technology is commonly used for such conversion, e. g., the most common electro-mechanical conversion proportional solenoid, DC servo motor, stepper motor, torque motor and moving coil motor. Proportional solenoid connects the electronic technology and the proportion hydraulic technology. The output is usually force and displacement in proportion to the input (current) produced by proportional solenoid.

B 液压先导级(Hydraulic pilot stage)

当液压系统的功率比较大时,要求比例控制元件必须提供足够大的驱动力。通过增加液压先导级,用电磁力控制先导级,然后用先导级的小流量控制功率级的大流量,从而去控制功率级的流阻。无需增大比例电磁铁的输出功率,又能保证比例控制元件的稳定性。

When the control power is relatively large, it is necessary for the proportional control component to produce enough driving force. Therefore hydraulic pilot stage, which can be controlled by the electromagnetic force, comes into consideration. The large flow of the power stage can be controlled by the small flow of the pilot stage, which means that the flow resistance of the power stage has been controlled. Without increasing the proportional solenoid output power, the stability of the proportional control components can also be ensured.

C 液压功率放大级(Hydraulic power amplifying stage)

它是连接先导级与功率级的中间环节。即利用先导级产生的液压功率去调节主功率级的流阻。针对不同的控制要求和控制参数,这个中间环节会采用不同的设计。

This stage is the connection between the pilot stage and the power stage, which means that the flow resistance of the main power stage is adjusted by the hydraulic power generated by the pilot stage. The design of such connection will be different due to different control requirement and parameters.

D 检测反馈元件(Detecting feedback component)

检测反馈元件是将液压执行器的位移、速度和力等反映运动状态的机械量通过传感器转化为电流、电压等电量的元件。通常是一些物理量传感器。

The detecting feedback component converts the mechanical quantities reflecting the states of movement, such as displacement, velocity and force of the hydraulic actuator into electric quantities such as electric current and voltage by sensors, which are usually physical sensors.

E 液压执行器(Hydraulic actuator)

液压执行器是将液压能转化为机械能的装置,目前主要指液压缸和液压马达。它们都是连接液压回路与工作机械的中间环节,不同的是液压缸产生直线运动并传递直线方向上的作用力,液压马达产生回转运动并输出转矩。

Hydraulic actuator converts the hydraulic energy into mechanical energy, such as hydraulic cylinders and hydraulic motors. Both of them are connection between the hydraulic circuit and the working machinery, but hydraulic cylinder moves linearly while hydraulic motor rotates.

3.2 电液比例压力控制阀(Electro-hydraulic proportional pressure control valve)

比例压力控制阀是用比例电磁铁控制液压传动系统整体或局部的流体压力,使系统压力与电控信号成比例变化的一种控制阀。具体地说,就是通过变化电信号的设定值,使工作系统的压力满足过程的需要。这种控制方式也称负载适应控制。

Proportional pressure control valve controls the fluid pressure of the whole or partial hydraulic system by proportional solenoid, making the system's pressure and electronic control signal change proportionally. Specifically, the pressure of the working system can be changed by the input signal to meet the requirements of the work cycle by setting the electrical signal. Such control is also called load-adapting system.

比例压力控制阀应用最多的是比例溢流阀和比例减压阀,有直动型和先导型两种。

The most popular proportional pressure valves are the proportional relief valves and the proportional pressure reducing valves. Both of them can be: direct- or pilot-controlled.

第3章 电液比例控制阀
Chapter 3　Electro-hydraulic Proportional Control Valve

3.2.1　比例溢流阀（Proportional relief valve）

比例溢流阀的结构，基本上与常规手动调节溢流阀相同。两者区别在于，比例溢流阀中用比例电磁铁取代常规阀中的调压弹簧调节装置。在先导式比例溢流阀中，还常配有手调安全阀。

The structure of the proportional relief valve is basically the same as the conventional manual adjusted relief valve. Difference between the two is that proportional relief valve replaces the spring-regulating device of the conventional valve with a proportional solenoid. A pilot-scale relief valve is often equipped with manual adjustment device.

比例溢流阀的功能较常规阀有明显的增强。在系统中不但可稳定系统压力为一定值，而且可以根据工况要求无级地快速改变系统压力。比例阀不用加二位二通电磁阀就具备了卸荷功能。比例溢流阀还可以根据需要构成闭环压力反馈控制。在与其他控制器件构成复合控制方面，例如 p-q 阀（压力控制与流量控制的组合）等，也显现出其结构紧凑、控制便利等优越性。

The function of the proportional relief valve is significantly improved compared with the conventional valve. In the system, proportional relief valve can not only stabilize the system pressure to a certain value but also change the system's pressure according to certain requirements quickly and variably. A proportional valve already has the unloading capability without adding 2-position 2-way solenoid valves. Proportional relief valve can also constitute a closed loop pressure feedback control when required. When it comes to complex integrity control with other components, such as p-q valve (pressure control and flow control in combination), the advantages of compact structure and simple control are also showed.

比例溢流阀是液压系统中重要的控制元件，其特性对系统的工作性能影响很大。它的作用为：

The proportional relief valve is an important control component for hydraulic power, its characteristics have an important impact on the system's working performance. The functions of proportional relief valve are as follows：

（1）构成液压系统的恒压源（Constituting constant pressure source of hydraulic system）。

比例溢流阀作为定压元件，当控制信号一定时，可获得稳定的系统压力；改变控制信号，可无级调节系统压力，且压力变化过程平稳，对系统的冲击小。

Using the proportional relief valve as constant-pressure components, stable system pressure can be obtained when the control signal is constant. Changing the control signal, the system pressure can be adjusted variably, and the adjustment is smooth, which

3.2 电液比例压力控制阀
(Electro-hydraulic proportional pressure control valve)

liminates the pressure shock.

（2）将控制信号置为零，即可获得卸荷功能。此时液压系统不需要压力油，油液通过主阀口低压流回油箱。

Setting the control signal at zero, the system can be unloaded. In such way, the pressurized oil is lead to the tank via the main valve port.

（3）比例溢流阀可方便地构成压力负反馈系统，或与其他控制元件构成复合控制系统。

The proportional relief valve can easily constitute a pressure negative feedback system, or a integrated control system combined with other control components.

3.2.1.1 直动型比例溢流阀(Directly operated proportional relief valve)

A 普通直动溢流阀 (Ordinary direct proportional relief valve)

直动型比例溢流阀结构及工作原理如图 3-2 所示。它是双弹簧结构的直动型溢流阀，与手调式直动型溢流阀功能完全相同。其主要区别是用比例电磁铁取代了手动的弹簧力调节组件。

The structure and operating principle of the direct proportional relief valve are shown in the Fig. 3-2. It is a direct proportional relief valve with double springs, which has the same function with the manual relief valve. But they are actuated by the proportional solenoid.

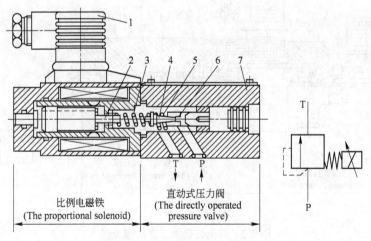

图 3-2 直动式比例溢流阀

(Fig. 3-2 Direct proportional relief valve)

1—插头 (Plug); 2—衔铁推杆 (Keeper pusher); 3—传力弹簧 (Force-transferring spring);
4—锥阀芯 (Conical valve core); 5—防振弹簧 (Anti-vibration spring); 6—阀座 (Valve seat);
7—阀体 (Valve body)

图 3-2 所示的直动式比例溢流阀的工作原理为：改变阀的电流会使衔铁推杆

2对传力弹簧产生的作用力按比例产生相应的变化,传力弹簧对锥阀芯的作用力也按比例产生相应地变化,从而按比例地改变了 P 口处的溢流压力,也就达到了通过改变比例阀的电流而按比例地改变 P 口溢流压力的目的。

The working principle of the direct proportional relief valve shown in Fig. 3-2 is that changing the current of the valve will proportionally change the force between the keeper pusher and the force-transferring spring accordingly, which will make a proportional change of the force between the force-transferring spring and the cone spool accordingly too. Then the relief pressure at port P will change proportionally. So the function of changing the relief pressure proportionally at port P by changing the input current of the valve is achieved.

B 带位置反馈的直动溢流阀（Direct relief valve with position feedback）

带位置反馈的直动溢流阀（图 3-3）包括力控制型比例电磁铁 4,以及由阀体 10、阀座 11、锥阀芯 9、弹簧 7 等组成的液压阀本体。输入电信号时,比例电磁铁 4 产生相应电磁力,通过弹簧 7 作用于阀芯 9 上。电磁力对弹簧预压缩,预压缩量决定了溢流压力。预压缩量正比于输入电信号,溢流压力正比于输入电信号,实现了对压力的比例控制。

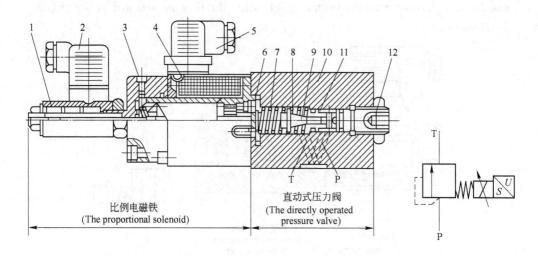

图 3-3 带位置反馈的直动溢流阀

(Fig. 3-3 Direct relief valve with position feedback)

1—位移传感器（Displacement sensor）; 2—传感器插头（Sensor plug）;
3—放气螺钉（Bleed screws）; 4—比例电磁铁（Proportional solenoid）;
5—线圈插头（Coil plug）; 6—弹簧座（Spring seat）; 7—调压弹簧（Regulator spring）;
8—防振弹簧（Anti-vibration spring）; 9—锥阀芯（Conical valve core）; 10—阀体（Valve body）;
11—阀座（Valve seat）; 12—调节螺塞（Adjustable plug screw）

3.2 电液比例压力控制阀
(Electro-hydraulic proportional pressure control valve)

The direct relief valve with position feedback (Fig. 3-3) is made up of the proportional solenoid of force control 4 and the hydraulic valve which is composed of the valve body 10, seats 11, conical valve core 9 and spring 7. Once receiving electrical signal, proportional solenoid 4 generates corresponding force acting on conical valve core 9 via spring 7. The relief pressure depends on the pre-compression which is provided by electromagnet strength. The pre-compression is in proportion with the input signal and the relief pressure is consequently in direct ratio with the input signal. The pressure control can be obtained proportionally.

之所以在动杆上加位移传感器，是因为电磁铁有水平吸力特性。即在给定电流后，动铁可以在工作行程内移动，而输出力不变。这样推杆推着的阀芯与对应的阀座之间构成可变液阻。加上与电磁铁动铁固联的位移传感器后，随时检测动铁位移并反馈至带 PID 控制单元的电控器，用反馈信号对输入信号的偏差值对电磁铁进行控制，使动铁继续移动，直至偏差值为零，构成动铁位移的闭环控制，使弹簧 7 得到与输入信号成比例的精确压缩量，使阀达到更小的磁环和更高的控制精度。

The reason for combining a displacement sensor to the moving pole is that the solenoid has the characteristic of constant suction, which means that the solenoid can move at any stroke with a constant output force when it is energised. So the valve core pushed by the push rod and the corresponding valve seat produce variable fluid resistance. The displacement of the electromagnetic moving iron is detected at any time by fixing the displacement sensor at the moving iron and the feedback is sent to the electric controller of the PID unit. The solenoid is controlled by the deviation of the feedback and the input. It will not stop moving until the deviation becomes zero. So the moving iron is controlled by the displacement-controlling close loop and compression of the spring 7 is precisely proportional to the input signal. So the valve will have a much smaller magnetic ring and higher control precision.

普通溢流阀采用不同刚度的调压弹簧改变压力等级。由于比例电磁铁的推力与电流成正比，比例溢流阀通过改变阀座 11 的孔径而获得不同的压力等级。阀座孔径小，控制压力高，流量小。

The common relief valve use adjusted spring with different stiffness to change the grade of pressure. The proportional relief valve achieves different grades of pressure by changing the diameter of passage in valve seat 11, due to the fact that the pushing force of the proportional solenoid is proportional to the current. The smaller the diameter is, the higher the control pressure can be, but correspondingly the flow is less as well.

调节螺塞 12 可在一定范围内调节溢流阀的工作零位。

Regulating plug screw 12 can regulate the initial position of the relief valve within a certain range.

直动型比例溢流阀在小流量场合下单独做调压元件，更多的是做先导型溢流阀或减压阀的先导阀。

The direct proportional relief valve can work alone under small flow as a pressure regulating element, but more often as a pilot stage for pilot operated relief or pressure reducing valve.

3.2.1.2　先导型比例溢流阀(Pilot operated proportional relief valve)

A　普通先导型比例溢流阀（Ordinary pilot operated proportional relief valve）

先导型溢流阀用比例电磁铁取代先导阀的调压手柄，便成为先导型比例溢流阀。

It will be a pilot operated proportional relief valve when using a proportional solenoid instead of the pressure regulating handle of the pilot valve of a pilot operated relief valve.

如图3-4所示，它属于带力控制型比例电磁铁的比例溢流阀。这种阀是在两极同心式手调溢流阀结构的基础上，由手调直动式溢流阀更换为带力控制型比例电磁铁的直动式比例溢流阀而来的。显然，除先导级采用比例压力阀外，其余与两极同心式普通溢流阀的结构相同，属于压力间接检测型先导式比例溢流阀。

As shown in Fig. 3-4, it is the proportional relief valve with a proportional force-controlling solenoid. This kind of valve is further developed by replacing the bipolar concentric manual structure of the manual direct-acting relief valve with force-controlling proportional solenoid. Obviously, the valve has the same structure with the bipolar concentric relief valve except that the pilot stage of the valve is a proportional relief valve. This valve is a kind of pilot operated proportional relief valve whose pressure is detected indirectly.

这种先导式比例溢流阀的主阀采用了两极同心式锥阀结构，先导阀的回油通过卸油口直接流回油箱，以确保先导阀回油背压为零。否则，如果先导阀的回油背压不为零（如与主回油口连接在一起），该回油压力就会与比例电磁铁的力叠加在一起，主回油压力的波动就会引起主阀压力的波动。

The main valve of this kind of pilot operated proportional relief valve adopts the structure of cone valve with a bipolar concentric design. The oil in the pilot valve is directly lead to the oil tank through the unloading port, which ensures that the back pressure of output port is zero. Otherwise, if the back pressure of the returned oil is not zero (if connected to the main drain port), the returned oil pressure will be superimposed together on the force of the proportional solenoid. The main returned oil pressure fluctua-

3.2 电液比例压力控制阀
(Electro-hydraulic proportional pressure control valve)

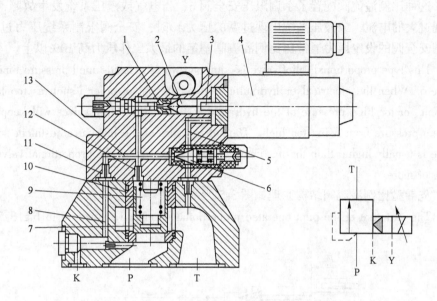

图 3-4 普通先导型比例溢流阀
(Fig. 3-4 Ordinary pilot operated proportional relief valve)
1—阀座(Valve seat); 2—先导锥阀(Pilot poppet valve); 3—衔铁推杆(Keeper pusher);
4—比例电磁铁(Proportional solenoid); 5—内泄油道(Internal leakage duct);
6—安全阀(Safety valve); 7—主阀芯(Main valve core); 8—先导油道(Pilot oil duct);
9—复位弹簧(Resetting spring); 10—主阀阀座(Seat of main valve);
11, 12—节流器(Orifice hole); 13—控制油道(Control oil duct)

tions can cause pressure fluctuation in the main stage.

主阀进油口压力作用于主阀芯 7 的底部，同时也通过控制油道 13（含节流器 11、12）作用于主阀芯 7 的顶部。当液压力达到比例电磁铁的推力时，先导锥阀 2 打开，先导油通过卸油口流回油箱，并在节流器 11、12 处产生压降，主阀芯 7 因此克服弹簧力上升，系统多余流量通过主阀口流回油箱，压力因此不会继续升高。

The main valve inlet pressure is applied to the bottom of the main valve core 7, and it also acts on the top of the main valve core 7 through the control passage 13 (including orifice 11, 12). When the fluid pressure reaches the pushing force of the proportional solenoid, the pilot conoid valve 2 will open. Then the pilot oil will return to the tank through the unloading port and the pressure drop will be caused at the orifice 11, 12. Thereby the main spool 7 rises since it overcomes the spring force and the excess flow of the system goes back into the oil tank through the main outlet port. Therefore, the pressure will not continue to rise.

这种比例溢流阀配置了手调限压安全阀6,当电气或液压系统发生故障(如出现过大的电流,或液压系统出现过高的压力)时,安全阀限制系统压力过高。手调安全阀的设定压力通常比比例溢流阀调定的最大工作压力高10%以上。

This type proportional relief valve is equipped with the hand-tuned pressure limiting valve 6. When the electrical or hydraulic systems are out of order (such as too large current, or too high pressure of the hydraulic system), the safety valves will keep the system pressure from being too high. The set pressure of the manual adjustment safety valve is usually higher than the maximum operating pressure of the proportional valve by 10% or more.

先导型比例溢流阀结构原理如图3-5所示。

The structure of the pilot operated proportional relief valve is shown in Fig. 3-5.

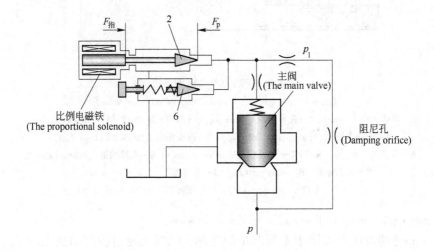

图 3-5 先导型比例溢流阀结构原理图

(Fig. 3-5 The structure of the pilot operated proportional relief valve)

先导型比例溢流阀下部与普通溢流阀的主阀相同,上部则为比例先导压力阀。该阀还附有一个手动调整的安全阀(先导阀)6,用以限制比例溢流阀的最高压力。

The lower part of the pilot operated proportional valve is the same as the main stage of the common relief valve, while the headpiece is the pilot proportional pressure valve. A safty valve (pilot valve) 6 which is manually regulated is affiliated to this valve to limit the maximum pressure of the relief valve.

B 带位置反馈先导型比例溢流阀 (Pilot operated proportional relief valve with position feedback)

带位置反馈先导型比例溢流阀结构如图3-6所示。它的上部为行程控制型直

3.2 电液比例压力控制阀
(Electro-hydraulic proportional pressure control valve)

动型比例溢流阀,下部为主阀级。当比例电磁铁2输入指令信号电流时,它产生一个相应的力经压缩弹簧4作用在锥阀5上。压力油经A输入主阀,并经主阀芯7的节流螺塞8到达主阀弹簧9腔,经通路a、b到达先导阀阀座6,并作用在锥阀芯5上。若A口压力不能使先导阀打开,主阀芯7的左、右腔压力保持相等,主阀芯7保持关闭;当系统压力超过比例电磁铁2的设定值,先导阀芯5开启,先导油经c从B口流回油箱。主阀芯右腔(弹簧腔)的压力由于节流螺塞8的作用下降,导致主阀芯7开启,则A口与B口接通回油箱,实现溢流。

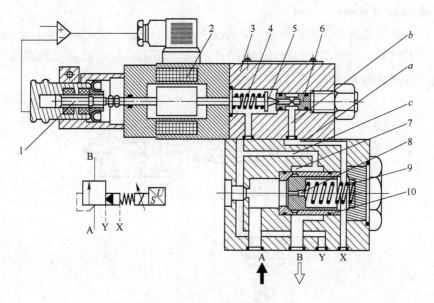

图 3-6 带位置调节型比例电磁铁的先导型比例溢流阀

(Fig. 3-6 Pilot operated proportional relief valve with the electromagnet of the location regulation)

1—位移传感器(Displacement sensor);
2—行程控制型比例电磁铁(Stroke-controlled proportional solenoid);
3—阀体(Valve body);4—弹簧(Spring);5—锥阀芯(Conoid valve core);
6—阀座(Valve seat);7—主阀芯(Main valve core);8—节流螺塞(Throttling plug screw);
9—主阀弹簧(Main valve spring);10—主阀座(Main valve seat)

The structure of the pilot operated proportional relief valve with position feedback is showed in Fig. 3-6. Its upper part is the stroke-controlled direct proportional relief valve, and the lower part is the main stage. When the solenoid 2 receives an input signal, it produces a corresponding force through the compressed spring 4 acting on the conoid valve 5. The pressure oil flows into the main valve. The oil reaches the chamber of the main valve spring 9 through the throttling plug screw 8 of the main valve spool 7, then it reaches the pilot valve seat 6 via path a, b, finally acting on the conical valve core 5. If the pilot valve seat 6 can not be opened by the pressure of port A, the main valve 7 will remain closed as the pressures of the left chamber and right chamber of the main valve spool are the same; when the system pressure exceeds the value set by the proportional solenoid 2, the pilot valve core 5 opens. The pilot oil flows back to the tank via c from port B. The pressure in the right chamber (spring chamber) of main valve spool will decline as the effect of the throttling plug screw 8, leading to the opening of the main valve 7, port A and port B are connected and pressurized oil flows back to tank, thus achieving the aim of pressure relief.

C 直接检测式比例溢流阀 (Proportional relief valve of direct detection)

直接检测式比例溢流阀结构如图 3-7 所示。测压面 a_0 检测 p_s，形成反馈力信号与衔铁的输出力 F_m 直接进行比较。

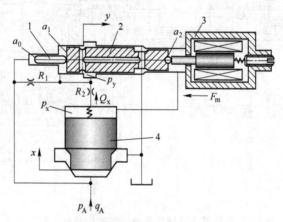

图 3-7 直接检测式比例溢流阀

(Fig. 3-7 Proportional relief valve of direct detection)

1—推杆 (Pusher); 2—先导阀阀芯 (Pilot valve spool); 3—比例电磁铁 (Proportional solenoid);
4—主阀阀芯 (Main valve core)

The structure of the proportional relief valve of direct detection is shown as Fig. 3-7. p_s detected by the pressure measuring surface a_0 produces the feedback signal which is compared directly with the output force F_m of the keeper.

当指令信号经比例放大器进行功率放大后，输给比例阀的比例电磁铁，比例电磁铁产生一个相应的输出力作用在先导阀阀芯 2 的右端。同时，压力油通过旁通道进入推杆左端，压力作用在其端面 a_1 上。推杆产生的力与比例电磁铁推力分别作用在阀芯 2 的两端，方向相反。

After amplified by the proportional amplifier, the command signal will be sent to the proportional solenoid of the proportional valve; then the proportional solenoid will generate a corresponding output force acting on the right side of the pilot valve spool 2.

3.2 电液比例压力控制阀
(Electro-hydraulic proportional pressure control valve)

At the same time, pressure oil flows into the left side of the push rod through bypass channel. The pressure of oil acts on surface a_1. The thrust generated by the push rod and the proportional solenoid respectively acting on both ends of valve core 2 in the opposite directions.

若 p_A 小于设定压力,先导阀芯 2 在比例电磁铁推动下处于左端,先导阀 2 的回油通道被关闭;主阀芯 4 两端的压力相等,在弹簧的作用下处于关闭状态。

If p_A is less than the preset pressure, the pilot valve spool 2 driven by the proportional solenoid will be located at the left side, the oil return channel of the pilot valve 2 will be closed. The pressure at both ends of the main valve core 4 is equal, thus the main valve spool will block the pilot connection.

若 p_A 升至大于比例电磁铁设定压力,先导阀芯 2 在 $p_A \times a_1$ 作用下向右移动,先导阀 2 的回油通道被打开,主阀上端压力油 p_x 通过 R_2 进入先导阀的回油通道,产生压力降,导致主阀上下压力不平衡而向上移动,压力油被维持在设定的压力值上。因为压力油 p_A 是直接通过推杆 1 与比例电磁铁相作用的,故称为直接检测式。

If the p_A is larger than the set pressure of the proportional solenoid, the pilot valve spool 2 driven by the force $p_A \times a_1$ will move to the right side, then the oil return channel of pilot valve 2 will open, and the pressure oil p_x in the top of the main valve will flow into the oil return channel of the pilot valve through R_2, resulting pressure drop which leads the difference between the upper and lower pressure of the main valve, which makes the main valve move up, and the oil pressure remains at the setting pressure value. The pressure oil p_A acts directly on the proportional solenoid through push rod 1, so it is called direct detection type.

3.2.2 比例减压阀 (Proportional pressure reducing valve)

比例减压阀是使阀的出口压力与进口压力的差连续地或按比例地随电信号而变化的压力控制阀,可实现按控制要求精确降低系统某一支路的油液压力,使同一系统有两个或多个不同压力。其减压原理是利用油液在某个地方的压力损失,使出口压力与进口压力的差值按比例变化,并保持出口压力恒定,故又称定值减压阀。

Proportional pressure reducing valve is the pressure control valve between whose outlet and inlet the pressure difference changes continuously or in proportion to the signal. It can precisely reduce the oil pressure of a branch according to the control requirements, so the same system can have two or more different pressures. By making use of the pressure loss somewhere, the pressure difference between outlet pressure and inlet pressure change proportionally according to the requirements of the electrical signal, and maintaining a constant outlet pressure. It is also known as the constant pressure reducing valve.

3.2.2.1 先导型比例减压阀(Pilot proportional pressure reducing valve)

先导型比例减压阀与先导型比例溢流阀工作原理基本相同。它们的先导阀完全一样，不同的只是主阀级。溢流阀采用常闭式锥阀，减压阀采用常开式滑阀，如图3-8所示。

The operational principles of the pilot pressure reducing valve and the pilot operated relief valve are basically similar. Their pilot valves are the same, while the main stage are different. The relief valve takes the poppet form, which is normally close, while the pressure reducing valve is always slide valve which usually switches on. It is just showed in the Fig. 3-8.

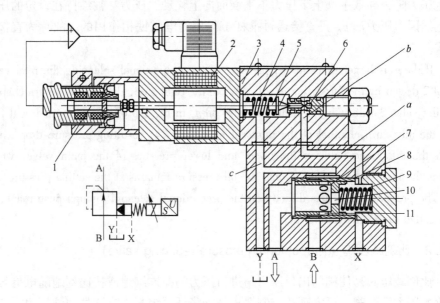

图 3-8 先导型比例减压阀

(Fig. 3-8 Pilot pressure reducing valve)

1—位移传感器（Displacement sensor）；
2—行程控制型比例电磁铁（Stroke-controlled proportional solenoid）；3—阀体（Valve body）；
4—弹簧（Spring）；5—先导锥阀芯（Pilot conoid valve core）；6—先导阀座（Pilot valve seat）；
7—主阀芯（Main valve core）；8—阀套（Valve sleeve）；9—主阀弹簧（Main valve spring）；
10—节流螺塞（Throttling plug screw）；11—减压节流口（Decompressing regulator orifice）

比例电磁铁接受指令电信号后，输出相应电磁力，经弹簧4将先导锥阀芯5

3.2 电液比例压力控制阀
(Electro-hydraulic proportional pressure control valve)

压在阀座 6 上。由 B 进入主阀的一次压力油,经减压节流口 11 后的二次压力油,经主阀芯 7 的径向孔经 A 口输出;二次压力油同时经阀芯 7 上的节流螺塞 10 至主阀芯弹簧腔(右腔)、通路 a、先导阀座 6 作用于先导阀芯 5 上。若二次压力不能使先导锥阀芯 5 开启,则主阀芯左、右两腔压力相等。在弹簧 9 的作用下,减压节流口 11 为全开状态,B→A 流向不受限制。当二次压力超过比例电磁铁设定值时,先导锥阀芯 5 开启,液压油经 c、Y 口泄回油箱。由于节流螺塞 10 的作用,主阀芯弹簧腔压力下降,主阀芯左、右两腔的压差使主阀芯克服弹簧 9 之作用,使减压节流口 11 关小,使二次压力降至设定值。为防止二次压力过高,可在 X 口接一手动式直动型溢流阀起保护作用。

When the proportional electromagnet receives the electrical directive signal, it can produce a corresponding force, which keeps the pilot poppet 5 at valve seat 6 through the spring 4. The original pressure oil that enters the main stage via B and the pressure oil that enters via decompressing regulator orifice 11 flow out from A through the radial orifice of the main valve spool 7. Meanwhile, the secondary pressure oil enters the spring chamber of the primary valve core (right), channel a and the poppet valve seat 6 acting on the pilot valve core 5 through the throttling plug screw 10 which is on the valve core 7. If the secondary pressure can not open the pilot conical valve core 5, the pressure in the left and right chambers of primary valve center would be equivalent. Because of the spring 9, the decompressing regulator orifice 11 is completely open, and the oil can flow from B to A without any restriction. If the secondary pressure exceeds the setting value of the proportional solenoid, the pilot valve core 5 will be open and the oil will run back to the tank via c and Y. Because of the throttling plug 10, the pressure which is in the spring chamber of the primary valve core descends, the differential pressure of left and right chamber of the primary valve core diminishes decompressing regulator orifice 11 and makes the secondary pressure get to the setting point by overcoming the spring 9. In order to limit the secondary pressure, a manual directly operated relief valve can be fitted on the mouth of X for protecting.

带限压阀的先导比例减压阀工作原理如图 3-9 所示。它的先导级是一个由力控制型比例电磁铁操纵的小型溢流阀,其主阀级也与手调式减压阀一样。事实上,限压阀 2 和主阀 3 就构成了一个先导手调减压阀,因此减压阀的调定压力值是由先导阀芯所处的位置决定的;而最高压力由限压阀 2 调定。

The operational principle of the pilot proportional pressure reducing valve with a pressure limiting valve is described as in Fig. 3-9. Its pilot-stage is a small proportional relief valve which is manipulated by the force-controlled proportional solenoid and its main valve is also the same as the hand-tuning reducing valve. In fact, the pressure limiting valve 2 and the main valve 3 constitute a pilot manual adjusted reducing valve. Therefore, the set pressure value of the reducing valve is determined by the location of pilot valve core; while its maximum pressure is adjusted by the pressure limiting valve 2.

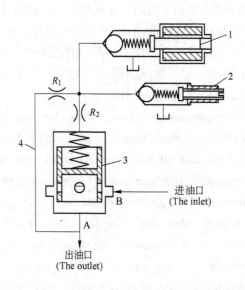

图 3-9 带限压阀的先导比例减压阀工作原理图
(Fig. 3-9 The operational principle of the pilot proportional pressure reducing valve with a pressure limiting valve)
1—比例溢流阀先导级 (The pilot grade of the proportional relief valve);
2—限压阀 (The pressure restrictive valve); 3—主阀 (Main valve);
4—先导油流道 (The fluid channel of pilot oil)

当阀接收到输入信号，比例电磁铁产生的电磁力直接作用在先导阀芯上。只要电磁力使阀芯保持关闭，先导油就处于静止状态。先导油从出口压力油（二次压力油）经通道 4 作用在主阀芯的上、下端面上。因主阀芯上下面积相等，所以主阀芯保持液压力平衡，一个很小的弹簧力保持主阀开启。当出口油的压力超过电磁力时，先导阀开启，先导油直接流回油箱。这导致在节流孔 R_1 处产生压力降，使主阀芯失去平衡而向上移动。这减小了油口 B 到油口 A 的通流面积（经过阀套和主阀芯上的径向孔），于是产生减压作用，在油口 A 处降为二次压力。主阀芯的调节作用，使油口 A 的压力保持在比例电磁铁的设定值上。

When the valve receives a input signals, the electromagnetic force generated by the proportional solenoid has a direct effect on the pilot valve core. As long as the electromagnetic force keeps the valve core closed, the pilot channel is cut. Pilot pressure oil

3.2 电液比例压力控制阀
(Electro-hydraulic proportional pressure control valve)

from export (secondary pressure oil) through channel 4 acts on the upper and down end of the main valve spool. As the upper area and the lower area of the main valve spool are equal, the main valve core is kept in balance still, and the small spring force keeps the main valve open. When the pressure of exported oil exceeds the electromagnetic force, the pilot valve is opened and the pilot oil directly runs into tank. This leads to a pressure drop generated at the orifice R_1, then the main valve spool is out of balance and moves upwards, which reduces the flow area (via the radial holes of the valve sleeve and the main valve core) from port B to port A, then the pressure is reduced to the secondary pressure at the port A. Due to the regulation of the main valve core, the pressure of port A can maintain the set pressure of the proportional solenoid.

3.2.2.2 直动式三通比例减压阀(Direct operated three-way proportional pressure reducing valve)

直动式三通比例减压阀是利用减压阀增大的出口压力来控制出油口和回油口的导通，达到精确控制出口压力，并保护执行元件的目的。直动式三通比例减压阀用作多级。

The direct operated three-way proportional pressure reducing valve uses the increased outlet pressure in the pressure reducing valve to control the connection between the outlet and tank port in order to achieve precise control of the outlet pressure and protect the actuator. The direct operated three-way proportional pressure reducing valve is often used in a multistage design.

直动式三通比例减压阀如图 3-10 所示。

The structure of the direct operated 3-way proportional pressure reducing valve is showed in Fig. 3-10.

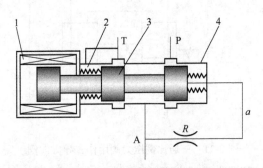

图 3-10 直动式三通比例减压阀

(Fig. 3-10 Directly operated three-way proportional pressure reducing valve)

1—比例电磁铁 (Proportional solenoid); 2—对中弹簧 (Centering spring);
3—阀芯 (Valve core); 4—阀体 (Valve body)

第 3 章 电液比例控制阀
Chapter 3　Electro-hydraulic Proportional Control Valve

无信号电流时，阀芯 3 在对中弹簧 2 作用下处于中位，P、T、A 各油口互不相通。比例电磁铁接收信号电流时，电磁力使阀芯 3 右移，P、A 接通，油口 A 输出的二次压力油输入到执行元件。二次压力油又经阀体通道 a 反馈到阀芯右端，作用于右端的油液压力与电磁力方向相反。二次压力与电磁力平衡时，滑阀芯 3 返回中位，A 口压力保持不变，并与电磁力成正比例。若对阀芯的作用力大于电磁力，阀芯移至左端，A 口与 T 接通，压力下降，直至新的平衡。三通比例减压阀可以控制二次压力油的压力和方向，成对使用时，用作比例方向阀的先导阀，如图 3-11 所示。

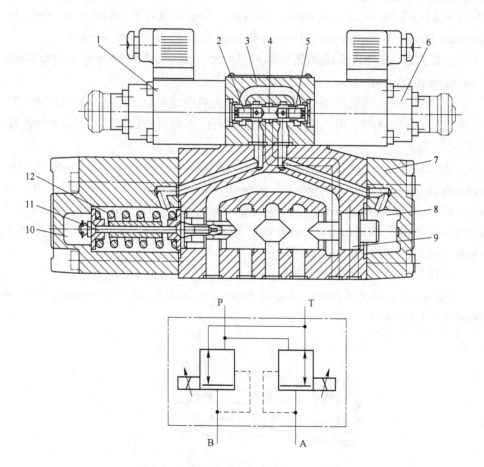

图 3-11　三通比例压力阀用作先导方向阀
(Fig. 3-11　Three-way proportional pressure reducing valve for the pilot directional valve)
1，6—比例电磁铁（Proportional solenoid）；2，5—压力测量（Pressure measurement）；
3—阀体（Valve body）；4—阀芯（Valve core）；7—主阀（Main valve）；
8，10—主阀腔（Main valve cavity）；9—主阀芯（Main valve core）；
11—连杆（Connecting rod）；12—弹簧（Spring）

3.2 电液比例压力控制阀
(Electro-hydraulic proportional pressure control valve)

When there is no input signal current, the valve spool 3 is located in the neutral position as a result of the centering spring 2, there is no connection between the port P, T and A. When the electromagnet receives signal current, the electromagnetic force pushes the valve core 3 towards right and connects P to A, and the reduced pressure oil flows from port A into the executive element. Then the secondary oil is fed back to the right of the valve core via the channel a of the valve body. The direction of the pressure acting on the right end is opposite to the direction of electromagnetic force. When the secondary pressure is equal to the electromagnetic force, the valve core 3 moves to the neutral position which keeps the pressure of the outlet A steady, and pressure is in direct proportion with the electromagnetic force. If the force which acts on the valve spool is higher than the electromagnetic force, the valve core is pushed towards left and connects A with T, so the pressure declines until getting a new balance. The three-way pressure reducing valve can control the pressure and the direction of the secondary pressure oil. When used in pairs, they might be the pilot valve of the proportional direction valve, as shown in the Fig. 3-11.

3.2.2.3 先导式三通比例减压阀(Pilot three-way proportional pressure reducing valve)

当调定放大器输入电压后，比例电磁铁输出电磁力，此时阀输出压力有一个相应值。若因某种干扰使出口压力降低，将引起先导阀芯向左移动，左边可变节流口增大，右边可变节流口减小，先导阀腔及主阀上腔压力上升，在上腔压力上升和下腔压力下降的共同作用下，主阀芯向下运动，主阀可变节流口开大，致使出口压力上升，这样就使输出压力保持在调定值。当输出压力增大超过调定值时（如用于动力负载时）先导阀芯向右移动，先导阀左边可变节流口减小，右边可变节流口增大，致使先导阀腔内压力和主阀上腔压力下降，主阀芯上移使进油口与出油口相通，此时相当于溢流阀，如图3-12所示。

After setting the input voltage of the enlarger, the proportional solenoid generates electromagnetic force and a corresponding output pressure value can be produced. The valve spool might move left if there are some interferences which can make the output pressure downwards. Then the left variable regulating orifice increases, while the right one decreases, which makes the pressure of the chamber of the pilot valve and the upper chamber of the primary valve increase. Under the joint effect of the ascending and descending of the pressure, the primary valve core moves down, the variable regulator orifice of the primary enlarges, so the pressure rises, which keeps the output pressure at the setting point. However, when the output pressure is higher than the set point (such as used it in the dynamic load), the pilot valve move towards the right and the left variable regulator orifice turns down, whereas the right one turn up, which makes the pressure of the chamber of the pilot valve and the upper chamber of the main stage decline, so the outlet connects with the inlet through the main valve core moving up, which acts as the relief valve, as shown in the Fig. 3-12.

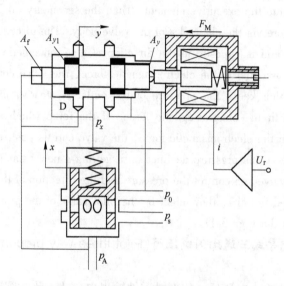

图 3-12　先导式三通比例减压阀工作原理图

(Fig. 3-12　The operational principle of the pilot three-way proportional pressure reducing valve)

通过对三通减压阀的工作原理的研究，得出以下推论：

By studying the operational principle of the three-way pressure reducing valve, the followings can be concluded：

（1）在稳态时，先导阀和主阀阀芯的力平衡方程。

At the steady state, the force balance equation between the pilot valve spool and the main valve spool can be established.

先导阀稳态时的力平衡方程为（The force balance equation of the pilot valve that is at steady state is）：

$$p_A A_f + p_x A_{y1} = F_M + F_{ky} + p_x A_y \tag{3-1}$$

式中　p_A——三通减压阀的输出压力（The outputting pressure of the three-way pressure reducing valve）；

p_x——先导阀腔内压力（与主阀上腔压力相等）（The inner pressure in the chamber of the pilot valve (that is equal to the pressure of the upper chamber））；

F_M——电磁力（The electromagnetic force）；

3.2 电液比例压力控制阀
(Electro-hydraulic proportional pressure control valve)

F_{ky} ——先导阀阀芯所受弹簧力(The spring force which acts on the pilot valve core);

A_f ——先导阀反馈推杆面积(The area of feedback push rod of the pilot valve);

A_{y1} ——先导阀阀芯左端面积(The left area of the pilot valve spool);

A_y ——先导阀阀芯右端面积(The right area of the pilot valve spool)。

主阀稳态时的力平衡方程为(The force balanced equation of the main valve at the steady state is):

$$p_A A = F_{kx} + p_x A \qquad (3-2)$$

式中 p_A ——三通减压阀的输出压力(The outputting pressure of three-way pressure reducing valve);

p_x ——先导阀腔内压力(与主阀上腔压力相等)(The inner pressure in the chamber of the pilot valve (that is equal to the pressure of the upper chamber));

F_{kx} ——主阀阀芯所受弹簧力(The spring force which acts on the main valve core);

A ——主阀阀芯上下腔的面积(The area of the main valve chamber)。

(2) 在输出压力 p_A 有变化时,假设 p_A 升高了 Δp_A,那么 p_x 随着 D 口的减小降低 Δp_x,那么则有以下的动态力平衡方程。

When the pressure p_A changes, if p_A gets the increment Δp_A, p_x decreases by Δp_x with the decline of outlet D, so there exists a dynamic force balanced equation.

先导阀动态时的力平衡方程为(The dynamic force balanced equation of the pilot valve is):

$$(p_A + \Delta p_A) A_f + (P_x - \Delta p_x) A_{y1} + m_y \frac{d\Delta y^2}{d^2 t} = F_M + k_y (y + \Delta y) + (P_x - \Delta P_x) A_y \qquad (3-3)$$

式中 Δp_A ——三通减压阀的输出压力的增量(The outputting pressure increment of the three-way pressure reducing valve);

Δp_x ——先导阀腔内压力增量(Pilot valve chamber pressure increment);

y ——先导阀弹簧原压缩量(D 口原有开度)(The original compression of the pilot valve spring (the original opening of D port));

Δy ——先导阀阀芯位移(The displacement of pilot valve core);

k_y ——先导阀弹簧刚度(The stiffness of pilot valve spring)。

主阀动态时的力平衡方程为(The dynamic force balance equation of the main valve is):

$$(p_A + \Delta p_A) A + m_x \frac{d\Delta x^2}{d^2 t} = k_x(x + \Delta x) + (p_x - \Delta p_x) A \tag{3-4}$$

式中 x ——主阀弹簧原压缩量（The original compression of the main spring）；

Δx ——主阀阀芯位移（The displacement of the main valve）；

k_x ——主阀弹簧刚度（The spring rigidity of the main valve）。

将式（3-1）代入式（3-3）可得（After substituting the equation (3-1) into the equation (3-3), the following can be concluded）：

$$\Delta p_A A_f - \Delta p_x A_{y1} + m_y \frac{d\Delta y^2}{d^2 t} = k_y \Delta y - \Delta p_x A_y \tag{3-5}$$

将式（3-2）代入式（3-4）可得（After substituting the equation (3-2) into the equation (3-4), the following can be concluded）：

$$\Delta p_A A + m_x \frac{d\Delta x^2}{d^2 t} = k_x \Delta x - \Delta p_x A \tag{3-6}$$

用增量代替变量，可得（Substituting incremental quantity for variable quantity, yields）：

$$p_A A_f - p_x A_{y1} + m_y \frac{dy^2}{d^2 t} = k_y y - p_x A_y \tag{3-7}$$

由阀的尺寸可知（According to the dimensions of the valve, it can be known）：

$$p_A A + m_x \frac{dx^2}{d^2 t} = k_x x - p_x A \tag{3-8}$$

由阀的设计尺寸可知：$A_f = A_{y1} - A_y$，则式（3-7）可化简为（According to the design dimensions of the valve, it can be found that $A_f = A_{y1} - A_y$, the equation (3-7) can be simplified as）：

$$p_A A_f - p_x A_f + m_y \frac{dy^2}{d^2 t} = k_y y \tag{3-9}$$

（3）D 口的流量增量方程为（The increment flow equation in port D）：

$$\Delta q_D = A_{y1} \frac{d\Delta y}{dt} + A \frac{d\Delta x}{dt} - A_y \frac{d\Delta y}{dt} = C_d W \Delta y \sqrt{\frac{2\Delta p_A}{\rho}} \tag{3-10}$$

式中 Δq_D ——D 口的流量增量（The incremental quantity of flow in the port D）；

C_d ——D 口的流量系数（The flowing coefficient of the port D）；

W ——伺服阀窗口的面积梯度（The area gradient of the servo valve port）。

用变量表示为（The above can be described by incremental quantity as）：

$$q_D = A_{y1} \frac{dy}{dt} + A \frac{dx}{dt} - A_y \frac{dy}{dt} = C_d W y \sqrt{\frac{2 p_A}{\rho}} \tag{3-11}$$

将式（3-10）线性化可得（Linearizing the equation (3-10)）：

3.2 电液比例压力控制阀
(Electro-hydraulic proportional pressure control valve)

$$q_D = K_q y - K_c p_A \tag{3-12}$$

公式中(In the equation):

$$K_q = \frac{\partial q_D}{\partial y} = C_d W \sqrt{\frac{2p_A}{\rho}} \qquad K_c = -\frac{\partial q_D}{\partial p_A} = -\frac{C_d W y \sqrt{\frac{2p_A}{\rho}}}{p_A}$$

对式(3-8)、式(3-9)、式(3-12)进行拉氏变换可得(By conducting the Laplace transform of the equation (3-8), (3-9) and (3-12), the following can be found):

$$P_A A + m_x X s^2 = k_x X - P_x A \tag{3-13}$$

$$P_A A_f - P_x A_f + m_y Y s^2 = k_y Y \tag{3-14}$$

$$A_f s Y + A s X = K_q Y - K_c P_A \tag{3-15}$$

联立以上三式,可得三通减压阀主阀芯位移变化量与输出压力变化量之间的传递函数:

The transfer function can be established by combining the above three equations, between the variable quantity of displacement and the outputting pressure of the spool of the three-way pressure reducing valve:

$$\frac{X}{P_A} = \frac{K_c A m_y s^2 - 2A_f^{\,2} A s + 2K_q A_f A - K_c k_y A}{(m_x A_f^2 - m_y A^2) s^3 + K_q m_x A_f s^2 + (k_y A^2 - k_x A_f) s + K_q k_x A_f} \tag{3-16}$$

液压缸上腔的压力是由三通减压阀调定的,那么忽略管路压降,则三通减压阀主阀芯位移变化量与输出压力变化量之间的传递函数也就是主阀芯位移变化量与液压缸上腔压力之间的传递函数。

The pressure in the upper chamber of hydraulic cylinder is set by the three-way pressure reducing valve. If the loss of pressure in the pipeline is neglected, the above transferring function is also the transferring function between the variable quantity of displacement of the primary valve spool and the pressure of the upper chamber of hydraulic cylinder.

由式(3-16)可见,当液压缸上腔的压力由于液压缸的位移或速度变化而产生扰动时,三通减压阀可以对其进行调节,也就是主阀芯的上下移动起到了稳定压力的作用。至于其响应情况,需对此系统进行仿真。

According to the equation (3-16), when the pressure of the upper chamber of hydraulic cylinder generates disturbance which is affected by the chance of the displacement and velocity of the hydraulic cylinder, the three-way pressure reducing valve can adjust it, that is to say, that the main valve core moves up or down can stabilize pressure. And a simulation is needed to evaluate the response characteristic.

3.3 电液比例方向控制阀 (Electro-hydraulic proportional directional control valve)

比例方向控制阀按输入信号的极性和幅值大小，同时对液压系统液流方向和流量进行控制，从而实现对执行器运动方向和速度的控制。在压差恒定条件下，电液比例方向阀的流量与输入电信号的幅值成正比，而流动方向取决于比例电磁铁是否受激励，是具有方向控制功能和流量控制功能的两参数控制复合阀。其外观与传统方向控制阀相同。

Proportional directional control valve according to the sign and amplitude of the input signal simultaneously controls the direction and the flow of the hydraulic system, thereby the aim to control the direction and speed of the actuator can be achieved. When the pressure difference is constant, the flow of the electro-hydraulic proportional directional valve is proportional to the input signal amplitude, while the flow direction depends on whether the proportional solenoid is actuated or not, so the valve is a two-parameter control composite valve which has direction control and flow control functions. It has the same appearance with traditional valve.

3.3.1 直动型比例方向阀 (Direct operated proportional directional valve)

直动型比例方向阀结构见图 3-13。阀体左、右两端各有比例电磁铁，阀体中阀芯 4 由对中复位弹簧 2、5 定位。滑阀 4 台肩上开有圆形和左右对称的节流槽。两电磁铁均不通电时，阀芯 4 在对中弹簧作用下处于中位。当电磁铁 1 有信号电流输入时，电磁力直接作用在阀芯上，比例电磁铁推力与相应的对中弹簧 5 的弹簧力平衡而定位在与输入信号成正比的位置上，从而使 P、B 连通，A、T 连通，并经相应节流槽节流。节流槽开口量的大小取决于输入电流信号的强弱。液流方向取决于哪个电磁铁接收信号。

The structure of the directly operated proportional directional valve is showed in the Fig. 3-13. There is a proportional solenoids on the both sides of the valve body, the valve spool 4 in the valve body is located by the centering spring 2 and 5. Circular symmetrical notches are placed on the step of the valve spool 4. When two electromagnets receive no signal, the valve core 4 stays in the center. However, if some signals are input into the electromagnet 1, the electromagnet force acts directly on the valve core and moves it to the input signals, which is the result of the force balance between the electromagnet and the spring 5. P and B, A and T are connected via the corresponding notch. The opening area of the notch depends on the input signal. The flow direction depends on which electromagnet receives the signal.

3.3 电液比例方向控制阀
(Electro-hydraulic proportional directional control valve)

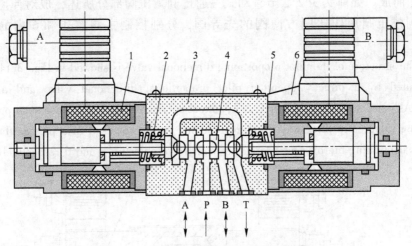

图 3-13 直动型比例方向阀（无位置控制）
(Fig. 3-13 Directly operated proportional directional valve (without position control))
1, 6—比例电磁铁 (Proportional solenoid); 2, 5—对中复位弹簧 (The centring and resetting spring); 3—阀体 (Valve body); 4—阀芯 (Valve core)

3.3.2 先导型比例方向阀(Pilot operated proportional directional valve)

直动型比例方向阀因受比例电磁铁电磁力的限制，只能用于小流量系统。在大流量系统中，过大的液动力将使阀不能开启或不能完全开启，应使用先导型比例方向阀。

Because of the restriction of the proportional solenoid force, the directly operated directional valve is only used in the system with small flow. However, in the large flow system, the hydraulic force becomes so huge that the valve can not open completely or at all, thus pilot operation is brought in.

先导型比例方向阀有两种：一种是以传统电液动方向阀为基础发展而成的，其先导阀是双向三通比例减压阀，主阀为液动式比例方向阀；二是在伺服阀简化基础上发展而成的，称作伺服比例方向阀或廉价伺服阀。

The pilot proportional directional valve have two kinds. One is developed from the traditional electro-hydraulic directional valve. Its pilot valve is reversible three-way proportional pressure reducing valve and the main valve is hydraulic proportional directional valve. And the other one is simplified from the servo valve, which is called the servo proportional directional valve or low priced servo valve.

3.3.2.1 普通先导式比例方向阀(Ordinary pilot proportional directional valve)

先导型比例方向阀结构如图 3-14 所示。它是由直动型比例方向先导阀和主阀叠加而成。如前文 3.2.2 节直动式三通比例减压阀部分所述，成对使用的三通比例减压阀用作比例方向阀的先导阀，分别控制主阀腔 4 和 6 内的油液压力。

The structure of the pilot proportional directional valve is showed in Fig. 3-14. It is constituted by a directly operated pilot proportional directional valve and a main valve. As is mentioned in section 3.2.2, direct operated three-way proportional pressure reducing valve, two three-way pressure reducing valves are in a pair as the pilot valve, controlling the oil pressure of the main valve chamber 4 and 6 respectively.

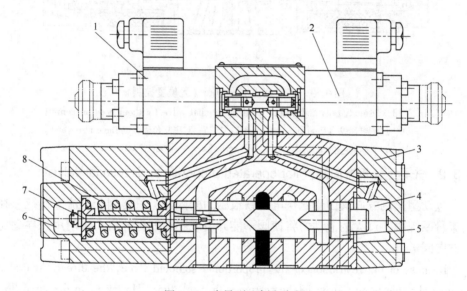

图 3-14　先导型比例方向阀

(Fig. 3-14　Pilot proportional directional valve)

1, 2—三通比例减压阀 (Control valve core)；3—主阀体 (Main valve body)；
4, 5—主阀腔 (Main valve chamber)；6—主阀芯 (Main valve core)；
7—连杆 (Connecting rod)；8—弹簧 (Spring)

此先导式比例阀实际是通过调节先导三通比例减压阀中的流阻来改变主阀芯两端面油液压力，并平衡左端的弹簧来实现对主阀阀芯位移的比例控制。自控制器的电信号，通过三通比例减压阀的比例电磁铁，成比例地转化为作用在先导阀芯上的力。与此作用力相对应，在先导阀的出口得到与之相应的油液压力，主阀腔 4 和 6 内油液压力分别作用于主阀芯的两个端面上，克服弹簧力推动主阀芯位移直到液压力和弹簧力平衡为止，从而实现比例方向阀的功能。

This pilot operated proportional valve is actually by adjusting the flow resistance in the pilot three-way pressure reducing valves to regulate the oil pressure in both ends of the main valve core, cooperation with the spring in the left end to achieve displacement

3.3 电液比例方向控制阀
(Electro-hydraulic proportional directional control valve)

control of the main valve spool. The electrical signal generated by the controller is proportionally converted into the driving force acting on the pilot valve core through the proportional electro-magnet of the three-way pressure reducing valves. Corresponding to this driving force, a specific oil pressure can be got in the outlet of the pilot valve. The pressure oil in main valve chamber 4 and 6 acting on both ends of the main valve spool, drives the main valve spool to move until the hydraulic pressure balances the spring force. This way, the function of proportional directional valve can be achieved.

主阀芯位移的大小,即相应的阀口开度的大小,取决于作用在主阀端面先导控制压力的高低,即相应地取决于输入先导阀两端控制器信号的大小。

The displacement of the main valve spool. namely the valve opening level, is determined by the pilot control pressure acting on the end face of the main valve spool, corresponding to the intensity of the control signal input to the pilot three-way pressure reducing valves.

3.3.2.2 位置反馈型先导式比例方向阀(Pilot proportional directional valve with position feedback)

这种阀由一个直动式的比例方向阀与一个零开口的液动阀叠加而构成,其结构原理如图 3-15 所示。该阀的先导阀工作原理与前面介绍的直动式闭环比例方

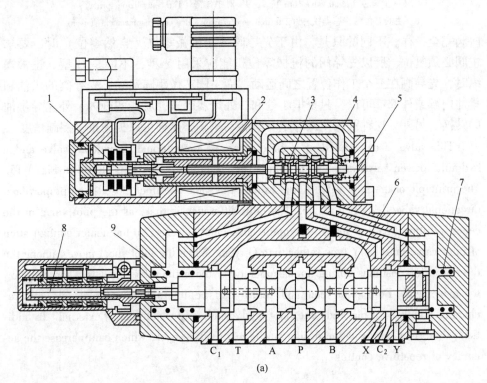

(a)

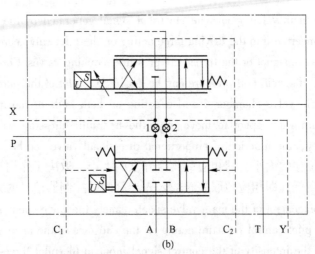

图 3-15 位置反馈型先导式比例方向阀

(Fig. 3-15 Pilot proportional directional valve with position feedback)

(a) 结构图 (Structure); (b) 原理示意图 (Schematic)

1—先导阀芯位移传感器铁芯 (The iron core of pilot valve core displacement sensor)

2—比例电磁铁 (Proportional solenoid); 3—先导阀芯 (Pilot valve core);

4—先导阀体 (Pilot valve body); 5—复位弹簧 (Resetting spring);

6—主阀芯 (Main valve core); 7, 9—对中弹簧 (The centralizing spring);

8—主阀芯位移传感器铁芯 (The iron core of main valve core displacement sensor)

向阀完全一样。不同的只是,由于先导阀处于机械零位(自然零位)时,要求主阀必须复位,所以先导阀的机械零位机能应采用 Y 型,不能是 O 型。正常操纵时,先导阀在三个工作位置之间运动,而主阀芯做跟随运动。先导阀的供油和排油内部或外部都可以,只要把 1 处和 2 处堵塞起来,就能实现内、外先导供油的转换。另外,这种阀的主阀芯上有防转动机构,这样可以提高重复控制精度。

This valve is constituted of a direct-operated proportional directional valve and a hydraulic driven valve with zero overlap. The structural principle is shown in Fig. 3-15. The principle of the pilot valve works exactly the same as direct closed-loop proportional directional valve described above. The only difference is that, as the pilot valve at the center position (natural zero), the main valve must be reset, so the center configuration of pilot valve must be Y type rather than O type. When the pilot spool moves among the three working positions during normal operation, the main valve spool is made to follow the movement. Both pilot feed and return can be internal or external. As long as channel 1 and 2 are blocked, the pilot feed switches from internal to external. In addition, there is an anti-rotation design on the main valve spool, which can increase the accuracy of repetitive control.

3.3 电液比例方向控制阀
(Electro-hydraulic proportional directional control valve)

3.3.2.3 整体式比例方向阀(Integral proportional directional valve)

(1) 从结构和控制特性上与普通比例阀相同,不同之处是将电子控制部分也集成在主阀或先导阀内部(图3-16)。在出厂前已做过仔细调整,使用前只需参考使用说明。在接线端子接上适当的电源电压,再做适当的工作设定,便可投入工作。它的工作原理与前面所述的直动式或先导式比例阀并无两样。其先导阀是一个直动式比例方向阀,主阀为双弹簧对中的液动阀。先导阀和主阀均带位置反馈,使阀芯的位置在较大的液动力干扰下仍能保持准确的位置。

The structure and the control characteristic of this valve are the same as the common proportional ones, the difference is that electronic control parts are also integrated in the inner part of the main valve or pilot valve (Fig. 3-16). Adjustment has been done carefully before sent out of the factory. The only thing that needs to be done before using is just reading the instructions. The terminals should be connected to the proper power supply voltage, and then make an appropriate work setting. The valve can be used. Its working principle has no difference with the previously described direct or pilot operated proportional valve. Its pilot valve is a direct operated proportional directional valve, and main valve is the double spring centered hydraulic valve. Pilot valve and main valve are equipped with position feedback, which makes the location of the valve remains accurate even under a larger fluid dynamic interference.

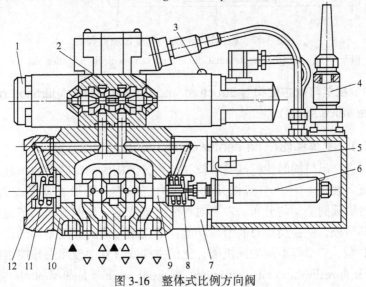

图 3-16 整体式比例方向阀
(Fig. 3-16 Integral proportional directional valve)
1, 3—放气螺钉(Bleed screw); 2—先导阀(Pilot valve); 4—出线口(Outlet port);
5—电控器(Electric controller); 6—位置传感器(Position sensor);
7—右端盖(The right end cap); 8, 11—对中弹簧(The centralizing spring);
9—主阀芯(Main valve core); 10—主阀体(Main valve body); 12—左端盖(The left end cap)

(2) 这种阀使用上可把一个隔离阀叠加在导阀和主阀之间作为安全阀使用,如图 3-17 所示。隔离阀断电时,主阀将保持在中位,不受先导阀的控制,起安全保护的作用。

This kind of valve can be used as a safety valve when a isolation valve is combined between the pilot valve and the main valve, as shown in Fig. 3-17. When out of power, the main valve will remain in the center position, free from the control of the pilot valve, which is the security part of the whole valve.

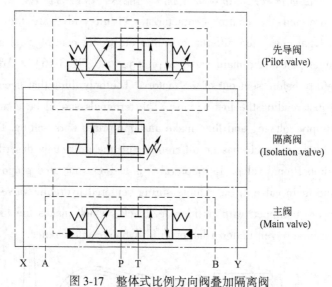

图 3-17 整体式比例方向阀叠加隔离阀

(Fig. 3-17 Integral proportional directional valve with the isolation valve)

3.3.3 中位机能与阀芯结构 (Structure and the center configuration of the valve spool)

3.3.3.1 中位机能 (The center configuration of the valve)

为了对进、出口同时执行准确节流,比例方向阀滑阀阀芯台肩圆柱面上开有轴向的节流(控制)槽。节流槽几何形状为三角形、矩形、圆形或其组合状。节流槽在台肩圆周上均匀分布、左右对称分布或成某一比例分布。节流槽轴向长度大于阀芯行程,使控制口总有节流功能。节流槽与阀套通过不同的配合可以得到 O 型、P 型、Y 型等不同的阀机能。比例方向阀有直动型和先导控制型。

In order to realize accurate restricting, an axial notch is located on the shoulder of the proportional directional valve spool. The notches have several shapes, such as triangle, rectangle, circular and their combination. The notches are well-distributed around the step and symmetrical or scatter in proportion. The axial length of the notch is longer

3.3 电液比例方向控制阀
(Electro-hydraulic proportional directional control valve)

than stroke distance of the valve core, which makes the control ports have control function constantly. The different functions of the valve, such as the type O, P, Y and so on, can be founded by differently combining the notch and the valve suit. The proportional directional valves have also the types of the directly operated and the pilot ones.

表 3-1 为阀的中位机能对照表(The Table 3-1 is comparison table of the valve's center comfiguration)。

表 3-1 阀的中位机能对照表
(Table 3-1 Comparison table of the valve's center comfiguration)

职能符号 (Function symbols)	机能符号 (Code symbols)	流通状态 (Circulation status)	应用 (Application)
(A B / P T)	O	P→B=q; A→T=q P→A=q; B→T=q	对称执行器或面积比接近1∶1的单出杆液压缸 (Symmetric actuator/area ratio close to 1∶1 single-rod hydraulic cylinder)
	O_1	P→B=q/2; A→T=q P→A=q; B→T=q/2	面积比接近2∶1的单出杆液压缸 (Area ratio close to 2∶1 single-rod hydraulic cylinder)
	O_1	P→B=q; A→T=q/2 P→A=q/2; B→T=q	
	O_1	P→B=q; A→T=q P→A=q; B→T=0	差动连接的单出杆液压缸 (Differential connection of single-rod hydraulic cylinder)
	PX	P→B=q; A→T=q P→A=q; B→T=q	对称执行器 (Symmetric actuator)
	YX	P→B=q; A→T=q P→A=q; B→T=q	对称执行器或面积比接近1∶1的单出杆液压缸 (Symmetric actuator/marca ratio close to 1∶1 single-rod hydraulic cylinder)
	YX_1	P→B=q/2; A→T=0 P→A=q; B→T=q/2	面积比接近2∶1的单出杆液压缸 (Area ratio close to 2∶1 single-rod hydraulic cylinder)
	YX_2	P→B=q; A→T=q/2 P→A=q/2; B→T=q	
	YX_3	P→B=q; A→T=q P→A=q; B→T=0	差动连接的单出杆液压缸 (Differential connection of single-rod hydraulic cylinder)

3.3.3.2 阀芯结构(Structure of valve spool)

阀芯的结构对阀的性能影响很大。图 3-18 所示为四种阀芯结构,它们的流量特性区别很大。特性曲线下部弯曲情况区别很大,原因是这种弯曲情况和节流阀口处各种不同形状的切槽有很大关系。这些曲线的下部都是平缓的曲线区段,这一点很重要:提高了分辨率。而曲线的上部弯曲是流量饱和在起作用。之所以流量达到饱和,是因为阀的节流面积不可能再变大了。

The valve properties depend on the structure of the valve spool significantly, there are four valve spool structures shown in Fig. 3-18, and their flow characteristics differs a lot. Bending of the lower part of the curves are quite different, because the bending have great relationship with various shapes of notches of the mouths in the throttle. The lower parts of these curves are gentle curve sections, which is very important. It enhances the resolution. The upper bend of the curve is the flow saturated at work. The reason why the flow is saturated is that the throttle valve area cannot be larger.

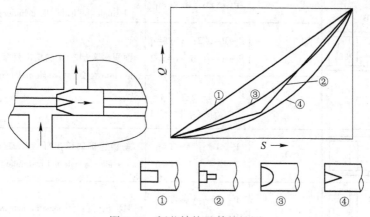

图 3-18 阀芯结构及其特性图

(Fig. 3-18 Valve core bodies and their characteristic diagram)

阀口打开的起始值,取决于阀芯上节流切槽的遮盖量。此遮盖量约为整个阀芯行程的20%,以保证原始位置一定的密封性。通过调整电放大器的零位,可以移动开口起始值。

The starting value opened by the valve port depends on overlap of the spool, which is about 20% of the entire spool stroke, to ensure the sealing of the center position. The initial value of the port can be changed by adjusting the zero position of the amplifier.

3.4 电液比例流量控制阀 (Electro-hydraulic proportional flow control valve)

电液比例流量控制阀的流量调节作用,都是通过改变节流阀口的开度(通流面积)来实现的,它与普通流量阀的主要区别是用比例电磁铁取代原来的手动调

3.4 电液比例流量控制阀
(Electro-hydraulic proportional flow control valve)

节机构,直接或间接地调节主流阀口的通流面积,并使输出流量与输入电信号成正比。节流阀的流量公式为

Electro-hydraulic proportional flow control valves regulate the flow by changing the throttle opening (flow area), and in contrary to the ordinary flow valve, a proportional solenoid is used to replace the manual adjustment mechanism. It regulates the flow area at the main stage directly or indirectly and makes the output flow proportional to the input signal. The flow formula of the throttle is

$$q = C_d A(x) \sqrt{\frac{2}{\rho} \Delta p} \tag{3-17}$$

式中,C_d 为阀口的流量系数,在紊流时近似为常数(The flow coefficient of the port, which in turbulent flow is approximately constant)。

由上式可见,控制通流面积可以控制通过阀口的流量,但是通过的流量还受到节流阀口的前后压差 Δp 等因素的影响。电液比例流量控制阀按其被控量是节流阀口的开度 x(或通流面积 $A(x)$)还是流量,可分为比例节流阀和比例流量阀,比例流量阀还可以分为二通型和三通型两种。

As shown in the previous equation, controlling the flow area can control the flow that pass through the valve port. The flow also depends on the pressure difference Δp of inlet and outlet of the throttle port and other factors. The electro-hydraulic proportional flow control valve can be divided into the proportional throttle valve and the proportional flow valve according to the amount of the opening x (or flow area $A(x)$) or flow. The proportional flow valve can also be divided into the two-way and the three-way type.

比例方向阀具有对进口和出口流量同时节流的功能,因此,它本质上是双路的比例节流阀。如果从外部加上压力补偿装置,就能使通过的流量与负载变化无关,具有调速阀的功能。

The proportional directional control valve has throttling function on both the inlet and outlet flows, therefore it is essentially a two-direction proportional throttle valve. If added with the pressure compensation device from the outside, it can eliminate the effect of load change, then the proportional directional control valve will have the function of the speed control valve.

3.4.1 利用方向阀作流量控制 (Directional valve used for flow control)

比例流量控制阀的流量调节作用在于改变节流口的开度。它是用电-机械转换器取代普通节流阀的手调机构,调节节流口的通流面积,使输出流量与输入信号成比例。

The flow regulation of the proportional flow valve is performed by changing the open amount. The area of flow can be adjusted by its electro-mechanical conversion device instead of manual regulating screw of common throttle restrictive valve, which makes the output flow proportional to the input signal.

比例流量阀分为比例节流阀和比例调速阀。直动型比例节流阀较少见。由于比例方向阀具有节流功能,实际使用中则以二位四通比例方向阀代替比例节流阀,如图 3-19 所示。

The proportional flow control valve can be divided into the proportional throttle restrictive valve and the speed regulating valve, but the directly acting throttle restrictive valve is used infrequently. As the proportional directional valve has the function of flow control, the four-way and two-position proportional directional valve is usually used to replace the proportional throttle valve, as showed in the Fig. 3-19.

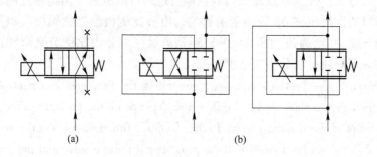

图 3-19 作比例节流阀时的四通比例阀的连接

(Fig. 3-19 The connection of four-way proportional valve used as a proportional throttle valve)
(a) 利用一个通道 (Use one channel); (b) 利用两个通道 (Use two channels)

3.4.2 节流阀 (Throttle valve)

当阀的输入电信号为零时,先导阀芯在反馈弹簧预压缩力的作用下,处于图 3-20 所示位置,即先导控制阀口为负开口,控制油不流动,主阀上腔的压力 p_c 与进口压力 p_s 相等。由于弹簧力和主阀芯上下面积差的原因,主阀口处于关闭状态。这时,无论进口压力 p_s 有多高,都没有流量从 A 口流向 B 口。

When no electrical signal is inputted, the pilot core is located in the place where is expressed in the Fig. 3-20 by the action of the advanced force of the feedback spring, namely, the mouth of the pilot valve is negative and the control oil is static. The pressure p_c in the upper chamber of the main valve is equal to the pressure p_s of the inlet. As the spring force exists and there is difference between the upper and under part area of the main valve spool, the main valve outlet is closed. Now there is no flow from A to B, no matter how high the inlet pressure p_s is.

3.4 电液比例流量控制阀
(Electro-hydraulic proportional flow control valve)

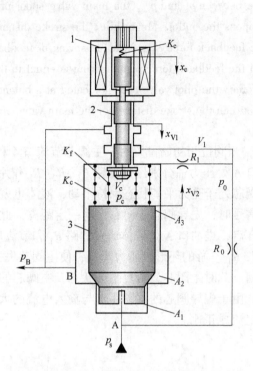

图 3-20 位置-力反馈型比例节流阀
(Fig. 3-20 Location - force feedback proportional throttle valve)
1—比例电磁铁(Proportional solenoid); 2—先导阀阀芯(Pilot valve core);
3—主阀芯(Main valve core)

当输入足够大的电信号时,电磁力克服反馈弹簧 K_f 的预压缩力,推动先导阀芯下移 x_{V1},先导阀口打开,控制油经过固定液阻 R_0 →先导阀口→主阀出口 B,沿油液流动方向压力有损失。故主阀上腔的控制压力 p_c 低于进口压力 p_s,在压差 p_s-p_c 的作用下,主阀芯产生位移 x_{V2},阀口开启。与此同时,主阀芯位移经反馈弹簧转化为反馈力 $K_f x_{V2}$,作用在先导阀芯上;当反馈弹簧的反馈力与输入电磁力达到平衡时,先导阀芯便稳定在某一平衡点上,从而实现主阀芯位移与输入电信号的比例控制。

When enough electrical signal is inputted, the electromagnetic force pushes the pilot valve spool down x_{V1} conquering the beforehand compression force of the feedback spring K_f, which opens the outlet of the pilot valve. The control oil enters the outlet B of the main valve via the fixed hydraulic resistance R_0 and the outlet of the pilot valve. As there is pressure loss along the flow direction of the oil, the control pressure p_c in the upper chamber of the main valve is lower than the pressure p_s of the inlet. For the action of differential pressure between p_c and p_s, the main valve spool produces the displacement of x_{V2}, which opens the outlet. Meanwhile, the stroke distance of the main valve core transfers into the feedback force $K_f x_{V2}$ via the spring of feedback and then acts on the valve core. When the feedback force of the spring is equal to the force of the inputting electromagnetic force, the pilot valve core is located at a balanced point. Thus, the proportional control between the stroke distance of the main valve and inputting electrical signal is finished.

如图 3-21 所示，当阀位于初始位置时，主阀芯节流口 4 和先导阀芯节流口 3 处于关闭状态，进口 A 处压力油分别经过油路 R_1、R_2、R_3 作用在主阀芯的三个正反端面上，此时主阀芯处于受力平衡状态保持不动，液体也就没有流动。当给比例电磁铁 1 一定电信号时，先导阀芯 2 向右移动一定距离，此时先导阀芯与主阀芯之间的节流口 3 开启，进油口 A 处液体通过油路 R_1 和节流口 3 经出口 B 流出，由于 R_1 的节流作用，R_1 下游的环形腔压力降低，使主阀芯失去平衡而向右移动，直至将节流口 3 关闭，此时主阀芯又恢复平衡状态，主阀芯的移动距离与先导阀芯的移动距离相等。由于先导阀芯的移动距离与输入电流的大小成正比，可知主阀芯的位移与输入电流成正比。

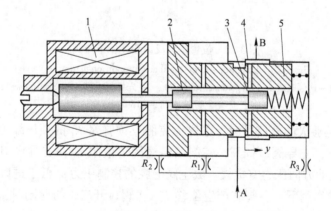

图 3-21 位置负反馈型比例节流阀

(Fig. 3-21 Proportional throttle valve with location negative feedback)

1—比例电磁铁 (Proportional solenoid); 2—先导阀芯 (Pilot valve core);
3—先导阀芯节流口; 4—主阀芯节流口; 5—主阀芯 (Main valve spool)

As is shown in Fig. 3-21, when the valve is at the initial position, the orifice 4 of the main valve core and the orifice 3 of the pilot valve core are closed. As the pressure acting on the three faces of the main valve core through the passage of R_1, R_2, R_3 is balanced due to the fact that no fluid moves in these spaces, and the main valve core does not move. When a certain electronic signal is given to the proportional electric-magnet

1, the pilot valve core 2 will move right by a corresponding distance. The orifice 3 will open which means the oil at A will move out of B through the pass of R_1 and orifice 3. The main valve core will move towards right until the orifice 3 is closed as the pressure of the ring cavity linked with R_1 drops due to the effect of throttle R_1. At this time, the main valve core is at the balancing state again and the moving distance of the main valve core is equal to the moving distance of the pilot valve core. As the moving distance of the pilot valve core is proportional to the electronic signal, the moving distance of the main valve core is proportional to the signal too.

3.4.3 比例调速阀 (Proportional flow control valve)

比例调速阀是在传统调速阀的基础上将其手调机构改用比例电磁铁而成的。它仍然是由压力补偿器（定差减压阀）和比例节流阀构成的，因为只有 A、B 两个主油口，又称为二通比例调速阀。比例调速阀的工作原理如图 3-22 所示。

The proportional flow control valve replaces the manual operated structure of the traditional speed control valve with proportional electromagnet. It is still made up of the pressure compensator (fixed differential pressure reducing valve) and the proportional restrictor. It is also called two-way proportional speed control valve, because it only has two main oil ports, A and B. The operating principle of the proportional speed control valve is shown as Fig. 3-22.

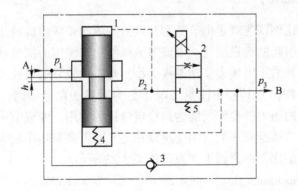

图 3-22 比例调速阀工作原理简图

(Fig. 3-22 Working schematic of proportional speed control valve)

1—定差减压阀 (Fixed differential pressure reducing valve);
2—比例节流阀 (Proportional restrictor); 3—单向阀 (Non-return valve); 4, 5—弹簧 (Spring)

图 3-22 中，压力补偿器的减压阀 1 位于节流阀 2 主节流口的上游，且与主节流口串联，减压阀阀芯由一小刚度弹簧 4 保持在开启位置上，开口量为 h；节流阀 2 阀芯也由一小刚度弹簧 5 保持关闭。比例电磁铁接受输入电信号，产生电磁

力作用于阀芯，阀芯向下压缩弹簧5，阀口打开，液流自 A 口流向 B 口。阀的开口量与控制电信号对应。行程控制型比例电磁铁提供位置反馈，可使其开口量更为准确。

In the Fig. 3-22, the pressure reducing valve 1 of the pressure compensator is located at the upstream of restrictor 2, and it is in series with the restrictor. The pressure reducing valve's cartridge is kept normally open with spring 4, and the opening amount is h. The throttle valve2's core is kept normally closed with spring 5. The proportional electromagnetperforms a electromagnetic force on the spool, when it receives the input electrical signal. And then the spool compresses downward the spring 5 to open the valve port. Meanwhile, the fluid flows from the port A to the port B. The opening amount of valve 2 depends on the electrical signals. The position feedback which is produced by stroke controlled proportional electro-solenoid can make the opening of the valve more accurate.

压力补偿器的功能是保持节流阀进出口压差 $\Delta p = p_2 - p_3$ 不变，从而保证流经节流阀的流量稳定。

Function of the pressure compensator is to keep inlet and outlet pressure of the throttle $\Delta p = p_2 - p_3$ unchanged, thus it could ensure the stability of flow regulation.

3.4.4 定差溢流型比例调速阀 (Pressure compensated flow bypass control valve)

定差溢流型比例调速阀是由节流阀与定差溢流阀并联而成的组合阀，它能补偿因负载变化而引起的流量变化。溢流节流阀能使系统压力随负载变化，没有调速阀中减压阀口处的压力损失，功率损失小，是一种较好的节能元件；但其流量稳定性略微差一些，尤其在小流量工况下更为明显。在进油路上设置溢流节流阀，通过溢流阀的压力补偿作用能达到稳定流量的作用。溢流节流阀也称为旁通调速阀。这种定差调速阀一般应用于速度稳定要求不高而功率较大的进口节流系统中。定差溢流型比例调速阀工作原理如图 3-23 所示。

Pressure compensated flow bypass control valve is a combination of a throttle valve and a pressure relief valve in parallel. It can compensate the flow changes caused by load changes. The relief valve can make the system pressure variable with respect to the load, and there is no pressure loss in the pressure reducing valve in contrast to flow control valve. Power loss is very low, so it is a better energy-saving component. But the flow stability is slightly worse, specially under condition of small flow, the relief valve pressure compensator located upstream can be used to stabilize the flow, therefore it is called pressure compensated flow bypass control valve. The fixed differential speed control valve is normally used in meter-in flow control system where the requirement of flow stability is not so strict but the power is large. The working principle of the fixed differential overflow proportional speed control valve is shown in the Fig. 3-23.

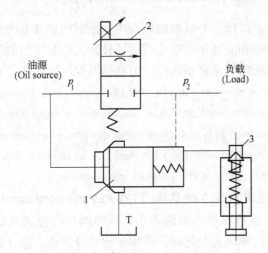

图 3-23 定差溢流型比例调速阀

(Fig. 3-23 Fixed differential relief proportional speed control valve)
1—定差溢流阀（Fixed differential relief valve）；2—比例节流阀（Proportional throttle valve）；
3—限压先导阀（Pressure limiting pilot valve）

3.5 压力补偿器（Pressure compensator）

压力补偿器是用于保持比例换向阀阀口两端压差一定的器件。补偿器的工作原理是通过调节给定弹簧的预紧力来调节比较机构的平衡状态。当输出压力、给定弹簧的力（或力矩、或位移）与输入压力（或力矩、或位移）平衡时，阀的开度保持不变，输出压力就维持不变。若输出压力发生变化，平衡状态被破坏，阀的开度就发生变化，最终输入量发生变化，从而使输出压力维持在给定弹簧设置的压力上，实现压力补偿。

Pressure compensator is the device used to maintain differential pressure between the up and down stream of proportional direction valve constant. The working principle of the compensator is that the balance of the comparator is kept by adjusting the spring pretension. When the output pressure, the given spring force (or torque, or displacement) and the upstream pressure (or torque, or displacement) are balanced, the valve opening remains unchanged and the downstream pressure remains unchanged. If the downs pressure changes, equilibrium will be broken. Then the valve opening and the final upstream will change, so the downstream pressure maintains at a given pressure with respect to the spring, and the pressure compensation is achieved.

3.5.1 进口压力补偿器 (Inlet pressure compensator)

在一个带二通进口压力补偿器的装置中，比例阀进口节流口的压降保持为常数，使负载压力波动和油泵压力变化得到了补偿，即泵压力的升高不会引起流量的增大。因而阀的额定流量必须按照压力补偿器的压力变化值来选择。

In a two-way inlet pressure compensator, the pressure drop across the throttle of the proportional valve remains constant. Thus, load pressure fluctuations and pump pressure changes are compensated. This also means that the flow is irrelevant of pump pressure. Therefore this valve must be selected with regard to its nominal flow in accordance with change of pressure difference of the pressure compensator.

3.5.1.1 二通进口压力补偿器(Two-way inlet pressure compensator)

在图 3-24 的二通进口压力补偿器中，调节阀口和检测阀口是串联的。当阀芯处于平衡位置而负载压力变化时，作用于检测口的压力降（进口压力减出口压力）将保持为常数。当弹簧很软，调节位移又很短时，弹簧力的变化也就很小，从而压力差近似为常数。只有当弹簧被进一步压缩时，调节阀芯才会调节阀口的过流截面变化。因此，只要负载压力增加，阀就能很好地起到调节作用。

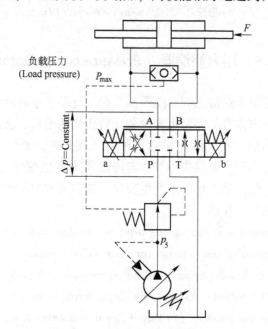

图 3-24 二通压力补偿器原理图

(Fig. 3-24 Schematic of two-way pressure compensator)

3.5 压力补偿器 (Pressure compensator)

In a two-way inlet pressure compensator shown in Fig. 3-24, the detection valve port are in series. When the spool is in balance and the load pressure changes, the pressure drop (inlet pressure minus outlet pressure) at the measuring orifice will remain constant. When the spring is very soft and the control stroke is short, the change in the spring force is only slight and therefore the pressure drop is almost constant. The control spool can only change the opening of the control orifice when the spring is further compressed. The flow regulation is regulated once the load pressure increases.

随着通过流量的增大，阀中的液流阻力就增大，此时只有相应增大外部压差，才能实现流量调节功能。

With the increase of flow through valves, the flow resistance increases. Then the outer pressure difference must also increase in order to achieve the flow control function.

3.5.1.2 三通进口压力补偿器(Three-way inlet pressure compensator)

三通压力补偿器相对于二通压力补偿器而言，尽管它对效率的改善不太明显，但在某些情况下，它能通过更换二通压力补偿器的控制阀芯而方便地构成。其负载取压点与二通压力补偿器相对应，分辨率和等流量特性与二通压力补偿器一样。这种补偿装置要和定量泵配套使用。

Relative to the two-way pressure compensators, the three-way pressure compensators can not efficiently increase the degree of efficiency. However, in some cases it can be produced relatively easily by changing the spool in two-way inlet pressure compensators. Their load application points correspond to those of the two-way inlet pressure compensators. The resolution capacity and uniform flow characteristics are identical to those of the two-way inlet pressure compensators. They are mainly used in conjunction with fixed pumps.

如图 3-25 所示，在使用三通进口压力补偿器时，调定的固定检测阀口 A_2 和由压力补偿器控制的调节阀口 A_1 是并联的。

As is shown in Fig. 3-25, when the three-way pressure compensator is incorporated, the fixed measuring orifice A_2 and the controlling orifice A_1 controlled by the pressure compensator are arranged in parallel.

调节阀口 A_1 在此作为到回油管路的出流口，在控制阀芯处于平衡位置而不考虑摩擦力和液动力时，可得：

The controlling orifice A_1 is the outlet of the oil return port. The following applies when the controlling spool is in balance and no friction and flow force is considered：

$$p_1 A_k = p_2 A_k + F \qquad (3\text{-}18)$$

$$\Delta p = p_1 - p_2 = \frac{F}{A_k} \approx 常数 \qquad (3\text{-}19)$$

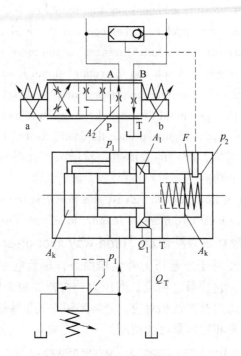

图 3-25 三通压力补偿器原理

(Fig. 3-25 Principle of the three-way pressure compensator)

这样在检测阀口上的压力差也就保持恒定,并得到一个与压力变化无关的流量 q。在使用二通压力补偿器时,油泵始终需提供由溢流阀调定的系统最高压力。

The pressure drop is also held a constant at the measuring orifice. Thereby flow q becomes independent of pressure changes. When using the two-way pressure compensator, the pump must constantly produce maximum pressure which was set by the relief valve.

与此相反,在使用三通压力补偿器时,其进口的工作压力仅需比负载压力高出检测阀口处的一个压降 Δp,因而其功率损失较小。当在比例阀中使用 W 型阀芯时,油液就以所设定的 Δp 压差循环于泵与油箱之间。

Contrary to the two-way pressure compensator, when the three-way pressure compensator is used, the working pressure needs to be greater than the actuator pressure only by the amount of the pressure drop Δp at the measuring orifice. As a result, the power loss is considerably less. If a W-spool is used in the proportional valve, the oil will work between the pump and tank at the set pressure difference Δp.

3.5.2 出口压力补偿器 (Outlet pressure compensator)

对于需改变负载方向的工作系统,使用进口压力补偿器是有条件的。可采用出口压力补偿器(图3-26)。这种出口压力补偿器按使用情况布置在执行器的一

3.5 压力补偿器 (Pressure compensator)

个或两个油口上。出口压力补偿器总是位于负载和阀之间，并保持从 A 或 B 到油箱的压差为常数。

Systems in which the direction of the applied load reverses, the use of inlet pressure compensator is severely restricted (Fig. 3-26). In such cases, an outlet pressure compensator is often used, arranged in one or in both connections of the actuator depending on the application. The outlet pressure compensator is always located between the load and the valve and maintains the pressure drop constant from A or B to the tank.

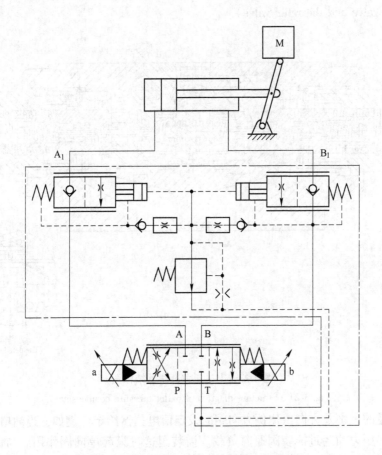

图 3-26 锥阀式出口压力补偿器应用油路

(Fig. 3-26 Application circuit of the outlet pressure compensator of poppet valve)

出口压力补偿器有 16、25 和 32 等规格，采用常规锥阀结构。与此同时，该补偿器兼容了可控单向阀功能。由于它不存在泄漏，所以这种功能特别适应于垂直负载的要求，并且在系统中可不用另外配置旁路单向阀。另外，锥阀阀芯很容易由相反方向的液流推开，可使流量在两个方向上流动。

Outlet pressure compensators of poppet design are available with sizes 16, 25 and

32, conventional poppet valve structure is applied. This arrangement therefore combines the compensator function and the function of the controllable check valve. As there is no leakage, the function is especially appropriate for the vertical loads. The by-pass single valve is not necessary in the system. Besides, as the poppet can be easily opened by the opposite direction flow, the flow can be in both directions.

如图 3-27 所示，出口单向截止型压力补偿器的基本构件为壳体、插装阀和限压阀（As described in the Fig. 3-27, the unit basically contains the housing, the cartridge valve and the relief valve）。

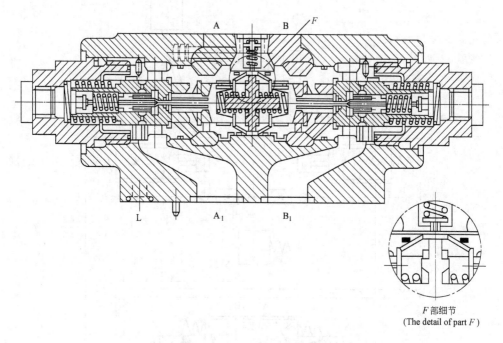

图 3-27 出口压力补偿器结构图

(Fig. 3-27 Structure diagram of outlet pressure compensator)

流量的大小及方向由比例方向阀的设定值电位器给定。例如，当油口 A 与泵相连时，压力油通过插装阀通向负载。插装阀这时只起单向阀作用。与此同时，从泵出口来的控制油，流经作为负载补偿流量调节器的开启控制阀芯，进入容腔。这一控制油在限压阀前形成一个压力，它通过控制阀芯的小孔作用到开启控制阀芯的 B 端。

The amount and direction of the flow are pre-set by the amplifier's signal potentiometer of the proportional directional valve. For example, if the pump is switched to port A, the fluid flows to the actuator via the cartridge valve. In this case, the cartridge valve functions as a check valve. At the same time, the flow of control oil is derived from the flow delivered by the pump and lead to the chamber via the control spool acting as a load compensating flow control valve. This flow of control oil builds up a pressure in upstream chamber of the relief valve which is applied to the B side of the control spool via the orifice.

此外,限压阀的出口与 T 口相连,控制阀芯克服在弹簧腔中的负载压力,把卸载锥阀打开;同时,卸载锥阀截断了弹簧腔与负载压力的通路。在弹簧腔中,通过卸载锥阀使其压力与比例方向阀前的 B 通道中的压力一致。这个压力还作用于液压缸环形腔一侧和开启控制阀芯的端面上。

In addition, the outlet of the relief valve is linked to the channel T. The control spool opens the pressure relief poppet against the load pressure applied in the spring chamber. At the same time, the pressure relief poppet closes off the link to the load pressure. In the spring chamber, the pressure is equal to the pressure in front of the proportional directional valve in channel B via the pressure relief poppet. This pressure also acts on the ring cavity side of the cylinder and the end face of the control spool.

由此,从 B 口经比例方向阀到 T 口的压力降为常数。这个压力降由控制阀口调节,数值上等于容腔的压力减去弹簧力的相应压力。弹簧的作用力是很小的。

The pressure drop from port B to port T via the proportional directional valve is therefore constant. This pressure drop is controlled by the valve port and represents the pressure which is formed by the pressure in chamber minus the spring force. The force of the spring is of no significance.

当通过比例方向阀把油口 B 和油泵相连时,在 A 通道中的插装阀组的功能与前述相似。

The cartridge valve in channel A functions as previously described when the proportional directional valve switches the pump to port B.

3.6 电液比例复合阀 (Electro-hydraulic proportional composite valve)

从广义上说,把两种以上不同的液压功能复合在一个整体上所构成的液压元件可称为复合阀。若其中至少有一种功能可以实现电液比例控制,这样的阀可称作电液比例复合阀。因此,比例复合阀有多种控制功能。

Generally speaking, hydraulic components which combine two or more different hydraulic functions into one can be called composite valves. When electro-hydraulic proportional control is incorporated, this valve can be called the electro-hydraulic proportional composite valve. Therefore, the proportional composite valve has a variety of control functions.

从上面的定义看,最简单的比例复合阀是比例方向阀,它复合了方向和流量控制两种功能。如果进一步把比例方向阀与定差溢流阀或定差减压阀组合,就构成了传统的比例复合阀,且参与复合的比例方向阀的联数可以不止一联。

From the definition above, the simplest proportional composite valve is the proportional directional valve, which combines two functions of the direction and flow control. When further combining proportional directional control valve with fixed differential pressure relief valve or fixed differential pressure reducing valve, it makes the traditional proportional composite valve. And the associated number participating in the compound of the proportional directional valve can be more than one. More than one proportional direction valve can be incorporated.

比例复合阀是多个液压元件的集成回路,具有结构紧凑,使用维护简单等特点,可用于对执行器的速度控制、位置控制要求连续有规律的调节场合。

Proportional composite valve is a integrated circuit composed of multiple hydraulic components, with the advantages of compact structure, low cost for service and maintenance and so on. The valve can be used to control the speed and the displacement of the actuator in regular control occasions.

常用的比例复合阀有压力补偿型复合阀和比例压力/流量复合阀等,其中后者又被称为 p-q 阀或比例功率调节阀(图3-28)。电-液比例压力/流量复合阀有时被称为比例功率调节阀。它是由先导式比例溢流阀与比例节流阀组成的一个复合阀。比例溢流阀的主阀同时在复合阀中兼作三通压力补偿器,为比例节流阀进行压力补偿,从而获得较稳定的流量。

The commonly used proportional composite valves include pressure compensation composite valves, proportional pressure / flow composite valves and etc. And the latter one is also known as p-q valve (Fig. 3-28), or the proportional power valve. Electro-hydraulic proportional pressure / flow composite valve is sometimes referred to the proportional power regulating valve. It is a composite valve composed by pilot operated proportional relief valve and proportional throttle valve. The main valve of the proportional relief valve is simultaneously used as the three-way pressure compensator in the composite valve, which supply the pressure compensation for the proportional throttle valve, resulting in more stable flow.

阀的详细符号如图3-29所示。由图可见,比例节流阀3的前后压差由三通压力补偿器2保持恒定,因而通过节流阀的流量仅取决于节流阀的开口面积,亦即取决于通入比例电磁铁的信号电流。三通压力补偿器同时又是先导式比例溢流阀的主阀芯级。当负载压力达到溢流阀1的调定压力时,阀芯2开启,保持进口压力 p_p 不变。有些 p-q 阀带有限压阀以确保系统安全。使用 p-q 阀的系统可以在

3.6 电液比例复合阀
(Electro-hydraulic proportional composite valve)

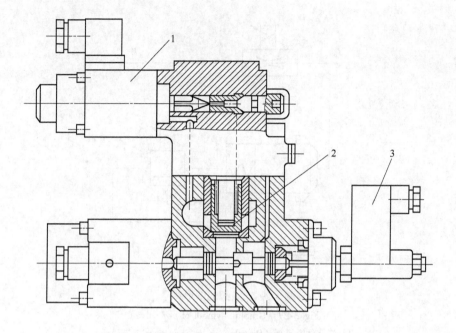

图 3-28 $p-q$ 阀结构图

(Fig. 3-28 Structure chart of p-q valve)

1—直动式比例溢流阀 (Directly operated proportional relief valve);
2—三通压力补偿器 (Three-way pressure compensator);
3—比例节流阀（带位置传感器）(Proportion throttle valve (with position sensor))

工作循环的不同阶段，对不同的多个液压执行器进行调速和调压，使系统得到大大的简化，同时控制性能也得到提高。

The specific valve symbol is shown in Fig. 3-29. The figure shows that the differential pressure upstream and downstream of the proportional throttle valve 3 is kept constant by the three-way pressure compensator 2. Therefore, the flow of the throttle valve only depends on the opening area of the throttle valve, i. e, the signal current sent to the proportional solenoid. The three-way pressure compensator is at the same time the main spool of the pilot proportional relief valve. When the load pressure reaches the pressure set by the relief valve 1, the spool 2 opens, so as to keep the inlet pressure p_p unchanged. Some p-q valves with pressure limiting valves are to ensure the safty of system. This p-q valve system can be applied in different stages of the work cycle, which can regulate the speed and pressure of different multiple hydraulic actuators. It will make the system much simplified, while the control performance is also improved.

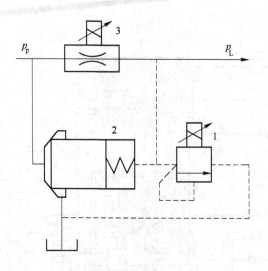

图 3-29 *p-q* 阀的图形符号

(Fig. 3-29 Graphic symbol of *p-q* valve)

1—直动式比例溢流阀（Directly operated proportional relief valve）；
2—三通压力补偿器（Three-way pressure compensator）；
3—比例节流阀（带位置传感器）（Proportion throttle valve (with position sensor)）

第4章 电液比例容积控制
Chapter 4 Electro-hydraulic Proportional Volume Control

4.1 容积泵的基本控制方法（The basic control method of capacity pump）

液压泵和液压马达都是液压系统中不可缺少的能源转换元件，为了节能的需要，发展了变量泵和变量马达。但由于通常的手动变量泵和变量马达只能适应一两种设定的工况，不能或很难适应在负载状态下进行连续比例变量或调压。在这些场合下会导致能量损失或性能下降，尤其在大功率及控制复杂、要求高的场合，更是如此。而比例式和伺服式容积控制的变量泵和马达能适应各种场合的要求，能够在工作状态下使液压参数作快速而频繁的变化，从而大大提高节能效果和控制水平。更重要的是采用了比例容积控制以后，可以通过电气技术进行各种补偿、校正、协调和适应控制，从而获得最大限度的性能提高和节能效果。

Both the hydraulic pump and hydraulic motor are indispensable energy conversion device to the hydraulic system. For the need of saving energy, the variable capacity pump and the variable capacity motor are developed consequently. However, as the usual manual variable capacity pump and variable motor can only adapt to one or two setting condition, and impossibly or hardly adapt to the continuous proportional variable quantity or pressure regulation under loaded condition. It will lead to energy loss or performance degradation in these situations, especially in the high-power, complex control and high requirement situations. The proportional and servo-controlled variable capacity pump and motor can adapt to the requirements to all kinds of conditions. Accordingly they are able to make the hydraulic parameters changing rapidly and frequently in any work state. So the energy-saving effect and control level are greatly improved. More importantly, after the adoption of the proportional capacity control, we can get the maximum performance improvement and energy-saving effect by a variety of compensation, calibration, coordination and adaptive control according to the electrical technology.

第4章 电液比例容积控制
Chapter 4 Electro-hydraulic Proportional Volume Control

比例变量泵和马达按其控制方式可分为比例排量、比例压力、比例流量和比例功率控制四大类。这四种类型都是在其相应手动控制的基础上发展起来的，所以有必要先来考察一下变量泵的各种控制方式。

Proportional variable capacity pump and motor can be divided into four types according to their control modes, and they are proportional displacement, proportional pressure, proportional flow and proportion power control. The four types are all developed on the basis of their respective manual control, so it is necessary to study different control modes of variable capacity pump firstly.

在液压控制系统中，节流控制，或称流阻控制，是实现各种控制功能的基本手段。它具有结构简单、成本低廉、容易调节及控制精度高等优点，但会产生压力降并导致能耗。对节流调速控制系统的功率损失研究表明，该系统的主要能量损失由两部分组成：即节流损失和溢流损失。上述能量损耗的原因究其本质有两个：流量不适应——过多的流量输入系统；压力不适应——供油压力大于工作压力，以补偿节流压降。解决这一问题的办法是采用容积调速控制，为此设计出了各种变量泵、变量机构及控制策略。

In the hydraulic control system, throttle control, or the so called flow resistance control is the basic method to achieve various control functions. It has the advantages of simple structure, low cost, easy to adjust and high control precision, but it will generate pressure drop and lead to energy loss. The study of power loss of the throttle control system has shown that the main energy loss of the system consists of two parts: the loss of throttle and overflow. The reasons for the energy loss have two aspects: Flow inadaptation—too much flow enters the system; Pressure inadaptation—supplying pressure is higher than the working pressure to compensate for the throttling pressure drop. The solution to this problem is to use capacity speed control. So, we have designed a variety of variable capacity pumps, variable mechanisms and control strategies.

4.1.1 流量适应控制 (Flow adaptive control)

由于节流调速系统既有节流损失，又存在溢流损失（图4-1），用数学式可表示为：

As the throttle speed system has both throttle loss and overflow loss (Fig. 4-1), it could be expressed by mathematical formulas as following:

$$\Delta P = \Delta p q_1 + p_p \Delta q \tag{4-1}$$

式中，ΔP 为液压回路总损失（Total loss of the hydraulic circuits）。

流量适应控制是指泵供给系统的流量自动地与系统的需要相适应，是为完全消除溢流损失而设计的。这种流量供给系统由于消除了过剩的流量，没有溢流损

4.1 容积泵的基本控制方法
(The basic control method of capacity pump)

失,提高了效率。流量适应控制的基本办法是采用变量泵。以下介绍几种较常见的流量适应控制变量泵。

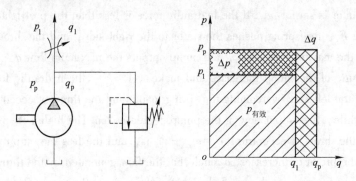

图 4-1 液压回路功率损失
(Fig. 4-1 Power loss of the hydraulic circuits)

Flow adaptive control is that the supply flow can automatically adapt to the needs of the system, and it is designed for the complete elimination of overflow loss. This flow supply system improves efficiency due to the elimination of surplus flow and overflow loss. The basic approach of flow adaptive control is to use variable capacity pump. The followings are some variable capacity pumps used for the flow adaptive control.

4.1.1.1 变排量型变量泵(Variable capacity pump of varying displacement)

A 自动调节式变量机构(限压式变量泵)(Automatic regulation variable mechanism (Pressure-limited variable capacity pump))

图 4-2 为外反馈限压式变量叶片泵结构示意图,由于配油盘上的吸油、压油窗口是关于泵的中心线对称的,所以压力油的合力垂直向上,可以把定子压在滚针轴承上。定子右边的柱塞与泵的压油腔相通。设柱塞面积为 A_x,则作用在定子上的液压力为 pA_x。当这个液压力小于弹簧的预紧力 F_s 时,弹簧把定子推向右边,此时的偏心距达到最大值 $e_{max} = e_0$,泵输出最大流量 q_{max}。当泵的工作压力升高使得 $pA_x > F_s$ 时,液压力克服弹簧力把定子向左推移,偏心距减小了,泵的输出流量也随之减小。压力越高,偏心距 $e_x = e_{max} - x$ 越小,泵输出的流量也越小。当压力增大到偏心距所产生的流量刚好能补偿泵的内部泄漏时,泵的输出流量为零。这意味着不论外负载如何增加,泵的输出压力不会再增高。这也是"限压"的由来。由于反馈是借助于外部的反馈柱塞实现的,故称为外反馈。

Fig. 4-2 shows the structural representation of external feedback pressure-limited variable vane pump. For the oil suction window and oil discharging window on the distributing plate are symmetrical with respect to the axis of the pump, the resultant of the pressure oil vertically upwards, so it can press the stator on the needle roller bearing. The plunger piston on the right side of the stator is connected with the high-pressure fluid chamber. The area of the plunger piston is set for A_x, so the hydraulic force effected on the stator is set for pA_x. If the hydraulic force is less than the pre-tightening force of the spring F_s, the spring pushes the stator to the right side, and the throw of eccentric reaches the maximum $e_{max} = e_0$, the pump outputs the maximum flow q_{max}. If the operating pressure of the pump increases and makes $pA_x > F_s$, the hydraulic force pushes the stator to the left side overcoming the spring load, and the throw of eccentric decreases, meanwhile, the output flow of the pump decreases too. The higher the pressure is, the smaller the throw of eccentric ($e_x = e_{max} - x$) is, and the less the output flow of the pump is. If the pressure increases so much that the flow generated by the throw of eccentric can exactly compensate the inside leakage of the pump, the output flow of the pump is zero. This means that no matter how the external load increases, the output pressure of the pump will not increase any more. This is the origin of "pressure-limited". Because the feedback is by means of external feedback plunger piston, it is called external feedback.

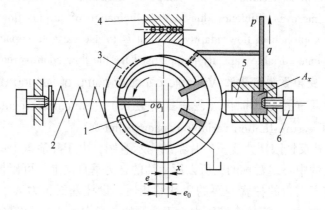

图 4-2 外反馈限压式变量叶片泵

(Fig. 4-2 External feedback pressure-limited variable capacity vane pump)
1—转子（Rotor）；2—弹簧（Spring）；3—定子（Stator）；4—滑块滚针支承（Slide block needle roller bearing）；5—反馈柱塞（Feedback plunger piston）；6—流量调节螺钉（Flow-regulated screw）

(1) 当 $p_B A_x < k_s x_0$ 时，定子相对于转子的偏向量最大，输出流量最大；

If $p_B A_x < k_s x_0$, the throw of eccentric that the stator is relative to the rotor reaches the maximum, and the output flow reaches the maximum；

(2) 当 $p_B A_x > k_s x_0$ 时，定子相对于转子的偏向量减小，输出流量减小；

If $p_B A_x > k_s x_0$, the throw of eccentric that the stator is relative to the rotor decreases, and the output flow decreases；

4.1 容积泵的基本控制方法
(The basic control method of capacity pump)

(3) 当 $p_B A_x = k_s x_0$ 时为转折点。p_B 为调定压力，k_s 为弹簧刚度，x_0 为弹簧的预压缩量。

If $p_B A_x = k_s x_0$, this is a turning point. Where, p_B is pressure, k_s is spring stiffness, x_0 is the pre-compression of spring.

设泵转子和定子间的最大偏心距为 e_{max}，此时弹簧的预压缩量为 x_0，弹簧刚度为 k_s，泵的偏心预调值为 e_0。当压力逐渐增大，使定子开始移动时，压力为 p_B，则有：

The maximum throw of eccentric between rotor and stator of the pump is set for e_{max}, and the pre-compression of spring is x_0, the spring stiffness is k_s, and the throw of eccentric of the pump is previously set for e_0. If the pressure increases gradually, the pressure that just makes the stator moved is set for p_B, so we can get:

$$p_B A_x = k_s (x_0 + e_{max} - e_0) \tag{4-2}$$

当泵压力为 p 时，定子移动了 x 距离，也即弹簧压缩量增加 x，这时的偏心量为：

If the pressure of the pump is p, and the distance that the stator moves is x, namely the compression of spring increases x, the throw of eccentric now is:

$$p A_x = k_s (x_0 + e_{max} - e_0 + x) \tag{4-3}$$

$$e = e_0 - x \tag{4-4}$$

$$p_B = \frac{k_s}{A_x} (x_0 + e_{max} + x) \tag{4-5}$$

图 4-3 所示为限压式变量泵，它的出口接有调速阀，构成了容积节流调速回路。该泵的工作原理是利用输出压力与参比弹簧反力的合力来推动定子移动，使偏心距改变，即改变排量，并使排量总在与反馈压力相应的平衡位置上。其输出的流量特性曲线 1 以及调速阀流量特性曲线 2 如图所示。最大输出流量 q_{max} 可由限制定子的最大偏心距来给定。泵的工作点由曲线 1 与曲线 2 的交点确定，在轻载时处于接近水平的一段上（拐点 B 之前），重载时处在斜线段上。图中斜线段的斜率由参比弹簧刚度 k_s 决定。设预压缩量为 x_0 和变量机构活塞面积为 A，则拐点处的压力 p_B 为：

The following Fig. 4-3 shows the pressure-limited variable capacity pump, and its outlet is connected with a speed regulating valve, which constitutes the capacity-throttle speed control circuit. The working principle of pump is that it uses the resultant force of output pressure and the reference spring reaction force to push the stator and change the throw of eccentric , namely change the displacement, and make the displacement at the balance position correspondent with the feedback pressure at all times. The output flow characteristic curve 1 and speed regulating valve flow characteristic curve 2 are shown in the figure. The maximum output flow q_{max} can be set by the maximum throw of eccentric that limits the stator. The operating point of pump is determined by the intersection point of the curve 1 and curve 2, which is at the subhorizontal section of the curve (before the inflection point B) with light load, at the slope segment with overload. The slope of the slope segment in the figure is determined by the reference spring stiffness k_s. The pre-compression is set for x_0 and the area of the variable mechanism piston is set for A, so the pressure p_B at the inflection point is:

$$p_B = k_s/A \tag{4-6}$$

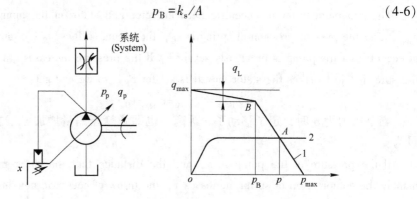

图 4-3 限压式变量泵原理图

(Fig. 4-3 Schematic diagram of pressure-limited variable capacity pump)

若系统压力小于 p_B，则表明输出压力不足以推动变量活塞，这时泵相当于一个定量泵，输出的流量为最大。流量随压力上升而减小是由泵的内泄漏所引起的。若系统压力超过拐点压力 p_B，则反馈弹簧被进一步压缩，定子朝偏心距减小的方向移动，输出流量减小，系统工作在斜线段上。泵的输出流量及压力为：

If the system pressure is less than p_B, it indicates that the output pressure is not able to push the variable piston, then the pump is equivalent to a constant displacement pump, and the output flow reaches the maximum. Flow decreases with the increase of pressure due to the interior leakage of the pump. If the system pressure is more than the inflection point pressure p_B, then the feedback spring is further compressed, and the stator moves towards the direction of reducing the throw of eccentric. The output flow decreases, and meanwhile, the system works at the slope segment. The output flow and pressure of pump are:

$$q = k_p(e_0 - x) - k_1 p_p \tag{4-7}$$

$$p_p = \frac{k_s(x_0 + x)}{A} \tag{4-8}$$

式中 k_p——泵常数，与泵的结构尺寸有关（Pump constant, related to the

4.1 容积泵的基本控制方法
(The basic control method of capacity pump)

structure and size of the pump);

e_0——最大偏心距(Maximum throw of eccentric);

k_1——泄漏系数(Leakage coefficient);

A——变量活塞面积(Area of variable piston);

x_0, k_s——参比弹簧预压缩量及刚度系数(Reference spring pre-compression and stiffness coefficient);

x——定子位移(Sttator displacement)。

当弹簧被压缩至允许的最大位移 x_{max} 时,泵的工作压力亦达到最大值 p_{max}

If the spring has been compressed to the maximum allowable displacement, the working pressure of the pump comes to its maximum p_{max}

$$p_{max} = \frac{k_s(x_0 + x_{max})}{A} \tag{4-9}$$

这时,泵的实际输出流量为零,全部流量仅用于补偿内部泄漏。

At this time, the actual output flow of the pump is zero, and all the flow is only used to compensate the internal leakage.

对上述限压式变量叶片泵的输出特性分析表明,它能基本上消除过剩的流量。能量损耗中没有式(4-1)中的第二项损失,获得了流量适应的性质。

The analysis of the output characteristic of pressure-limited variable capacity vane pump shows that it can basically eliminate the surplus flow. There is no the second loss in formula (4-1) in the energy loss, so it obtains the character of flow adaptive.

必须注意到,上述流量适应性的获得是以系统供油压力的额外增加为代价的,即式(4-1)中右面第一项的 Δp 被增大,它最后被消耗在调速阀的补偿环节上。

It must be noted that the above flow adaptive is at the cost of extra addition of the system supply pressure, that is, the first item Δp on the right side of formula (4-1) is increased, then it is consumed at the compensation link of the speed governing valve.

为了减小 Δp 的提高,应使特性曲线中的倾斜段有近似垂直的性质,即应具有恒压特性。由于倾斜段的斜率仅决定于参比弹簧的刚度,因此,为获得恒压特性,宜采用较软的参比弹簧。但由于限压式变量泵是直接反馈,降低其弹簧刚度不大可能。

In order to reduce the increase of Δp, the slope segment of the characteristic curve should be nearly vertical. That is, it should have a constant pressure characteristic. As the slope of the inclined section is only determined by the reference spring stiffness, so we should adopt a more soft reference spring in order to obtain a constant pressure characteristic. However, as the pressure-limited variable capacity pump is direct feedback, the spring stiffness is unlikely to decrease.

B 机液伺服变量机构（Mechanical and hydraulic servo variable mechanism）

机液伺服变量机构是在手动伺服变量机构的基础上发展起来的。在这种变量方式中，变量活塞跟踪先导阀芯的位移而定位，可见变量活塞的行程与先导阀芯的行程相等。下面介绍的机液伺服式变量机构和电液伺服式变量机构，都是在国产 CY14-1 型手动伺服变量泵的变量机构的基础上，增设比例控制电-机械转换部件而构成的。所以先要了解 CY14-1 型手动伺服变量机构的工作原理。

Mechanical and hydraulic servo variable mechanism is developed on the basis of manual servo variable mechanism. In the variable method, the variable piston is located by following the shift of pilot spool. Visibly, the stroke of variable piston is the same as that of pilot spool. The mechanical-hydraulic servo variable mechanism and electro-hydraulic servo variable mechanism that will be introduced next are both composed by adding the proportional electronic-mechanical switch control components to the domestic CY14-1 type manual servo variable capacity pump. So we should understand the working principle of CY14-1 type manual servo variable mechanism first.

图 4-4 所示是一种常用的机液伺服变量机构，它是由一个双边控制滑阀和差动活塞缸组成的位置伺服控制系统。伺服系统的供油取自泵本身。活塞小端腔 1（直径 D_2）常通来自液压泵的压力油，而大端腔 7 中的油压力受滑阀 5 的位置控制。当输入一个位移信号 Δx 以后，滑阀上部的开口打开，高压油便流入上腔 7（直径 D_1）。上腔面积比下腔大，所以合力向下，使活塞产生对 Δx 的跟随运动 Δy，直到 $\Delta y = \Delta x$ 为止（此时上部的阀口重新关闭）。这时斜盘已产生了一个倾角增量 $\Delta \alpha$，使流量发生正比于输入信号 Δx 的变化。当拉杆上移时，大腔经通道 8 接通回油腔 9，活塞在小腔端推力作用下上移，运动规律与下移时相同。

The Fig. 4-4 is a common mechanical-hydraulic servo variable mechanism, it's a position servo control system constructed with a bilateral control sliding spool valve and a differential piston-cylinder. The pump serves the oil to the servo system itself. The small side cavity 1 of piston (the diameter is D_2) is usually connected with the pressure oil from the hydraulic pump, and the oil of big side cavity 7 is controlled by the position of sliding spool valve 5. As a shift signal Δx is input, the upper port of the sliding spool valve will be opened and the high pressure oil will be injected into the upper cavity 7 (the diameter is D_1). The upper cavity area is bigger than that of the lower cavity, so the resultant force is downward and drives the piston to move a distance Δy following the Δx until $\Delta y = \Delta x$ (the upper valve port is closed again at this time). Meanwhile, the swash plate has generated a dip angel increment $\Delta \alpha$, which will make the flow change proportional to the input signal Δx. When the bar is moved upward, the bigger cavity is

4.1 容积泵的基本控制方法
(The basic control method of capacity pump)

connected with the oil return cavity 9 bypass the channel 8, the piston moves upward driven by the thrust of the smaller cavity, and the principle is the same as that of moving downward.

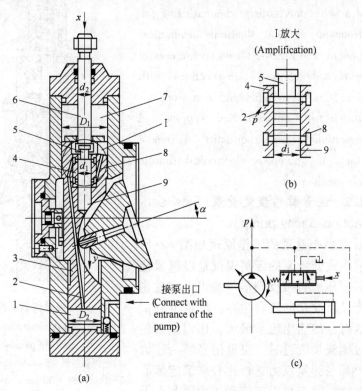

图 4-4 机液伺服变量机构

(Fig. 4-4 The mechanical-hydraulic servo variable mechanism)

(a) 伺服机构结构图 (The servo mechanism construction drawing);
(b) 滑阀放大图 (Sliding spool valve enlarged drawing);
(c) 液压伺服机构职能符号 (The function symbol of hydraulic servo mechanism)

1—活塞小端油腔 (Piston small side cavity); 2—先导阀进油通道 (The inlet channel of pilot valve);
3—差动活塞 (Differential piston); 4—阀套 (Valve pocket); 5—滑阀 (Sliding spool valve); 6—拉杆 (Bar);
7—活塞大端油腔 (Piston big side cavity); 8—大端油腔控制油通道
(The control oil channel of big side cavity); 9—回油腔 (Oil return cavity)

C 电液伺服变量机构 (The electro-hydraulic servo variable mechanism)

伺服电动机输出的是转角,而伺服滑阀需要的是直线位移。因此,利用伺服电动机作电-机械转换元件时,需要加上一个螺旋机构。图 4-5 所示就是这样的一种控制方式。这种变量泵控制精度很高,工作可靠,能与电气自动化系统共同工作,实现遥控比例变量。这种泵实际上就是电液伺服变量泵。

Chapter 4 Electro-hydraulic Proportional Volume Control

The output of servo electromotor is rotation angle, while what the servo sliding spool valve needs is straight-line displacement. Therefore, we need to add a screw mechanism when utilizing the servo electromotor as the electronic-mechanical switch component. The Fig. 4-5 shows such a control method. The variable capacity pump controls with high precision. It works reliably and can work together with electric automatization system and achieves proportional variable quantity by remote control. In fact, the pump is electronic-hydraulic servo variable mechanism.

4.1.1.2 流量敏感型变量泵(Flow-sensitive variable capacity pump)

流量敏感型变量泵的工作原理如图4-6所示。它与限压式变量泵的区别仅仅是以流量检测信号代替了压力直接反馈信号。当没有过剩流量时,流过液阻R的流量为零,控制压力p_0也为零。这时泵的输出流量最大,作定量泵供油。当有过剩流量流过时,流量信号转为压力信号p_0,然后与弹簧反力进行比较来确定偏心距。适当选择液阻R可以把控制压力限制在低压范围,从而使参比弹簧的刚度减小。由于过剩的流量先经过顺序阀,而顺序阀的微小变动就能引起调节作用,故这种流量敏感型变量泵同时具有恒压泵的特性。其压力-流量特性曲线如图4-6所示,显然这种泵的拐点压力及最大压力均由顺序阀的手调机构调节确定。当工作压力低于调定压力时,它是个定量泵。工作压力高于调定压力时,按流量适应变量泵工作,其出口压力基本上与流量无关。

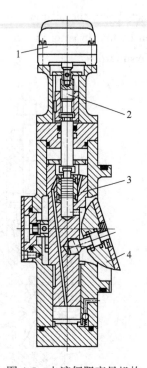

图 4-5 电液伺服变量机构
(Fig. 4-5 The electronic-hydraulic servo variable mechanism)
1—伺服电机 (Servo electromotor);
2—螺旋机构 (The screw mechanism);
3—变量机构 (Variable mechanism);
4—斜盘机构 (Swash plate mechanism)

The working principle of the flow-sensitive variable capacity pump is shown in the Fig. 4-6. The only difference between it and the pressure-limited variable capacity pump is that it replaces the direct pressure feedback signal with the flow detection signal. If there is no surplus flow, the flow flowing through the fluid resistance R is zero, and the control pressure p_0 is zero too. At this time, the output flow of the pump reaches the maximum, and the pump supplies the oil as a constant capacity pump. If there is surplus flow flowing through it, the flow signal is transformed into pressure signal p_0, which then compares with the spring reaction force to determine the throw of eccentric. The appropriate selection of fluid resistance R can limit the control pressure to a low-pressure range, thus reduce the stiffness of the reference spring. Since the surplus flow flows through the sequence valve firstly and a small fluctuation of the sequence valve can cause the regulatory function. So this kind of flow-sensitive variable capacity pump also has the characteristic of constant pressure pump. The pressure-flow characteristic curve is shown in Fig. 4-6, evidently, the inflection point pressure and the maximum pressure of the pump are both determined by manual adjustment mechanism of the sequence valve. When the working pressure is lower than the set pressure, it is a constant capacity pump. And when the working pressure is higher than the set pressure, it works as an adaptive flow variable capacity pump and its outlet pressure has nothing to do with the flow.

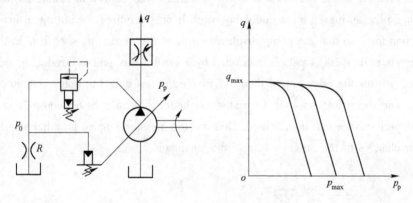

图 4-6 流量敏感型变量泵

(Fig. 4-6 Flow-sensitive variable capacity pump)

4.1.1.3 恒压变量泵(Constant pressure variable capacity pump)

恒压变量泵是指当流量作适应调节时压力变动十分微小的变量泵。

Constant pressure variable capacity pump is a variable capacity pump whose pressure change is very small when the flow is adaptively adjusted.

从前两节的分析中可知，要获得恒压的特性，必须尽量采用弱刚度的弹簧作为参比对象，和先导式溢流阀一样，用压差来与弹簧力平衡，其恒压性能就比直动式的好得多。恒压变量泵也是利用这一原理来获得压力浮动很小的恒压性能的。

According to the analysis of former two sections, we must adopt the weak stiffness spring as the reference object to obtain the characteristic of constant pressure. Just like the pilot overflow valve, it uses differential pressure to make a balance with the spring force. So its performance of constant pressure is much better than the direct-action overflow valve. Constant pressure variable capacity pump also uses this principle to obtain the constant pressure performance of small fluctuation.

图 4-7 给出了恒压变量泵的工作原理和性能曲线。在泵中设置两个控制腔,利用两端的压力差来推动变量机构做恢复平衡位置的运动。这样弹簧刚度就可以较小,只要求弹簧能够克服摩擦力,使泵在无压时能处于最大排量的状态即可。其弹簧腔由一小型的三通减压阀控制。实质上是利用了先导控制原理,把检测环节(减压阀的弹簧)和调节环节(变量机构)分开,以获得较好的恒压变量性能。即流量从零至最大的范围变化时,压力变化甚小。

Fig. 4-7 shows the working principle and performance curve of the pump. Two control cavities were set in the constant pressure variable capacity pump and it use the differential pressure of the two sides to drive the variable mechanism to restore the balance position. So the spring stiffness could be lower. It only requires the spring to overcome the friction force so that the pump displacement is at the maximum when it is in the no-pressure state. Its spring cavity is controlled by a small 3-way pressure reducing valve. In essence, it uses the pilot control theory to divide the test unit (pressure reducing valve spring) and the regulatory unit (variable mechanism). So a better constant pressure variable performance can be obtained. That is, the pressure change is rather small when the flow changes in the range of zero to the maximum.

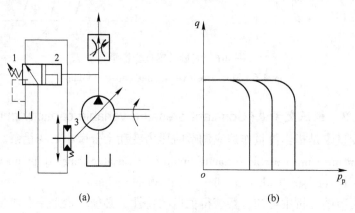

图 4-7 恒压变量泵原理图及其特性曲线

(Fig. 4-7 Schematic diagram and characteristic curve of the constant pressure variable capacity pump)

(a) 恒压变量泵原理图 (Schematic diagram of the constant pressure variable capacity pump);
(b) 恒压变量泵特性曲线 (Characteristic curve of the constant pressure variable capacity pump)
1—调压弹簧 (Pressure-regulator spring); 2—三通减压阀 (Three-way pressure reducing valve);
3—变量机构 (Variable mechanism)

4.1 容积泵的基本控制方法
(The basic control method of capacity pump)

泵的工作原理是：当输出压力比减压阀 2 的调定压力小时，减压阀全开不起减压作用。这时，变量机构 3 液压力平衡，在复位弹簧作用下处在排量最大位置。当泵的输出压力等于或超过调定压力时，减压阀阀芯向左移动，使供油压力和变量机构下腔部分与油箱接通。因此下腔压力降低，使变量机构失去平衡。上腔液压力推动变量机构下移压缩复位弹簧，直至重新取得力平衡。与此同时，变量机构的移动使流量作适应变化。

The principle of pump is that when the output pressure is less than the set pressure of pressure reducing valve 2, the pressure reducing valve is opened throughly and can not reduce the pressure. At this time, the hydraulic pressure of variable mechanism 3 is in balance and it is at the maximum displacement position under the impact of the reset spring. If the output pressure of the pump is equal to or more than the set pressure, the valve spool moves to the left, then the supply pressure and the below cavity of the variable mechanism is connected with the tank. Therefore, because of the pressure of below cavity decreasing, the variable mechanism is not in balance. The hydraulic force of the above cavity pushes the variable mechanism and compresses the reset spring until the force balance is obtained again. Meanwhile, the flow changes adaptively due to the move of variable mechanism.

图 4-7b 所示为获得的恒压力曲线，图中垂直段的斜率由参比弹簧的刚度决定。调节三通减压阀的弹簧 1，改变其压缩量就可以方便地定出其特性曲线上转折点的位置，实现对该泵的压力调节。图中水平段由泵的最大输出流量决定。显然，合理地使用恒压变量泵比限压式变量泵有更佳的节能效果。这种泵也很容易制成比例压力变量泵，只要用比例电磁铁取代调节弹簧 1 便可。对于要求恒压的地方，该泵很有实用价值。图 4-8 所示为恒变量泵的能量平衡图。

The Fig. 4-7b is the obtained constant pressure curve, the slope of the vertical segment in the figure is determined by the stiffness of reference spring. We can conveniently obtain the position of the inflection point of its characteristic curve by adjusting the spring 1 of the three way reducing valve and changing its compression. The pump pressure regulation has been accomplished. The horizontal segment in the figure is determined by the maximum output flow of the pump. Obviously, rational usage of the pump has a better energy-saving performance than the pressure-limited pump. This kind of pump is also easily turned into a proportional pressure variable capacity pump, as long as we replace the adjustable spring 1 with proportional solenoid. The pump is very practical to the condition that requires constant pressure. Fig 4-8 shows the energy balance diagram of constant pressure variable capacity pump.

图 4-9 所示为一种新型的恒压控制式 PV 泵。

The Fig. 4-9 is a new constant pressure control pump called PV.

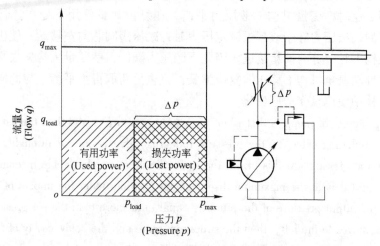

图 4-8 恒压变量泵的能量平衡图

(Fig. 4-8 The energy balance diagram of constant pressure variable capacity pump)

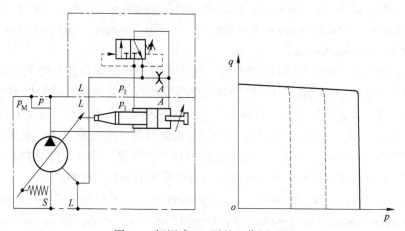

图 4-9 恒压式 PV 泵的工作原理图

(Fig. 4-9 The working principle diagram of constant pressure type PV pump)

与图 4-1 比较可知，泵损失的能量有了明显减少，原因是泵的出口与变量机构连接，泵可做流量适应，减少了溢流损失。

Compared with Fig. 4-1, we can see that the loss energy of pump decreases obviously. The reason is that the outlet of pump is connected with variable mechanism, and the pump can make flow adaption and reduce the overflow loss.

泵的输出压力小于调定压力时，变量机构处于最大流量状态。泵的输出压力超过调定时，控制阀左移，油液进入变量机构右腔，推动变量机构左移，直至达到新的平衡点，同时流量作适应变化。

4.1 容积泵的基本控制方法
(The basic control method of capacity pump)

When the output pressure of pump is lower than the set pressure, the variable mechanism is in the state of the maximum flow. When the output pressure is over the set pressure, the control valve spool moves left, and the oil flows into the right cavity of variable mechanism, then drives the variable mechanism left until it reaches a new balanced point, meanwhile the flow changes adaptively.

4.1.2 压力适应控制 (Pressure adaptive control)

压力适应控制是指泵供给系统的压力自动地与系统的负载相适应，完全消除多余的压力。压力适应控制可以由定量泵供油系统来实现。这对于负载速度变化不大，而负载变化幅度大且频繁的工况，具有很好的节能效果。在定量泵压力适应系统中，当供油量超过需要量时，多余的流量不像定压系统那样从溢流阀中排走，也不像限压变量泵那样会迫使系统压力升高来减少输出流量，而是从非恒定的流阻排出。此流阻的阻值随着执行机构的负载和速度不同而变化。常见的压力适应控制系统有恒流源系统、旁路节流系统和负载压力反馈回路。下面只介绍与比例变量泵关系最为密切的负载压力反馈压力适应控制回路。

Pressure adaptive control refers that the pump pressure supplies to the system automatically adapting to the load of the system to eliminate the surplus pressure completely. Pressure adaptive control can be achieved by the system that the constant capacity pump supplies the oil. It has a perfect energy-saving effect on the working condition of the little load speed changing but the load change amplitude is large and frequent. In the constant capacity pump pressure adaptive system, if the oil supplying is in excess of requirement, the surplus flow is not drained in the same way as the overflow valve in the constant system, and does not make the pressure of the system increase to reduce the output flow as it does in the pressure-limited variable capacity pump. It is drained from the non-constant flow resistance. The value of the flow resistance changes with the distinction of the load and speed of the actuator. Common pressure adaptive control system includes constant flow source system, bypass throttle system and load pressure feedback circuit. Here, we only introduce the pressure adaptive control circuit with load pressure feedback, which is closest to the proportional variable capacity pump.

4.1.2.1 定差溢流型压力适应控制(Constant differential relief pressure adaptive control)

这种回路只需使溢流阀的弹簧腔接受负载压力反馈信息，构成压差控制溢流阀（或称定差溢流阀），就能使溢流阀失去恒阻抗流阻的性质，成为一种与负载压力相适应的非恒定流阻，从而使系统供油压力与负载需要相适应，实现节能。

This circuit only makes the spring cavity of the overflow valve to accept the feedback of load pressure and constitutes a differential pressure control overflow valve, or it is called constant differential overflow valve, which can make the overflow valve lose the characteristic of constant flow resistance and make it a non-constant flow resistance corresponding with the load pressure, so it makes the system oil supply pressure adaptto the need of load and obtain energy conservation.

若将节流阀的出口压力作为反馈压力,实质上就构成了一个溢流节流型调速阀(图4-10b)。压力适应控制虽然有利于节能,但系统的压力总是追随负载的变化而变化,使系统的泄漏和容积效率等对系统的影响不稳定,这对系统的刚性、平稳性不利。如果使定差溢流阀的进口压力点与取出反馈压力点尽可能地包围更多的液压元件,可以提高系统的刚性和速度的稳定性。

If the output pressure of the throttle valve acts as the feedback pressure, it constitutes an overflow throttle speed regulating valve actually (Fig. 4-10b). Although the pressure adaptive control is good to energy conservation, the system pressure always changes with the change of the load, which makes the leakage and volumetric efficiency of the system have unstable effect on the system, and it is negative to the stiffness and stationarity of the system. If the inlet pressure point of the constant differential overflow valve and the feedback pressure point cover more hydraulic components as many as possible, the stiffness and the speed stability of the system will be improved.

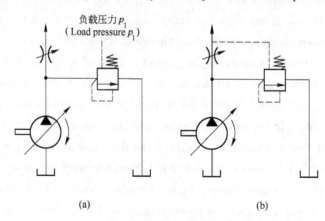

图 4-10 定差溢流型压力适应控制回路

(Fig. 4-10 Constant differential relief pressure adaptive control circuit)

(a) 使用定差溢流阀 (With constant differential overflow valve);
(b) 使用溢流节流型调速阀 (With relief throttle type speed control valve)

4.1.2.2 流量敏感型稳流量变量泵(The flow-sensitive steady flow variable capacity pump)

把定差溢流阀用于变量泵时,可以解决由于压力追随内泄漏和容性引起的不

4.1 容积泵的基本控制方法
(The basic control method of capacity pump)

稳定问题。把定差溢流节流阀的溢流信息作为变量泵的变量信号,向减小流量方向变化,就可以完全消除溢流量,还能保证输出流量不受外载的影响,构成所谓的稳流量变量泵。

The adoption of constant differential overflow valve in the variable capacity pump can solve the problem of instability caused by leakage and volumetric efficiency which follows load pressure. If we regard the overflow information of the constant differential relief throttle as the variable signal of the variable capacity pump, and change it toward the direction of reducing the flow, it can completely eliminate the overflow and ensure that the output flow will not be influenced by external load, which is called steady flow variable capacity pump.

图 4-11 为流量敏感型稳流量变量泵的工作原理图。节流阀 R_1 与定差溢流阀构成溢流节流阀,节流阀的前后压差 $\Delta p = p_p - p_1$ 被定差溢流阀固定,所以进入系统的流量不受负载影响,只由节流阀的开口面积来决定。泵的出口压力 p_p 追随负载压力 p_1 变化,两者相差一个不大的常数 Δp。所以它是一个压力适应动力源,特点是以流量检测信号代替原来的压力直接反馈。在工作压力适应调节时,过剩的流量从溢流阀流走。由固定流阻 R_2 检测,并转为压力信号。然后反馈到变量柱塞的弹簧腔,与弹簧力一起和泵出口压力进行比较。由于节流阀 R_1 的截面积比 R_2 大得多,且主节流孔后的液容也比 R_2 后的液容大得多,因此,同样的流量变化,Δp_2 要比 Δp 的变化大得多。因 R_2 很小,使溢流阀的溢流量保持在很小的水平上。通过定差溢流阀流量的微小变化能引起压力 p_1 的较大变化,故溢流量的微小变化就能引起泵的动作。所以这种泵被称为流量敏感型泵。实质上此泵同时具有压力适应与流量适应的性质。

The figure below shows the working principle of flow-sensitive steady flow variable capacity pump. The throttle valve R_1 and the constant differential overflow valve constitute the overflow throttle valve, and the input and output differential pressure of the throttle valve $\Delta p = p_p - p_1$ is fixed by the constant differential overflow valve. So the flow that enters into the system will not be influenced by the load, and it is only determined by the opening area of the throttle valve. The output pressure of the pump p_p changes along with the load pressure p_1, the difference between them Δp is a small constant. So it is a pressure adaptive power source. Its characteristic is to replace the original direct pressure feedback with flow detection signal. When the working pressure is regulated adaptively, the surplus flow flows away through the overflow valve. Then it is detected by the fixed flow resistance R_2 and is converted into a pressure signal. Then it sends feedback to the spring cavity of the variable plunger and compares with the pump output pressure together with the spring force. As the area of the throttle valve R_1 is much larger than that of R_2. As the liquid capacity behind the main orifice is much larger than that behind R_2, the change of Δp_2 is much larger than that of Δp with the same flow changing. The overflow quantity of the overflow valve is kept to a very small level because R_2 is very small. A small flow changing through the constant differential overflow valve can cause large change of pressure p_1. So a slight change of the overflow will cause a response of the pump, then the pump is called flow-sensitive type. In essence, the pump has the characteristics of pressure adaptive as well as the flow.

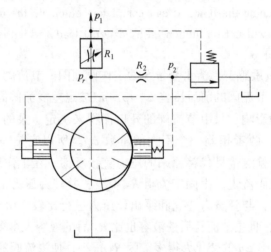

图 4-11 流量敏感型稳流量变量泵

(Fig. 4-11 The flow-sensitive steady flow variable capacity pump)

4.1.3 功率适应控制 (Power adaptive control)

液压功率由压力与流量的乘积来决定。无论是流量适应系统或压力适应系统，都只能做到单参数适应，因而都是不够理想的能耗控制系统。功率适应动力源系统，即压力与流量两参数同时正好满足负载要求的系统，才是理想的能耗控制系统。它能把能耗限制在最低的限度内。事实上，上一节中介绍的流量敏感型稳流量变量泵就具有压力与流量的适应能力。采用这种泵的系统具有功率适应性。

The hydraulic power is determined by the product of pressure and flow. Both the flow adaptive control system and the pressure adaptive control system can only achieve single parameter adaptation, thus they can't be ideal power control systems. Power adaptive dynamic source system is an ideal power control system that both parameters of pressure and flow simultaneously meet the requirement of the load. It can limit the energy consumption to a minimum range. In fact, the flow-sensitive steady flow variable capacity pump introduced in the last section has both pressure adaption and flow adap-

4.1 容积泵的基本控制方法
(The basic control method of capacity pump)

tion. The system using this kind of pump has power adaptability.

图 4-12 所示为一种差压式稳流量变量叶片泵。泵的变量机构由在定子两边的两个柱塞缸组成。定子的移动靠两个柱塞缸上的液压作用力之差克服弹簧 4 的作用力来实现。所以这种泵称为差压式变量泵。这种泵的工作情况为：节流阀 2 的出口压力，即负载压力反馈到右边柱塞弹簧腔，泵的出口压力，即节流阀的进口压力反馈到左柱塞腔。由于左右柱塞承压面积相等，故压差 Δp 为：

Fig. 4-12 shows a differential pressure steady flow variable vane pump. The variable mechanism of the pump is formed with two plunger cylinders at both sides of the stator. The movement of the stator is driven by the hydraulic difference force of the two plunger cylinders that overcomes the force of spring 4. So this pump is called differential pressure variable capacity pump. The working condition of the pump is: the outlet pressure of the throttle valve 2, that is, the load pressure sends feedback to the plunger spring cavity on the right. The outlet pressure of the pump, that is, the inlet pressure of the throttle valve sends feedback to the plunger cavity on the left. As the two plungers have the same bearing area, so the differential pressure Δp is:

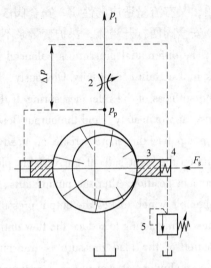

图 4-12 差压式稳流量变量泵
(Fig. 4-12 Differential pressure steady flow variable capacity pump)
1, 3—左右变量柱塞 (Left and right variable plunger);
2—节流阀 (Throttle valve); 4—参比弹簧 (Reference spring); 5—安全阀 (Safety valve)

$$\Delta p = p_p - p_1 = \frac{F_s}{A} \tag{4-10}$$

式中　F_s——弹簧力 (Spring force);
　　　A——承压面积 (Pressure area)。

这个压差由弹簧力来平衡，使泵在某一偏心距下工作。显然，Δp 的稳定精度与参比弹簧的刚度直接相关。如果 F_s 变化不大，则 Δp 近似不变，泵的输出流量也变化不大。故这种泵具有稳流量的特性。泵启动前，变量柱塞右端的弹簧使定子保持在最大排量位置。泵启动后，节流阀 2 限制了泵的输出流量通过，使泵出口压力增高，这导致参比弹簧动作使流量减小，直到节流阀限定的流量值。由于负载压力大小的信息被反馈到柱塞缸的弹簧腔，这构成了压力负反馈。负载压力变化时，会引起泵的压力作适应调节。可见，采用此泵的系统既有压力适应的性质，又有流量适应的性质。安全阀5的作用是当 p_1 升高到安全阀的调定压力时，安全阀便打开溢流，保护变量泵不受损害。

The differential pressure is balanced by the spring force, and it makes the pump work in a certain eccentricity. Obviously, the stability accuracy of Δp is directly related to the stiffness of the reference spring. If the F_s doesn't change much, the Δp doesn't change approximately, and the output flow of the pump doesn't change much either. So this pump has the characteristic of steady flow. Before the pump starts to work, the spring on the right side of the variable plunger makes the stator at the maximum displacement position. After the pump starts, the throttle valve 2 limits the output flow of the pump, which makes the output pressure of the pump increasing, and leads the reference spring acting to reduce the flow until the limit flow of the throttle valve. As the information of the load pressure is transmitted to the spring chamber of the variable plunger cylinder, it constitutes a pressure negative feedback. When the load pressure changes, the pump pressure will make corresponding adjustment adaptively. It can be seen that the system adopting this pump has both pressure adaptive and flow adaptive characteristics. The function of safety valve 5 is that if the p_1 increases to the set pressure, the safety valve will open to protect the variable capacity pump from damage.

图4-13所示为一种较完善的功率适应动力源。其特点是泵的变量机构不是靠负载反馈信号直接控制，而是通过一个三通减压阀作为先导阀来控制，因而具有先导控制的许多优点，特别是灵敏度高，动特性好。主节流阀的压差被减压阀的参比弹簧固定，因此，通过主节流阀的流量仅由其开口面积决定。改变开口面积可使流量-压力特性曲线上下平移。调节减压阀的弹簧预压缩量可以改变拐点的压力，即可使垂直段左右移动，这样便可适应不同的工况要求。

In Fig. 4-13, there is a relatively perfect adaptive power source. Its characteristic is that the variable mechanism of the pump is not controlled by the load feedback signal directly, but controlled by a three-way pressure reducing valve as a pilot valve. Therefore, it possess many advantages of the pilot control, especially the high sensitivity and good dynamic characteristic. The differential pressure of the main throttle valve is fixed by the reference spring of the reducing valve, so the flow of the main throttle valve is only determined by the opening area. Changing the opening area can make the characteristic curve of the flow and pressure shift up and down. Regulating the pre-compression of the spring of the sliding spool valve can change the pressure of the inflection point, which is to make the vertical segment move right and left. So it can adapt to different working condition requirements.

4.1 容积泵的基本控制方法
(The basic control method of capacity pump)

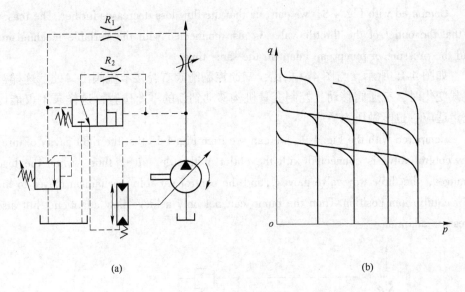

图 4-13 功率适应变量泵
(Fig. 4-13 Power adaptive variable capacity pump)
(a) 原理图 (Principle diagram);
(b) 流量-压力特性曲线 (Characteristic curve of flow and pressure)

图 4-14 所示为一种负载敏感型 PV 泵。
The Fig. 4-14 is a load sensitive PV pump.

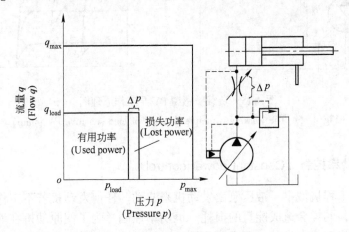

图 4-14 负载敏感变量泵的能量平衡图
(Fig. 4-14 The energy balance diagram of load sensitive variable capacity pump)

与图 4-8 比较可知，泵的流量损失进一步减小，原因是节流阀的出口也连接到变量机构处，泵的压力同样也能作适应。

Compared with Fig. 4-8, we can see that the flow loss decreases further. The reason is that the outlet of the throttle valve is also connected with the variable mechanism, and the pressure of pump can adapt in the same time.

如图 4-15 所示,与图 4-9 相比,三通控制滑阀右腔接节流阀出口处,这样当负载变化时,三通阀移动,控制变量机构移动到新的平衡位置,这样泵不仅能做流量适应,也能做压力适应。

Compared with the Fig. 4-9, we can see from Fig. 4-15 that the right cavity of three way control valve is connected with the outlet of throttle valve, thus, when the load changes, the three way valve moves, and the control variable mechanism moves to the new equilibrium position. Then the pump can not only achieve flow adaption, but also pressure adaption.

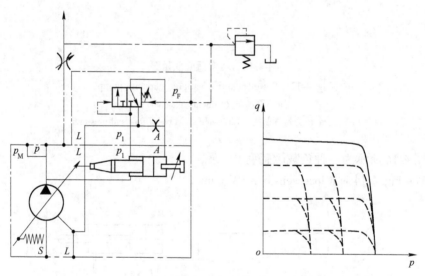

图 4-15 负载敏感型 PV 泵工作原理图
(Fig. 4-15 The working principle diagram of load sensitive PV pump)

4.1.4 恒功率控制 (Constant power control)

在许多工程机械中,液压泵由发动机来驱动。任何发动机若不工作在高效率的最佳工况,同样会造成能量的损耗。恒功率控制是为了使原动机在液压泵的任何工况下都能工作在最佳工况下而设计的,从而减少原动机的能耗。

In many construction machineries, the hydraulic pump is driven by the engine. If the engine does not work in the high-efficiency optimum working condition, it will also cause the loss of energy. Constant power control is designed to make the prime motor work in the best working condition whatever the hydraulic pump working condition is. So

4.1 容积泵的基本控制方法
(The basic control method of capacity pump)

it can reduce the energy losing of the prime motor.

恒功率控制是使泵的总输出功率具有与负载无关、维持恒定的性质。因液压功率 P 为压力 p 与流量 q 的乘积，即 $P=p \cdot q$，所以精确的恒功率控制中，流量与压力的关系是压力升高时，流量随压力呈双曲线规律减小。如图 4-16b 所示。

Constant power control has the characteristic of making the total output power of the pump has nothing to do with load and keep the power constant. As the power P is the product of pressure p and flow q, that is $P=p \cdot q$. So in the precise constant power control, the relationship between flow and pressure is that if the pressure increases, the flow decreases with pressure in the hyperbolic law. It is shown in Fig. 4-16b.

图 4-16a 为一种恒功率控制器的结构原理图。控制器中有两根套在一起的弹簧，低压时一根较软的弹簧起作用，这时泵的输出对应于 bc 段，高压时两根弹簧同时接触变量活塞。这时弹簧刚度变大，其特性曲线对应 cd 段。两段合起来就能近似实现恒功率控制。

In the Fig. 4-16a, there is a structural schematic diagram of a constant power controller. There are two springs in the controller set together. If the pressure is low, the soft spring works, the output of the pump corresponds to the bc segment. While if the pressure is very high, the two springs contact the variable piston at the same time, and the stiffness of the spring becomes larger, meanwhile the characteristic curve corresponds to the cd segment. Combining the two together, it will be able to achieve the constant power control.

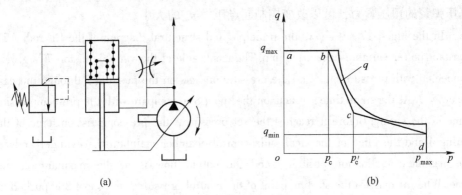

图 4-16 恒功率控制变量泵

(Fig. 4-16 Constant power control variable capacity pump)

(a) 恒功率控制器的结构原理图 (Structural schematic diagram of constant power controller);

(b) 恒功率控制器特性曲线 (Characteristic curve of constant power controller)

为了能调节变量控制的起点 b，控制腔与一小溢流阀串接，且并联一个流阻产生可变压差。这个压差与弹簧的合力用来推动变量活塞。调节溢流阀的溢流压力可改变变量起点压力 p_c。在低压段，溢流阀未开启，泵处于最大排量位置，这时是定量泵状态。当压力升高到溢流阀设定压力后，变量活塞开始移动并压缩弹簧，使流量随压力呈双曲线减小。

In order to adjust the starting point b of variable control, the control chamber cascades a small overflow valve and connects in parallel with a flow resistance to get a variable differential pressure. The resultant force of the differential pressure and the spring is used to push the variable piston. Adjusting the overflow pressure of the overflow valve can change the pressure p_c of the variable starting point. In the low-pressure section, the overflow valve is closed, and the pump is at the maximum displacement position. At this time the pump is a constant capacity pump. When the pressure increases to the set pressure of the overflow valve, the variable piston will begin to move and compress the spring so that the flow decreases with pressure in the hyperbolic law.

图 4-17 为力士乐型 A7V 恒功率变量泵的原理及结构图。变量活塞 7 小端常通压力油，压力油经流道 2 作用在推杆 5 上，并作用在控制阀 8 的阀芯上与调压弹簧的预压缩力比较。在达到预调压力之前，弹簧 4 的预压缩力及小端控制腔上的作用力使缸体保持在最大倾角位置。这时输出流量为最大。当到达控制起点压力时，阀芯 8 被推杆推动，使油腔 a 与 b 接通，这时差动活塞带动缸体向倾角减小的方向移动，直至变量活塞上的差动作用力与弹簧 4 上的压缩力（或弹簧 4 与弹簧 6 的合力）平衡为止。这时输出的流量随压力的增加而减小，其斜率由弹簧的组合刚度确定，流量-压力曲线如图 4-16b 中曲线 abcd 所示。调节螺钉 9，改变调压弹簧的预压紧力，可以改变控制起点压力 p_c 的大小。

In the Fig. 4-17, it shows the principle and structural diagram of the Rexroth A7V constant power variable capacity pump. The small side of the variable piston 7 is always connected with pressure oil, and the pressure oil acts on the pushrod 5 through the passageway 2. At the same time, it acts on the spool 8 to compare with the pre-compression force of the spring. Before it reaches the set pressure, the pre-compression force of the spring 4 and the force of the small side control chamber maintain the cylinder body at the maximum inclination position, and the output flow is at the maximum at this time. When it reaches the starting point of the control pressure, the spool 8 is pushed by the pushrod so that the chamber a and chamber b is connected. Then the differential piston does not drive the cylinder body to the direction of inclination reduction until the differential force of the variable piston is equivalent to the compression force of the spring 4 (or the resultant force of spring 4 and spring 6). Then the output flow decreases with the pressure increasing. The slope is determined by the combination stiffness of the spring. The flow-pressure curve is shown as the *abcd* segment in Fig. 4-16b. Adjusting

4.1 容积泵的基本控制方法
(The basic control method of capacity pump)

the screw 9 to change the pre-compression of the pressure regulating spring can change the control starting point pressure p_c.

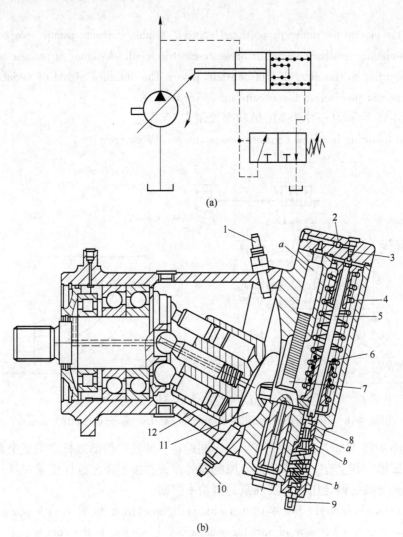

图 4-17 A7V 恒功率泵的原理及结构图
(Fig. 4-17 Schematic and structural diagram of the A7V constant power pump)
(a) 原理图 (Schematic diagram); (b) 结构图 (Structural drawing)
1—q_{min} 限位螺钉 (Limit screw of q_{min}); 2—控制油道 (Control oil duct); 3—柱塞 (Plunger);
4——级弹簧 (Primary spring); 5—推杆 (Puncher); 6—二级弹簧 (Secondary spring);
7—变量活塞 (Variable capacity piston); 8—控制阀 (Control valve);
9—控制起点压力调节螺钉 (Control starting point adjustment screw);
10—q_{max} 限位螺钉 (Limit screw of q_{max}); 11—配流盘 (Thrust plate); 12—缸体 (Cylinder body);
a—小腔 (常通压力油) (Small cavity (Open to pressure oil)); b—大控制腔 (Large control cavity)

在电液比例控制的变量泵中，还可以利用流量-电反馈或压力-电反馈的信号，按恒功率的算法进行处理和运算。求出变排量的信号，可以准确地实现恒功率控制。

In the electro-hydraulic proportional control variable capacity pump, we can also use flow-electric feedback signal or pressure-electric feedback signal to process and operate according to the algorithm of constant power. The obtained signal of variable displacement can precisely achieve constant power control.

图4-18所示为一种新型的恒功率控制式PV泵。

The following is a new constant power control PV pump.

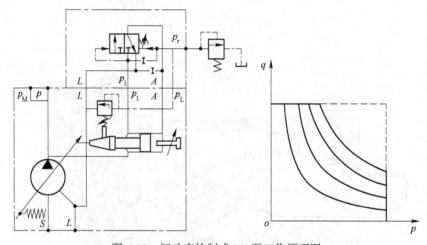

图4-18 恒功率控制式PV泵工作原理图

(Fig. 4-18 The working principle diagram of constant power control PV pump)

由图4-18可知，与前面的PV泵不同的是，变量机构活塞杆左端成类似凸轮轨迹的凹形，其上有一根杆与先导控制阀的弹簧直接相连，这样变量机构运动时能控制先导阀的调定压力，实现泵的恒功率控制。

We can see from the Fig. 4-18 that what is different from the above PV pump is that the shape of rod's left side of variable mechanism is similar to the concave of cam locus. It is equipped with a bar which directly connected with the spring of pilot control valve. So the variable mechanism can control the set pressure of pilot valve when it is working to achieve the constant power control of pump.

4.2 比例排量变量泵和变量马达（Proportional displacement variable pump and variable motor）

在一定转速下，变量泵吸入和马达输出的流量正比于排量。而排量是由变量

4.2 比例排量变量泵和变量马达
(Proportional displacement variable pump and variable motor)

机构的位置决定的,因此,电液比例排量控制本质上是个电液比例位置控制系统。它利用适当的电-机转换装置,通过液压放大级对变量泵或马达的变量机构进行位置控制,使其排量与输入信号成正比。这样的控制称为电液比例排量控制。

At a definite rotating speed, the output flow or suction flow of the variable capacity pump or variable motor is proportional to the displacement. The displacement is determined by the position of the variable mechanism. So the electro-hydraulic proportional displacement control is essentially an electro-hydraulic proportional position control system. It utilizes an appropriate electro-mechanical conversion equipment and controls the position of the variable mechanism of the variable capacity pump or variable motor by a hydraulic amplifier so that the displacement is proportional to the input signal. This is called electro-hydraulic proportional displacement control.

比例排量调节变量泵的输出流量能在负荷状态下跟随输入信号作连续比例变化,而泵的压力由外部负载决定。由于泵的容积效率随工作压力的上升而下降,使这种泵的流量随负载而变化,得不到精确地控制。但由于它能在工作状态下用电气遥控变量,故因容积效率下降而减小输出的流量,可以通过电气控制的办法,增加或减小输入控制信号,使排量作小量的变化而得到补偿。

The output flow of the proportional variable capacity pump can change continuously and proportionally following the input signal under loaded conditions. The pump pressure is determined by the external load. As the volumetric efficiency decreases with the increase of the working pressure, it makes the flow change with the load and can not achieve precise control. However, it can achieve electric remote control variable under working condition, so the decreased output flow because of the decrease of the volumetric efficiency can increases or decreases by electric control signal. This is realized through making the displacement change a little.

虽然比例容积控制只是用比例电磁铁和先导阀来取代手调变量机构,但这不仅是操作方式的不同,更重要的是能够利用电气控制来进行各种压力补偿、流量补偿以及功率的适应控制,从而使控制水平大为提高。

Although the proportional volume control is just using the proportional solenoid and pilot valve to replace the manual variable mechanism, it is more than the difference of operation modes, what's more important is that it can carry out a variety of pressure compensation, flow compensation and power adjustment control by electricity control so as to improve the control level remarkably.

图 4-19 为几种常见的比例排量调节方案。比例排量调节的方式按反馈信号的类型分为三种,即位移直接反馈式、位移-力反馈式和位移-电反馈式。图中压

第4章 电液比例容积控制
Chapter 4 Electro-hydraulic Proportional Volume Control

力油输入处可以接内部或外部的压力油。比例阀可以是直动式的,有时需要先导式的,使其获得足够大的推力。也可用机械力代替控制力,必要时可以用手动,这时就相当于手动变量式变量泵。变量缸内设有对中弹簧,以便在无信号时回到无流量状态。图 4-19a 所示为一种单向变量机构,图 4-19b 和 4-19c 可以用于双向变量。

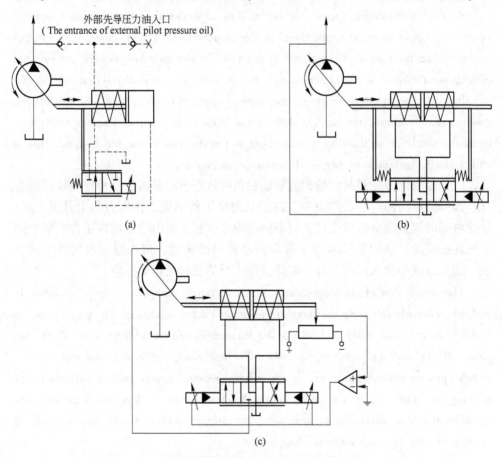

图 4-19 电液比例排量调节

(Fig. 4-19 Electro-hydraulic proportional displacement adjustment)
(a) 位移直接反馈式 (Direct shift feedback); (b) 位移-力反馈式 (Shift-force feedback);
(c) 位移-电反馈式 (Shift-electricity feedback)

From the Fig. 4-19 we can see several common proportional displacement adjustment schemes. The proportional displacement adjustment modes are divided into three types according to the feedback signal, the direct shift feedback, shift-force feedback and shift-electricity feedback. The position P where the pressure oil enters in the figure can get access to internal or external pressure oil. The proportional valve in the figure can be direct-acting, sometimes pilot-operated in need, so as to obtain a large

4.2 比例排量变量泵和变量马达
(Proportional displacement variable pump and variable motor)

enough thrust. The control force can be replaced by mechanical force. If necessary, it can be replaced by hand control. At this time it is the same as the manual variable type variable capacity pump. There is a alignment spring in the variable cylinder so that it returns to non-flow state when there is no signal. Fig. 4-19a shows a one-way variable mechanism, but Fig. 4-19b and 4-19c can be used to bi-directional variable.

4.2.1 位移直接反馈式比例排量调节 (Shift direct feedback proportional displacement adjustment)

位移直接反馈式比例变量泵是在手动伺服变量泵的基础上增设比例控制电-机械转换部件而构成的。在这种变量方式中，变量活塞跟踪先导阀芯的位移而定位。可见变量活塞的行程与先导阀芯的行程相等。如图4-4和图4-5所示的变量机构，即属于位移直接反馈。图4-20所示比例减压阀控制式的变量机构也属于位移直接反馈式。

The shift direct feedback proportional variable capacity pump is constructed on the basis of the manual servo variable capacity pump by adding proportional control electro-mechanical switch components. In such a variable method, variable piston locates by following the shift of the pilot spool. It's seen that the stroke of variable piston is equal to that of pilot spool. As the variable mechanism showed by the above Fig. 4-4 and 4-5, both of them are shift direct feedback. The proportional pressure reducing control variable mechanism showed by the Fig. 4-20 is also shift direct feedback.

4.2.1.1 比例减压阀控制的比例排量调节变量泵 (Proportional displacement regulating variable capacity pump controlled by the proportional pressure reducing valve control)

采用比例减压阀输出的液压油来推动伺服阀的阀芯，实现电液比例排量调节，这样比例电磁铁的行程就不受变量活塞的行程限制。图4-20所示比例排量变量泵使用的就是这种形式的比例排量调节机构。它是在手动伺服变量泵上增设电液比例减压阀2和操纵缸3而构成的。其控制精度及灵敏度虽不如电液伺服变量泵，但其抗污染能力强，价格低廉，工作可靠，对许多机械的远程控制是很理想的。

It utilizes the hydraulic oil from the proportional pressure reducing valve to drive the spool of servo valve, and achieves the electronic-hydraulic proportional displacement regulation. So the stroke of proportional solenoid is not limited by the variable piston. The following figure shows that the proportional displacement variable capacity pump utilizes this kind of proportional displacement regulation mechanism, it's a manual servo variable capacity pump added with a electronic-hydraulic proportional pressure reducing valve 2 and actuating cylinder 3. Its control precision and sensibility are not as good as electronic-hydraulic servo variable capacity pump, but its antipollution capacity is stronger, its price is lower and it operates reliably. So it's ideal to many mechanical remote control.

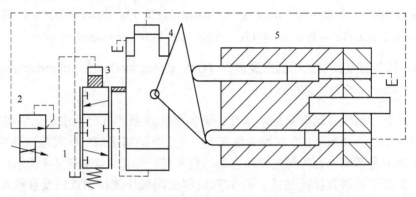

图 4-20 电液比例排量调节变量机构

(Fig. 4-20 Electro-hydraulic proportional displacement regulation variable mechanism)
1—液动伺服阀（Hydraulic servo valve）；2—比例减压阀（Proportional pressure reducing valve）；
3—操纵缸（Actuating cylinder）；4—变量机构（Variable mechanism）；5—柱塞泵（Plunger pump）

4.2.1.2 位移直接反馈型比例排量变量泵的特性分析（The characteristic analysis of the shift direct feedback proportional displacement variable capacity pump）

位移直接反馈比例排量变量泵的特性是指输入信号电流或当量电流 ΔI 变化时，流量 q 的响应特性。从上面比例排量泵的结构原理分析中可知，它的结构由三个主要环节组成，就是电-机械位移转换机构、伺服变量机构及泵主体。在简化的条件下，除伺服变量机构外，都可简化为比例环节，则泵的特性可由以下的基本方程来描述。

The characteristic of the shift direct feedback proportional displacement variable pump is the response characteristic of the flow q when the input signal current or equivalent current ΔI changes. It can be seen from the above analysis of the structural principle of the proportional displacement pump that its structure is composed of three main links, which are electrical-mechanical shift switch mechanism, servo variable mechanism and main body mechanism of the pump. In the simplified conditions, it is all simplified to the proportional link except for the servo variable mechanism, and the characteristic of the pump can be described by the following basic equations.

A　电-机械转换机构方程（Electrical-mechanical switch mechanism equation）

$$\Delta I = K_c \Delta x \tag{4-11}$$

式中　ΔI——输入信号电流或脉冲数（Input signal current or pulses number）；
　　　K_c——电-机械转换常数（Electrical-mechanical switch constant）；
　　　Δx——滑阀阀芯位移（Shift of the slide valve spool）。

4.2 比例排量变量泵和变量马达
(Proportional displacement variable pump and variable motor)

B 伺服变量机构特性方程 (Characteristic equation of the servo variable mechanism)

伺服变量机构实质上是一个由双边滑阀控制的差动缸组成的动力机构，它的工作原理如图 4-21 所示。该图可供分析其特性用。假设阀为理想零开口三通滑阀，对应于阀芯位移和阀压降变化的流量变化能瞬时发生，并忽略所有的泄漏。各物理量的正方向如图所示，对控制腔应用连续性方程，得

The servo variable mechanism is essentially a power mechanism composed of differential cylinder controlled by bilateral slide valve. The working principle is shown in the Fig. 4-21. The figure can be used to analyze its characteristic. Assume that the valve is a three way slide valve that has ideal zero opening,

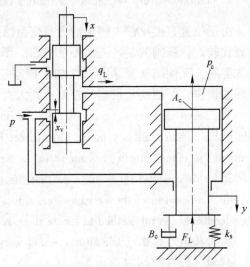

图 4-21 位移直接反馈比例排量调节机构简图
(Fig. 4-21 The simplified schematic of the shift direct feedback proportional displacement regulation mechanism)

and the flow changes instantaneously following with the spool shift and the valve pressure drop ignoring all the leakage. The positive direction of the all physical quantities are shown in the figure, and using continuity equation of the control chamber, we can get

$$q_L = K_q x_v - K_c p_c \tag{4-12}$$

$$q_L = A_c \frac{dy}{dt} + \frac{V_0}{\beta_e} \frac{dp_c}{dt} \tag{4-13}$$

式中 K_q——伺服阀的流量增益 (The flow gain of the servo valve);

K_c——伺服阀的流量压力系数 (The flow-pressure coefficient of the servo valve);

p_c——变量活塞控制腔大腔控制压力 (The big chamber control pressure of the variable piston control chamber);

x_v——阀芯开口量 (The amount of the spool opening);

A_c——变量活塞大端面积 (The big side area of the variable piston);

V_0——控制腔容积 (The capacity of the control chamber);

β_e——油液的体积弹性模数 (The volume elastic modulus of the oil);

y——变量机构输出位移 (The output shift of the variable mechanism)。

C 泵的流量方程 (The flow equation of the pump)

$$q = K_2 \tan\alpha \tag{4-14}$$

4.2.2 位移-力反馈式比例排量调节变量泵（Shift-force feedback proportional displacement regulation variable capacity pump）

由于位置直接反馈中的反馈量是变量活塞的位移，它必须采用伺服阀来进行位置比较。伺服阀的制造工艺要求很高。而且为了驱动伺服阀，往往需要增加一级先导级，又使结构复杂化，增加了成本。采用位移-力反馈的控制方式来使变量活塞定位，可使变量机构的控制得到大大的简化。

The servo valve must be adopted for the position comparison as the shift of the variable piston is the feedback in the shift direct feedback method. However, the requirements of the manufacturing technique of the servo valve are very high, and a pilot is needed for driving the servo valve so that the structure is complicated which increases the cost. The control of the variable mechanism can be simplified by adopting the shift-force feedback control method to locate the position of the variable piston.

结构及工作原理（The structure and working principle）

我国引进生产的力士乐型A7V斜轴式轴向柱塞泵就是一种位移-力反馈式的比例排量调节变量泵，如图4-22所示。

The Rexroth A7V axial plunger pump that China has imported is a shift-force feedback proportional displacement regulation variable capacity pump, and it is shown in the Fig. 4-22.

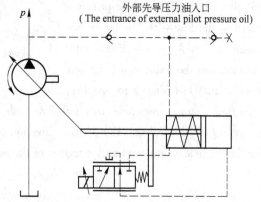

图4-22 位移-力反馈式比例排量调节变量泵原理图
(Fig. 4-22 The schematic diagram of the shift-force feedback proportional displacement regulation variable capacity pump)

泵的排量与输入控制电流成正比，可在工作循环中无级地按程序来控制泵的排量。如图4-23所示，无控制电流时，该泵在复位弹簧的作用下返回原位；当有信号电流时，比例电磁铁10产生推力，通过调节套筒9和推杆7作用于控制阀上。当电磁铁10产生的推力足以克服起点调节弹簧3和反馈弹簧8的预压缩力的总和时，控制阀4阀芯位移使 a、b 控制腔接通。控制活塞5从最小排量位置向增大排量的方向移动，实现变量。与此同时，在移动中不断压缩反馈弹簧，直至弹簧上的压缩力略大于电磁力时，先导阀芯开始关闭，并最终完全关闭，实现位移-力反馈，使活塞定位在与输入信号成正比的新位置上。

The displacement of the pump is proportional to the input control electric cur-

4.2 比例排量变量泵和变量马达
(Proportional displacement variable pump and variable motor)

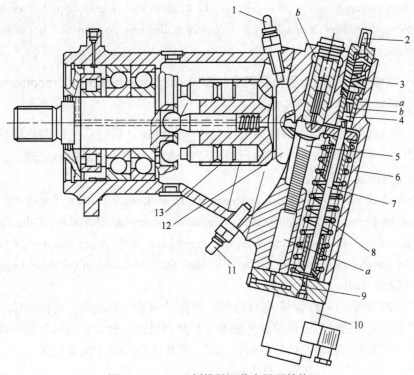

图4-23 A7V比例排量调节变量泵结构
(Fig. 4-23 The structure of the A7V proportional displacement regulation variable capacity pump)
1—最小流量限位螺钉(Minimum flow limit screw); 2—调节螺钉带护罩(Regulation screw with a fender);
3—控制起点调节弹簧(Control origin regulation spring); 4—控制阀(Control valve);
5—控制活塞(Control piston); 6—端盖(End cover); 7—推杆(Push rod); 8—反馈弹簧
(Feedback spring); 9—调节套筒(Regulation bush); 10—比例电磁铁(Proportional solenoid);
11—最大流量限位螺钉(Maximum flow limit screw); 12—配流盘(Port plate); 13—缸体(Cylinder body);
a—变量活塞小腔(常通压力油)(Small cavity of the variable piston (always connected with pressure oil));
b—变量活塞大腔(控制腔)(Large cavity of the variable piston (control cavity))

rent. The displacement of the pump can be controlled steplessly according to the program in the working cycle. As is shown in Fig. 4-23, when there is no control electric current, the pump returns to the initial position effected by the reset spring. And when there is a signal current, the proportional solenoid 10 produces a thrust and acts on the control valve by control sleeve 9 and push rod 7. The shift of the spool 4 connects control cavity a with control cavity b if the thrust produced by 10 can overcome the resultant force of original regulation spring 3 and feedback spring 8. The variable is achieved when the control piston 5 moves from minimum displacement position to the maximum displacement position. Meanwhile, the feedback spring is compressed continuously during the motion until the compressive force is somewhat bigger than the electromagnetic force, then the spool of the pilot valve begins to close and closes completely eventually and achieve shift-force feedback, which locates at the new position that is proportional to the input signal.

4.2.3 位移-电反馈型比例排量调节 (Shift-electronic feedback proportional displacement regulation)

位移-电反馈式排量调节把带有排量信息的斜轴（斜盘）的倾角或变量活塞的直线位移转换成电信号，并把此信号反馈到输入端由电控器进行处理，最后得出控制排量改变的信息。

Shift-electronic feedback displacement regulation turns the inclination of the clino-axis (swash plate) with displacement information or straight-line shift of the variable piston into electrical signal, and send the signal back to the input channel to be processed by the electrical controller. Then it obtains the information of the control displacement variable finally.

图4-24所示为一种位移-电反馈型比例排量调节的原理图。机构的位移或倾角，代表了流量的信息，由位移传感器2检测和反馈。所以它实际上是间接的流量反馈，反馈的实际值与给定值进行比较，得出需要的排量调节信息。

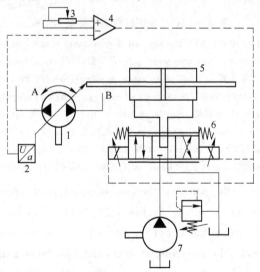

图 4-24 位移-电反馈型比例排量调节原理图
(Fig. 4-24 Shift-electronic feedback proportional displacement regulation)
1—双向变量泵 (Bilateral variable capacity pump); 2—位移传感器 (Shift sensor);
3—输入信号电位器 (Input signal potentiometer); 4—电控器 (Electrical controller);
5—双向变量机构 (Bilateral variable mechanism); 6—三位四通比例换向阀
(Four-port three-position proportional control valve); 7—先导油泵 (Pilot pump)

4.2 比例排量变量泵和变量马达
(Proportional displacement variable pump and variable motor)

The Fig. 4-24 shows the schematic diagram of a shift-electronic feedback proportional displacement. The shift or inclination angle of the mechanism represents the flow information, which is detected and fed back by the displacement sensor 2. So it is actually an indirect flow feedback, and demanding displacement regulation information will be obtained by comparing the actual value of the feedback with the given value.

因为是双向变量机构，需要采用辅助先导泵向比例阀供油，使在主泵排量为零的位置下变量机构仍能可靠工作。变量机构由弹簧对中，控制失效时，通过比例阀的中位节流口实现油液体积平衡而回零。

Because of the bilateral variable mechanism, it needs an auxiliary pilot pump to supply oil for the proportional valve, which keeps the variable mechanism operating reliably even if there is no displacement of the main pump. The variable mechanism is aligned by spring. Variable mechanism utilize the mid-position throttle port to achieve the balance of oil volume and backs to zero when control fails.

该系统流量随压力升高有较大的负偏差，这是由于泵的泄漏流量 Δq_L 引起的。可见，比例排量调节虽然可以通过电气信号来控制泵的输出流量，但因负载变化以及泄漏等因素引起的流量变化无法得到抑制和补偿。所以，比例排量调节的流量得不到精确地控制。当需要精确控制时，就要采用比例流量调节变量泵。

The system flow has a large negative bias as the pressure increasing, which dues to the leakage flow Δq_L of the pump. It can be seen that although the proportional displacement regulation could control the output flow of the pump by electrical signal, yet the flow change caused by load change, leakage and other factors can't be inhibited and compensated. So, the flow of proportional displacement regulation can't be controlled precisely. The proportional flow regulation variable capacity pump will be adopted when precise control is in need.

4.2.4 一种新型比例排量控制式 PV 泵 (A new proportional displacement control PV pump)

新型比例排量控制式 PV 泵的结构见图 4-25，工作原理见图 4-26。

The structure of a new proportional displacement control PV pump is shown in Fig. 4-25, and its working principle diagram in Fig. 4-26.

这实际上是一种位移-电反馈式比例排量调节泵，变量机构左端接收反馈信号，再控制三通阀的比例电磁铁的输入，从而控制泵的排量的改变。

In fact, it is a shift-electronic feedback proportional displacement regulation pump, the left side of variable mechanism picks up the feedback signal, and then controls the proportional electromagnet input of three-way valve, thus controls the change of the pump displacement.

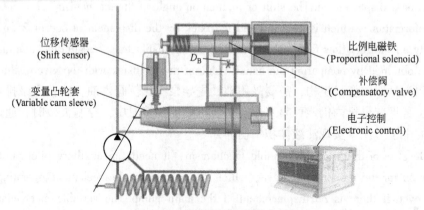

图 4-25 比例排量控制式 PV 泵
(Fig. 4-25 Proportional displacement control PV pump)

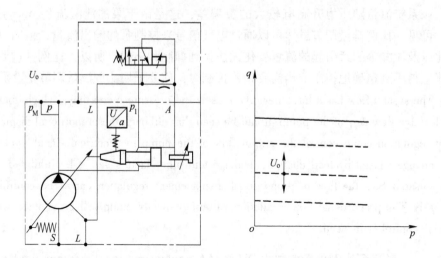

图 4-26 比例排量控制式 PV 泵工作原理图
(Fig. 4-26 Working principle diagram of proportional displacement control PV pump)

4.3 电液比例压力调节型变量泵（Electro-hydraulic proportional pressure regulation variable pump）

4.3.1 工作原理及结构（Working principle and structure）

电液比例压力调节型变量泵是一种带负载压力反馈的变量泵，其特点是利用负载压力变化的信息作为泵本身变量的控制信号。泵的出口压力代表了负载大小

4.3 电液比例压力调节型变量泵
(Electro-hydraulic proportional pressure regulation variable pump)

的信息。当它低于设定压力时，像定量泵一样工作，输出最大流量。当工作压力等于设定压力时，它按变量泵工作。这时泵输出的流量随工作压力的变化下降很快。任何工作压力的微小变化都将引起泵较大的流量变化，从而对压力提供了一种反向变化的补偿。当系统的负载压力大于设定压力时，泵的输出流量迅速减小到仅能维持各处的内、外泄漏，且维持设定压力不变。可见这种变量泵具有流量适应的特点，在泵作流量适应调节时，工作压力变动很小，基本保持不变。泵的最大供油能力由其最大排量决定，而泵的出口压力由比例电磁铁的控制电流来设定。这种泵完全消除了流量过剩，而输出压力又可根据工况随时重新设定，因而有很好的节能效果。这种泵是流量适应的，无需设置溢流阀，但应加入安全阀来确保安全。

Electro-hydraulic proportional pressure regulation variable pump is a variable pump with pressure feedback of the load, its characteristic is to use the information of load pressure change as the control signal of the pump itself. The outlet pressure of the pump represents the information of the load. When the pressure is lower than the set pressure, it works as a constant pump, and its output flow is the maximum. When the working pressure is equal to the set pressure, it works as a variable pump. The output flow of the pump decreases greatly with the change of the working pressure. Any tiny change of the working pressure will cause great flow change of the pump, which provides the pressure with a reverse compensation. When the loading pressure of the system is greater than the set pressure, the output flow of the pump decreases greatly just to meet the internal and external leakage and maintains the set pressure. Visibly, the pump has the characteristic of flow adaption, and when the pump carries on flow adaption regulation, the working pressure changes very small and nearly keeps unchanged. The maximum providing oil capacity of the pump is determined by its maximum displacement, and the outlet pressure of the pump is set by the control electricity of the proportional electromagnet. As the pump eliminates excess of flow completely, and the output pressure can be reset according to the working condition at any time, so it has a good energy-saving effect. This kind of pump is flow-adaptive, and doesn't have to set relief valve, but a safety valve is needed to ensure safety.

比例压力泵是在4.1.1节中所介绍的恒压变量泵和流量敏感型变量泵的基础上开发出来的。事实上，用比例溢流阀代替手调溢流阀，它就成了一种直接控制式的比例压力调节型的变量泵，如图4-27所示。如果用比例顺序阀代替先导控制中的手调机构就成了先导控制式的比例压力变量泵，如图4-28所示。由于直接控制式压力调节泵的参比弹簧刚度较大，泵的特性有较大的调压偏差。

The proportional pressure pump is developed based on the constant pressure variable pump and flow-sensitive variable pump which is shown in the section 4.1.1. In fact, by replacing the manual relief valve with proportional relief valve, it will become a direct control proportional pressure regulation variable pump which is shown in the Fig. 4-27. If replacing the manual mechanism of pilot control with proportional sequence valve, it will become a pilot control proportional pressure variable pump which is shown in the Fig. 4-28. As the stiffness of the reference spring is very large, the characteristic of the direct control pressure regulation pump has a big pressure regulation deviation.

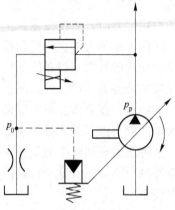

图 4-27 比例压力泵

(Fig. 4-27 Proportional pressure pump)

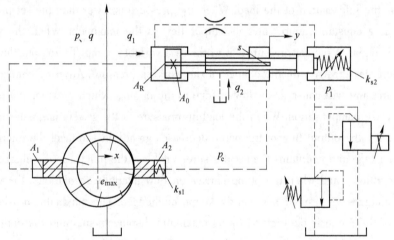

图 4-28 先导式比例压力调节变量叶片泵半结构原理图

(Fig. 4-28 Semi-structural principle diagram of the pilot proportional pressure regulation variable vane pump)

4.3.2 先导式比例压力调节变量叶片泵的特性分析 (Characteristic analysis of pilot proportional pressure regulation variable vane pump)

变量叶片泵的定子环由两个控制柱塞保持在一定的位置上，较大的一个控制柱塞由一软弹簧支持，其有效面积为小柱塞的两倍。弹簧的推力应在零压力时足以克服各种摩擦力，把定子推至最大偏心位置，因此泵启动时能给出最大流量。直到输出压力达到设定的拐点压力为止，它按定量泵工作，其后便按变量泵工作。

4.3 电液比例压力调节型变量泵
(Electro-hydraulic proportional pressure regulation variable pump)

The stator ring of the variable vane pump is kept at some position by two control plungers, and the bigger plunger is supported by a soft spring, whose valid area is twice of the smaller one. The thrust of the spring could overcome all kinds of frictions when the pressure is zero so as to push the stator to the maximum deviation position. So the pump can supply the maximum flow when it starts. It doesn't work as the variable displacement pump until the output pressure reaches at the set pressure inflection point, then it works as a variable pump.

当液压反馈力 p 小于设定压力 p_0 和弹簧 k_{s2} 的预紧力时,控制阀芯关闭(如图 4-28 所示),大柱塞腔不通回油,定子处在极左位置。这时泵的输出流量为

When the hydraulic feedback force p is less than the set pressure p_0 and the pretightening force of spring k_{s2}, the spool of the control valve will close (shown in the Fig. 4-28), and the big plunger chamber is not connected with return oil, then the stator is located at ultra-left position. The output flow of the pump is

$$q = k_q e_{\max} - k_1 p \qquad (4-15)$$

式中 k_q ——泵的流量常数(单位偏心距产生的理论流量),其值由泵的几何参数决定(The flow constant of the pump (the flow in theory generated by unit deviation), the value is decided by the geometry parameter of the pump);

　　　k_1 ——泵的泄漏系数(The leakage coefficient of the pump);

　　　$e_{\max}$ ——定子的最大偏心距(The maximum deviation of the stator)。

当液压反馈力 p(工作压力)大于设定压力 p_0 和弹簧顶紧力的合力时,控制阀芯右移并在 s 处开启,使大腔与回油相通,控制压力 p_c 下降,定子向右移动,偏心距变小,叶片泵改变流量,并保持压力不变。

When the hydraulic feedback force p (working pressure) is greater than the resultant force of the set pressure p_0 and the jacking force of the spring, the control spool moves right and opens in the position s, and it connects the bigger chamber with oil return, then, the control pressure p_c decreases, and the stator moves to right so that the deviation turns smaller, and the flow of vane pump changes, which keeps the pressure constant.

4.3.2.1 动态特性(Dynamic characteristic)

根据上述变量叶片泵的工作原理,可写出泵与流量、工作压力有关的诸数学式。为简便,忽略有关的库仑摩擦以及油液压缩性等影响。

According to the working principle of the above variable vane pump, we can get many mathematic expressions about the flow and working pressure of the pump. The relevant Coulomb frictions and the effect of the compressibility of the oil are neglected for simplification.

第4章 电液比例容积控制
Chapter 4 Electro-hydraulic Proportional Volume Control

A 泵的输出流量方程（The output flow equation of the pump）

$$q = k_q(e_{max} - x) - k_1 p - \frac{V_p}{\beta_e}\frac{dp}{dt} \tag{4-16}$$

式中　x——定子从零位算起的位移（The shift of stator from the zero position）；

　　　V_p——泵出口处及管道容积（The volume of the pump outlet and the pipe）；

　　　β_e——油液的体积弹性模数（The volume elasticity module of the oil）。

B 定子受力平衡方程（The force balance equation of the stator）

$$pA_1 - p_c A_2 = M\frac{d^2 x}{dt^2} + B_c\frac{dx}{dt} + k_{s1}x + F_f + F_s \tag{4-17}$$

式中　A_1，A_2——变量泵小端和大端控制柱塞截面积（The area of the small and big side control plunger of the variable pump）；

　　　M——定子移动的质量（The moving mass of the stator）；

　　　B_c——定子移动时的黏性阻力系数（The viscosity resistance coefficient when the stator moves）；

　　　k_{s1}，F_s——定子复位弹簧刚度及预紧力（The stiffness of the stator reset spring and pre-tightening force）；

　　　F_f——定子摩擦阻力（The friction resistance of the stator）。

C 通过节流口 S 和阻尼孔的连续性方程（The continuity equation through choke S and orifice）

$$q_1 = C_{d1}A_0\sqrt{\frac{2}{\rho}(p - p_c)} \tag{4-18}$$

$$q_2 = C_{d2}Wy\sqrt{\frac{2}{\rho}p_c} \tag{4-19}$$

$$q_1 = q_2 \tag{4-20}$$

式中　q_1——通过节流孔 A_0 的流量（The flow through the orifice A_0）；

　　　A_0，C_{d1}——节流孔 A_0 的过流面积及其流量系数（The flow area of the orifice A_0 and the flow coefficient）；

　　　p_c——变量泵大腔控制压力（The control pressure of the big chamber of the variable pump）；

　　　q_2，C_{d2}——通过节流口 3 的流量及其流量系数（The flow through orifice 3 and the flow coefficient）；

　　　W——控制阀的面积梯度（The area gradient of the control valve）；

　　　ρ——油液的密度（The density of oil）。

4.3 电液比例压力调节型变量泵
(Electro-hydraulic proportional pressure regulation variable pump)

D 控制阀芯的力平衡方程 (The force balance equation of the control valve spool)

$$pA_R - p_0 A_R = m\frac{d^2 y}{dt^2} + B_0 \frac{dy}{dt} + k_{s2}(y + y_0) \tag{4-21}$$

式中 A_R ——控制阀芯面积 (The area of the control valve spool);
p ——设定压力 (Set pressure);
m ——阀芯质量 (The mass of spool);
B_0 ——阀芯黏阻系数 (The viscosity resistance coefficient of the spool);
k_{s2} ——调节弹簧刚度 (The stiffness of the regulation spring);
y_0 ——弹簧预压缩量 (The pre-compression of spring)。

忽略比例溢流阀的动态影响,仅认为设定压力 p_0 和输入电流 i 成比例,设 K_e 为电流-压力增益,则输入方程为

Ignoring the dynamic effect of the proportional relief valve, the set pressure p_0 is only thought to be proportional to the input current i, and K_e is set as the flow-pressure gain, then the input equation is

$$p_0 = K_e i \tag{4-22}$$

以上各方程就反映了比例压力泵的动态特性。对以上各方程取增量,线性化后求拉普拉斯变换,可得如下各方程

The above equations reflect the dynamic characteristics of proportional pressure pump. Abstracting increment from the above equations and linearizing them, then using Laplace transformation and we can get the following equations

$$q = -K_q X - \left(k_1 + \frac{V_p}{B_t}s\right)p \tag{4-23}$$

$$pA_1 - p_c A_2 = (Ms^2 + \beta_c s + k_{s1})X \tag{4-24}$$

$$K_q Y = -(K_{c1} + K_{c2})p_c \tag{4-25}$$

式中,$K_q = \frac{\partial q_2}{\partial y} = C_{d2}W\sqrt{\frac{2}{\rho}p_c}$; $K_{c2} = -\frac{\partial q_2}{\partial p_c} = \frac{C_{d2}Wy\sqrt{\frac{1}{\rho}p_c}}{2p_1}$;

$$K_{c1} = -\frac{\partial q_1}{\partial p_c} = \frac{C_{d1}A_0\sqrt{\frac{1}{\rho}(p - p_c)}}{2(p - p_c)};$$

$$p - p_0 = \frac{1}{A_R}(ms^2 + B_0 s + k_{s2})Y \tag{4-26}$$

$$p_0 = K_e i \tag{4-27}$$

其中,ω_v、ω_p 分别为控制阀和变量泵定子的固有频率;δ_v、δ_p 分别为控制阀和变量泵定子的无因次阻尼系数;ω_1 为泵的出口容积腔的频率(参见图 4-29)。

Among other things, ω_v and ω_p are respectively the natural frequency of the control valve and the variable pump stator; δ_v and δ_p are respectively the non-dimensional damping coefficient of the control valve and the variable pump stator; ω_1 is the frequency of the outlet volume chamber of the pump (See Fig. 4-29).

从而得到传递函数方框图见图 4-29 (So the block diagram of the transfer function is shown in Fig. 4-29).

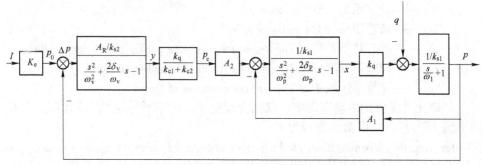

图 4-29 传递函数方框图

(Fig. 4-29 The block diagram of the transfer function)

4.3.2.2 静态特性(Static characteristic)

静态特性是指泵的流量与压力、功率与压力及效率与压力之间的关系。相应的静态特性方程为

The static characteristic refers to the relation between flow and pressure, power and pressure, efficiency and pressure. The corresponding static characteristic equation is

$$q = k_q(e_{\max} - x) - k_L p \tag{4-28}$$

$$pA_1 - pA_2 = k_{si}x + F_s + F_f \tag{4-29}$$

$$F_f = K_y p$$

式中 K_y ——与压力有关的摩擦系数 (A friction coefficient relative with pressure)。

$$K_q y = -(K_{c1} + K_{c2}) p_c \tag{4-30}$$

$$p - p_0 = \frac{k_{s2}}{A_R}(y + y_0) \tag{4-31}$$

式中 y_0 ——控制阀的弹簧预压缩量 (The spring pre-compression of the control valve)。

由以上四个式子,消去中间变量,求得泵的静态特性方程:

From the above four equations, we can get the static characteristic equation of the pump by eliminating the intermediate variables:

$$q = \frac{k_q}{k_{s1}}\left[F_s + \frac{k_q A_R A_2}{k_{s2}(K_{c1}+K_{c2})}(p_0 + \frac{k_{s2}}{A_R}y_0) + k_{s1}e_{\max}\right] - \frac{k_q}{k_{s1}}\left[A_1 \pm A_y + \frac{k_q A_R A_2}{k_{s2}(K_{c1}+K_{c2})} + \frac{k_{s1}K_L}{k_q}\right]p$$

$$\tag{4-32}$$

4.3 电液比例压力调节型变量泵
(Electro-hydraulic proportional pressure regulation variable pump)

上述方程须在控制阀芯取得力平衡后才适用,因此它对应特性曲线的倾斜段(图 4-30),而式(4-28)对应泵的不变量段的特性。整条特性曲线如图 4-30 中 p-q 曲线所示。曲线中 C 点的压力称为设定压力 p_C(拐点压力),利用式(4-28)和式(4-32)的流量相等可求出 p_C。

The above equation is applicable only when the force of the control valve spool is balanced, so it corresponds to the slope segment of the characteristic curve (Fig. 4-30), the equation (4-28) corresponds to the characteristic of invariant section of the pump. The entire characteristic curve is shown as the p-q curve in the Fig. 4-30. The pressure of point C in the curve is called set pressure p_C (inflection pressure), and we can figure out p_C by using the flow equation between equation (4-28) and (4-32).

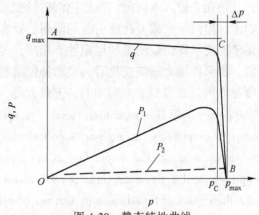

图 4-30 静态特性曲线
(Fig. 4-30 Static characteristic curve)
q —流量压力曲线(Flow pressure curve);
P_1 —最大输出时的功率(The maximum output power);
P_2 —零行程时的功率输出(The power output at zero stroke)

$$p_C = \frac{F_s + \dfrac{k_q A_R A_2}{k_{s2}(K_{c1}+K_{c2})}(p_0 + \dfrac{k_{s2}}{A_R}y_0)}{A_1 \pm A_y + \dfrac{k_q A_R A_2}{k_{s2}(K_{c1}+K_{c2})}} \quad (4\text{-}33)$$

从图中可见,在 $p < p_C$ 的区域内,功率特性与定量泵相同,在 $p > p_C$ 区域内,输出功率急剧下降,直至到达零行程功率曲线。因此,该泵用于保压系统时,具有很好的节能效果。

As shown in the figure, in the area of $p < p_C$, the power characteristic is the same as the fixed displacement pump. In the area of $p > p_C$, the output power decreases sharply until it reaches the zero stroke power curve. So if the pump is used in the pressure maintaining system, it will have a good energy-saving effect.

4.4 电液比例流量调节型变量泵(Electro-hydraulic proportional flow regulation variable pump)

比例流量调节型变量泵是一种稳流量型的自动变量泵,它的输出流量与负载无关,且正比于输入电信号。虽然比例排量变量泵的输出流量也有正比于输入信号的功能,但由于负载变化所引起的泄漏,液容的影响无法得到自动补偿,因此排量调节控制不能保证流量的稳定性。相反,比例流量调节型变量泵以流量为控制对象,在泵作压力适应变化时,自动补偿流量的变化,维持稳流量。但由于它的恒流量性质是靠容积节流实现的,流量大时,其节流损失不容忽视。

The proportional flow regulation variable pump is a steady flow and automatic variable pump. Its output flow has nothing to do with the load and is proportional to the input electric signal. Although the output flow of the proportional displacement variable pump is also proportional to the input signal, yet because of the leakage caused by the change of load, the impact of liquid capacity can not be compensated automatically. So the displacement regulation control can't ensure the stability of flow. On the contrary, the proportional flow regulation variable pump regards the flow as control object, and when the pump is making pressure adaption, it compensates the change of flow automatically and keeps the stability of flow. But as its characteristic of constant flow is achieved by capacity throttle, if the flow is very large, the throttle loss can't be ignored.

图 4-31 所示为比例流量调节型变量泵的职能符号和半结构原理图。将该图与流量敏感型稳流量变量泵相比较可知,除用比例节流阀代替手调节流阀外,其他是完全相同的,所以它们的工作原理也完全相同。

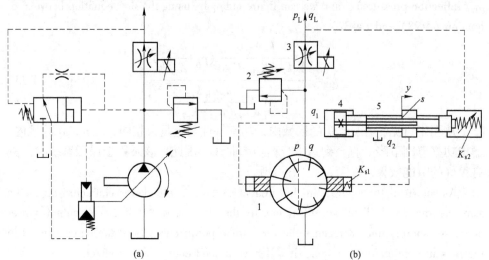

图 4-31 比例流量调节变量泵

(Fig. 4-31 The proportional flow regulation variable pump)

1—变量叶片泵(Variable vane pump);2—安全阀(Safety valve);

3—比例节流阀(Proportional throttling valve);

4—固定阻尼孔(Fixed damping hole);5—控制阀(Control valve)

4.4 电液比例流量调节型变量泵
(Electro-hydraulic proportional flow regulation variable pump)

Fig. 4-31 shows the functional symbols and semi-structural principal diagram of the proportional flow regulation variable pump. Comparing the diagram with the flow-sensitive steady flow variable pump, we can see that they are completely the same except for replacing the manual throttling valve with the proportional throttling valve, so the operating principles are also exactly the same.

4.4.1 稳流量调节控制原理 (Steady flow regulation control principle)

如图 4-31 所示。在设定的安全压力范围内，泵的全部流量通过由电气设定的节流孔 3。对于任意设定开口量，压降与通过的流量成正比。因此，该压降是泵流量的一个度量信息。该压降由泵的控制装置保持恒定。控制阀芯 5 两端作用着比例节流孔 3 的进出口压力。出口压力与阀芯右端弹簧腔相通。弹簧通常调定在某一压缩量下，这一弹簧力决定了通过比例节流孔的压降。该压降应使控制阀在控制边处有 0.05mm 左右的开口量，使其保持少量溢流。通过调整这一压差，对任何输入的电信号，都可获得准确的流量控制。

As is shown in the above figure, in the range of set safety pressure, all the flow of the pump passes through the throttle hole 3 which was set by electrical component. For any amount of set opening, the pressure drop is proportional to the passing flow. Therefore, the pressure drop is a measuring information of the pump flow. The pressure drop is kept constantly by the control device of the pump. The inlet and outlet pressure of the proportional throttle hole 3 acts on the two sides of control spool 5. The outlet pressure is connected with the right side spring cavity of the spool. The spring is always set at a certain amount of compression, and the spring force determines the pressure drop through the proportional throttle hole. The pressure drop should make the control valve own an opening at the amount of approximately 0.05mm at the control side S and keep a small amount of overflow. By regulating the differential pressure, we can achieve an accurate flow control for any input electrical signal.

控制阀芯在正常工作时，在约 0.1mm 的行程内不断调整位置，力图保持平衡。当比例节流口关小或负载力 p_L 下降时，入口压力 p 便有大于出口压力 p_L 与弹簧力之和的趋势，阀芯就有向右移动使溢流量增大的倾向。其结果是大柱塞腔部分卸压，使泵的出流减小，直至调节阀芯恢复到原来的位置，保持节流孔压差不变。反之过程类似。

When the control spool is working normally, it adjusts the position continuously within the stroke of about 0.1mm to keep balance. When the proportional throttle hole is turned down or load force p_L drops, the inlet pressure p is likely to tend to be greater than the resultant of outlet pressure p_L and spring force. The spool has the trend of moving right and increasing the overflow. As a result, the partial pressure of the big plunger cavity overflows and the output flow of the pump decreases until the regulating spool comes back to its original position and keeps the differential pressure through the throttle hole unchanging. The reverse process is similar.

4.4.2 泵的特性分析 (Analysis of pump characteristics)

比例流量调节型变量泵的动、静态特性分析与压力调节型相仿，可以利用其结果做相应的修改而获得。其主要差别是以压差反馈代替压力反馈。比例流量调节的输入方程为

The dynamic and static characteristic analysis of the proportional flow regulation variable pump is similar to the pressure regulation type, so it can be achieved by modifying the corresponding results. The main difference is to replace the pressure feedback with differential pressure feedback, the input equation of proportional flow regulation is

$$A = K_e i \tag{4-34}$$

式中 A ——比例节流孔通流面积 (The flow area of proportional throttle hole);

K_e ——电流面积转换系数 (The area conversion coefficient of current)。

通过节流阀的流量方程 (The flow equation through the throttle valve)

$$q_L = KA\sqrt{p - p_L} = KK_e i\sqrt{\Delta p_L} \tag{4-35}$$

式中 K ——节流阀常数 (The throttle valve constant)。

线性化得到 (Linearization to be)

$$\Delta q_L = K_q \Delta i + K_c \Delta p_L \tag{4-36}$$

式中 K_q ——流量增益 (Flow gain);

K_c ——流量-压力系数 (Flow-pressure parameter)。

另外，通过节流口 s 及阻尼孔 4 的流量连续方程式 (4-25) 应做些修改，因为此时泵的出口压力不能被认为是常数。对式 (4-18)、式 (4-19) 线性化后代入式 (4-20)，得

In addition, the flow continuity equation (4-25) of the throttle hole s and damping hole 4 should be modified a little, because the outlet pressure of the pump at this time can't be considered as a constant. By linearizing the equation (4-18) and (4-19), then substituting them to the equation (4-20), we can obtain

$$K_q \Delta y = K_{c1} \Delta p - (K_{c1} + K_{c2}) \Delta p_c \tag{4-37}$$

考虑到溢流量 q_1 对输出流量的影响，出口连续方程为 (Taking into account the impact that the overflow q_1 makes on the output flow, the outlet continuity equation is)

$$q_L = q - q_1 = q - K_{c1}(\Delta p - \Delta p_c) \tag{4-38}$$

利用以上式子，比例流量调节变量叶片泵的方框图如图 4-32 所示。

4.4 电液比例流量调节型变量泵
(Electro-hydraulic proportional flow regulation variable pump)

Making use of the above equations, the block diagram of proportional flow regulation variable vane pump is shown in the Fig. 4-32.

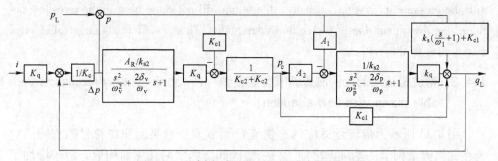

图 4-32 比例流量调节型变量叶片泵方框图

(Fig. 4-32 The block diagram of proportional flow regulation variable vane pump)

由图 4-32 可以看出，负载压力 p_L 作为干扰量影响着流量调节精度。依具体情况，可对方框图进行适当的简化，得出实用的传递函数并进行动态特性的分析。

It is seen from the Fig. 4-32 that the load pressure p_L influences the flow regulation precision as a disturbance variable. The block diagram can be simplified appropriately in specific conditions so as to obtain a practical transfer function and carry on the dynamic characteristic analysis.

比例流量调节变量泵的静态特性如图 4-33 所示（The static characteristic of the proportional flow regulation variable pump is shown in the Fig. 4-33）

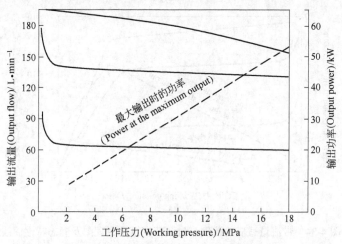

图 4-33 比例流量调节变量泵的静态特性

(Fig. 4-33 The static characteristic of proportional flow regulation variable pump)

从特性曲线中可以看到，随着压力升高不存在截流压力，即泵不会回到零流量处，所以这种泵不具流量适应性，因而系统必须设有足够大的溢流阀。

It can be seen from the characteristic curve that there is not stop-check pressure with the pressure increasing, namely, the pump will not come back to the zero-flow position, so this pump doesn't have flow adaptability. Thus, sufficiently large relief valve must be set in the system.

4.4.3 带流量适应的比例流量调节型变量泵 (Proportional flow regulation variable pump with flow adaption)

图 4-34 所示为带流量适应的比例流量调节型变量泵的职能符号原理图。它是在比例流量调节的基础上增加了一个溢流阀 2 和阻尼孔 4 而构成。由于增加了这些元件，当工作压力达到溢流阀 2 的调定压力后，控制阀 3 就失去了保持恒压

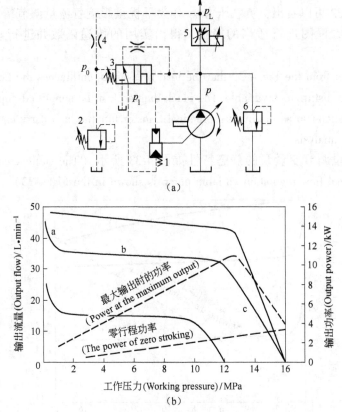

图 4-34 带流量适应的比例流量调节变量泵原理图及静态特性曲线
(Fig. 4-34 The schematic diagram and static characteristic curve of proportional flow regulation variable pump with flow adaption)

1—变量机构 (Stroking mechanism); 2—溢流阀 (Relief valve); 3—控制阀 (Control valve);
4—固定阻尼孔 (Fixed damping hole); 5—比例节流阀 (Proportional throttle valve); 6—安全阀 (Safety valve)

4.4 电液比例流量调节型变量泵
（Electro-hydraulic proportional flow regulation variable pump）

差的性质。溢流后，阀 3 的弹簧腔压力被固定在溢流阀调定压力上；当系统压力继续升高时，阀芯 3 便左移，并最终使泵的大控制腔压力为零。在小控制腔液压力的推动下，变量泵处于零流量位置。

The Fig. 4-34 shows the functional symbol schematic diagram of proportional flow regulation variable pump with flow adaption. It is constituted based on the proportional flow regulation adding with a relief valve 2 and orifice 4. As it is added with the above elements, when the working pressure reaches the set pressure of relief valve 2, the control valve 3 will lose the characteristic of constant differential pressure. After overflow, the spring cavity pressure of the valve 3 is fixed at the set pressure of the relief valve. If the system pressure continues to increase, the spool 3 will move left and ultimately make the big control cavity pressure of pump become zero. Under the drive of hydraulic force of small control cavity, the variable pump is at zero flow position.

压力和流量是两个最基本的液压参数，它们的乘积就是液压功率。前面介绍的几种比例泵中，都只有一个参数是可以任意设定的，另一个参数需要手调设定或自动适应。自动适应控制虽然有它的优点，但并非总是最佳选择。例如，比例压力变量泵有流量适应能力，但它在作流量适应时，其代价是供油压力要额外提高；或者比例流量泵在作压力适应时，要损失一定的流量。所以，从能耗控制的观点来看，它们并非最优。

Pressure and flow are two basic hydraulic parameters, and hydraulic power is the product of pressure and flow. For the proportional pumps mentioned above, only one parameter can be set randomly and another parameter needs to be set manually or adapted automatically. Although the automatic adaptive control has its advantages, yet it's not always the best choice. For example, the proportional pressure variable pump has flow adaptive ability, but when it carries on the flow adaption, it is at the cost of additional increasing oil pressure, or the proportional flow pump will lose some flow under pressure adaption. So they are not the optimum in the view of energy consumption controlling.

比例压力和流量调节型变量泵被称为复合比例变量泵，能充分地利用电液比例技术的优点。它控制灵活、调节方便，在大功率系统中节能十分明显。

Proportional pressure and flow regulation variable pump is called combined proportional variable pump, and it can take advantage of electric-hydraulic proportional technology mostly. It controls flexibly, regulates conveniently and saves energy obviously in the large power system.

复合比例变量泵的复合控制功能是通过信号处理来实现的。这样两个基本参数的控制信号并非完全独立，有时要利用一个算出另一个，或限制另一个的取值范围。

The compound control function of the compound proportional variable pump is realized by signal process. Thus, the two control signals of basic parameter are not completely independent. Sometimes, it's essential to figure out one or limit the value range of one by another one.

电液比例压力和流量调节型变量泵大致可分为压力补偿型和电反馈型两种。压力补偿型是以容积节流为基础，由一个变量泵加上比例节流阀构成，并由一个特殊的定差溢流阀（称为恒流控制阀）和一个特殊的定压溢流阀（称恒压控制阀）对变量机构进行控制，实现压力和流量的比例控制。

The electro-hydraulic proportional pressure and flow regulation variable pump can be approximately divided into two types of pressure compensation and electrical feedback. Basing on the volume throttle, the pressure compensation type is constructed of a variable pump and a proportional throttle valve, and the stroking mechanism is controlled by a special fixed differential relief valve (called as constant flow control valve) and a special fixed pressure relief valve (called as constant pressure control valve) so as to achieve pressure and flow proportional control.

电反馈型的变量泵可以避免使用比例节流阀，是纯粹的容积调速。它需要利用流量和压力传感器对被控制流量和压力进行检测和反馈，构成闭环控制系统，因而有更好的控制精度和节能效果。

The variable pump of electrical feedback type can refrain from the proportional throttle valve as it is purely volume speed regulation. It needs making use of flow and pressure sensors to monitor the flow and the pressure and give feedback to the system, which constitutes a closed loop control system, so it has a better control precision and energy saving effect.

4.5　电液比例压力和流量调节型变量泵（Electro-hydraulic proportional pressure and flow regulation variable pump）

4.5.1　压力补偿型比例压力和流量调节（Pressure compensation proportional pressure and flow regulation）

从上节所示的特性曲线图中可知，带流量适应的比例流量调节型变量泵是一种比例流量调节加手动压力调节的变量泵。把其中的手调溢流阀 2 换成比例溢流阀，便构成了一种比例压力和流量复合调节型变量泵。其原理图和特性曲线见图 4-35。当比例溢流阀 2 不溢流时，是比例流量调节，节流阀 5 的前后压差被特殊

4.5 电液比例压力和流量调节型变量泵
(Electro-hydraulic proportional pressure and flow regulation variable pump)

的定差溢流阀 3 固定。这时泵的输出流量决定于弹簧 k_{s2} 的预紧力（决定压差）与节流阀 5 的输入电流（决定开口面积）。当比例溢流阀溢流时，定差溢流阀 3 失去定压差的性质，其阀芯被泵的出口压力推向左，并使泵的流量变少，这时是比例压力调节。当负载压力继续升高时，变量机构回到零行程位置，最终导致截流。

We can see it from the characteristic curve shown in the last section that the proportional flow regulation variable pump with flow adaption is a proportional flow regulation and manual pressure regulation variable pump. So a proportional pressure and flow compound regulation variable pump is constituted by replacing the manual relief valve 2 with the proportional relief valve. The schematic diagram and characteristic curve are shown as Fig. 4-35. It is proportional flow regulation when the proportional relief valve 2 does not overflow, and the differential pressure through the throttle valve 5 is fixed by a special fixed differential relief valve 3. At this time, the output flow of the pump is determined by the pre-compression of spring (determining differential pressure) and the input current of throttle valve 5 (determining opening area). When the proportional relief valve overflows, the fixed differential relief valve 3 loses the characteristic of fixed differential pressure, and its spool is pushed towards the left by the outlet pressure of pump so that the flow of pump decreases. It is proportional pressure regulation at this time. If the load pressure increases continuously, the stroking mechanism comes back to the zero-stroke position and leads to interception at last.

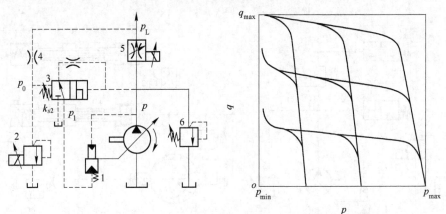

图 4-35 共用一个压力补偿阀的 p-q 调节的原理及特性曲线图
(Fig. 4-35 The principle and characteristic curve diagram of p-q regulation sharing a pressure compensation valve)

1—变量机构 (Stroking mechanism); 2—溢流阀 (Relief valve); 3—控制阀 (Control valve); 4—固定阻尼孔 (Fixed damping hole); 5—比例节流阀 (Proportional throttle valve); 6—安全阀 (Safety valve)

第 4 章　电液比例容积控制
Chapter 4　Electro-hydraulic Proportional Volume Control

这种泵具有较大的流量调节偏差和压力调节偏差。造成这种偏差的原因主要是压力和流量调节共用一个控制阀 3 所致。为了克服上述缺点，可以采用两个压力补偿阀，分别进行压力和流量的调节。

This kind of pump owns a rather large regulative deviation of the flow and pressure. The main reason is that the pressure regulation and flow regulation share a control valve 3. For overcoming the above shortcomings, we can use two pressure compensation valves to regulate pressure and flow respectively.

这种泵实际上就是前面介绍过的比例压力调节和流量调节的组合，其职能符号及半结构图如图 4-36 所示。其中各控制阀可以以叠加阀的形式，叠加在泵体上。该泵的调节特性得到了明显改善。它的流量与压力在额定值范围内可随意设定来满足不同的工况。但在作压力调节时存在滞环，在低压段也存在一个最小控制压力和流量不稳定阶段（图 4-37）。

This pump is actually a combination of the proportional pressure regulation and flow regulation which has been introduced before, and its functional symbol and semi-structure diagram are shown in the Fig. 4-36. The included control valves can be superimposed on the pump by the form of stacked valve. The regulation characteristic of the pump has been improved obviously. Its flow and pressure can be set randomly in the rated range so as to meet different working conditions. The hysteresis will exist when it regulates pressure, and there is a minimum control pressure and flow instable segment in the low pressure segment (Fig. 4-37).

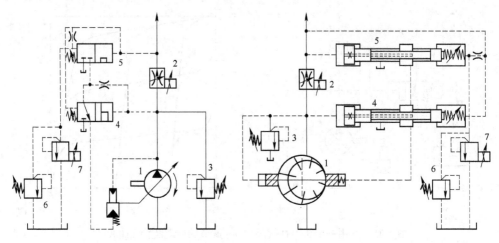

图 4-36　比例压力和流量调节型变量泵职能符号及半结构图

(Fig. 4-36　The functional symbols and semi-structure diagram of proportional pressure and flow regulation variable pump)

4.5 电液比例压力和流量调节型变量泵
(Electro-hydraulic proportional pressure and flow regulation variable pump)

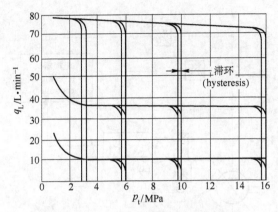

图 4-37 $p-q$ 调节变量泵的静态特性

(Fig. 4-37 The static characteristic of $p-q$ regulation variable pump)

4.5.2 电反馈型比例压力和流量调节 (Electrical feedback proportional pressure and flow regulation)

流量和压力-电反馈型比例调节是把流量信息和压力信息通过传感器转换成相应的电信号，并反馈到输入端。它代表实际输出值与给定值进行比较。任何偏差都由电控器进行处理，得出控制排量改变的信号，通过改变排量来控制输出的流量与压力来满足要求（图 4-38）。与压力补偿型调节相比，它可以取消比例节流阀，因而消除了节流损失，是完全的比例容积控制。压力的检测通常利用压力传感器直接检测（图 4-39）。而流量的直接检测较为困难，可以利用流量传感器来直接检测，也可以对变量机构的位移或倾角做间接检测。

Flow and pressure electrical feedback proportional regulation is to convert the flow information and pressure information into corresponding electrical signal by sensor and then gives feedback to the input port. It represents the comparison of actual output value and given value. Any deviation is totally processed by the electrical controller and obtaining the signal that controls displacement change to control the output flow and pressure so as to meet the requirements by changing displacement (Fig. 4-38). Compared with the pressure compensation type, it can cancel the proportional throttle valve, so it relieves the throttle losses, and it is a completely proportional volume control. The detection of pressure is usually directly detected by the pressure sensor (Fig. 4-39). However, the direct detection of flow is a little difficult so far. And it can be directly detected by flow sensor or indirectly detected by detecting the shift or inclination of the stroking mechanism.

电反馈型比例压力和流量调节的特点包括（The characteristics of electrical feedback proportional pressure and flow regulation include）：

第 4 章 电液比例容积控制
Chapter 4　Electro-hydraulic Proportional Volume Control

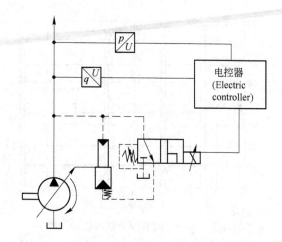

图 4-38　直接检测式电反馈 $p-q$ 调节变量泵原理图
(Fig. 4-38　The schematic diagram of direct detective electrical feedback $p-q$ regulation variable pump)

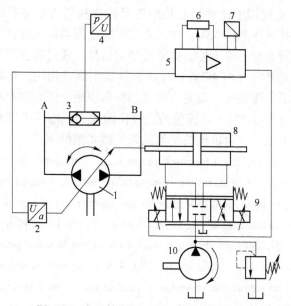

图 4-39　电反馈型 $p-q$ 调节变量泵原理图
(Fig. 4-39　The schematic diagram of electrical feedback $p-q$ regulation variable pump)
1—双向变量轴向柱塞泵（Bidirectional variable axial plunger pump）；2—位移传感器（Displacement sensor）；
3—梭阀（Shuttle valve）；4—压力传感器（Pressure sensor）；5—电控器（Electronic controller）；
6—输入信号电位器（Input signal potentiometer）；7—直流稳压电源（DC regulated power supply）；
8—双向变量机构（Bidirectional stroking mechanism）；9—三位四通比例换向阀
（Three-position and four-way proportional directional valve）；10—先导压力油源（Pilot pressure oil source）

4.5 电液比例压力和流量调节型变量泵
(Electro-hydraulic proportional pressure and flow regulation variable pump)

（1）能同时输出流量反馈（间接检测）和输出压力反馈，因而可以依靠电控器进行 $p-q$ 函数的运算处理，实现复杂的多种控制功能。

It can output flow feedback (indirect detection) and pressure feedback at the same time, so it can carry out function processing of $p-q$ function and achieve complicating multiple control functions by the electronic controller.

（2）由于是双向变量机构，所以必须采用辅助先导油源向比例阀供油，以保证泵在排量为零的情况下不使控制及反馈失效。

As it is bidirectional stroking mechanism, for making sure the control and feedback success under the situation that the displacement is zero, it's essential to adopt auxiliary pilot oil source to supply oil to the proportional valve.

（3）图 4-40 所示为该泵的静态特性（The Fig. 4-40 is the static characteristic of the pump）.

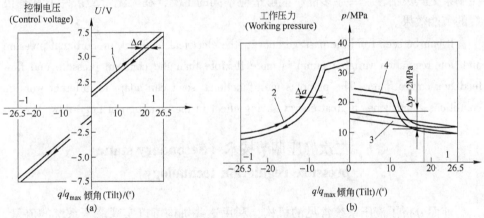

图 4-40　位移及压力电反馈比例 $p-q$ 调节变量泵的特性曲线

(Fig. 4-40　The characteristic curve of shift and pressure electrical feedback proportional p-q regulation variable pump)

1—比例排量调节特性曲线（The characteristic curve of proportional displacement regulation）；
2—恒功率调节曲线（Constant power regulation curve）；3—恒压力调节曲线
（Constant pressure regulation curve）；4—Q-N-P 型调节曲线（Q-N-P type regulation curve）

其中曲线 1 为比例排量调节时控制电压与缸体倾角的关系曲线。由于先导阀芯的力平衡、阀口遮盖量、摩擦力等的影响，曲线上可见有 ±10%~15% 的死区，因而需要起始控制电压 u_0 来克服死区的影响。图中滞环 $\Delta\alpha_{max}$ 约为最大倾角的 3%。重复精度约为 2%。曲线 2 是通过对反馈的倾角位移信号及压力信号实行运算后实现的恒功率控制的情况，曲线 3 是实现恒压控制时的压力与倾角之间的关系曲线。由曲线可见，在做恒压力调节时，压力的变化值为 $\Delta p = 2\text{MPa}/26.5°$。曲线 4 是 Q-N-P 型调节特性。

The curve 1 is the relationship curve of the control voltage and cylinder body inclination when proportional displacement regulates. Because of many influencing factors, for example, the force balance of the pilot spool, the amount of valve port cover, the friction and so on, there is a visible dead zone about ±10%~15%, so the start control voltage u_0 is needed to overcome the effect of the dead zone. The hysteresis $\Delta\alpha_{max}$ in the figure is about 3% of the maximum inclination. The replication precision is about 2%. The curve 2 is the constant power control situation by implementing the operation of the feedback inclination shift signal and pressure signal. The curve 3 is the relationship curve of pressure and inclination when constant pressure control is achieved. It can be seen from the curve that the pressure changing value is $\Delta p = 2\text{MPa}/26.5°$ when it is regulating for a constant pressure. The curve 4 is Q-N-P type regulation characteristic.

由以上可见，电反馈型比例压力和流量调节变量泵比普通的压力和流量反馈控制泵更具灵活性。一泵多能，能够适应不同的工况，在大流量的条件下具有很好的节能效果。

It can be seen from the passage above, the electrical feedback proportional pressure and flow regulation variable pump is more flexible than the common pressure and flow feedback control pump. The pump is multifunctional so it can adapt to different working conditions, and it has a great energy-saving effect in the condition of large flow.

4.6 二次静压调节技术 (Secondary static pressure regulation technology)

前面分析了液压系统常见的机构，利用这些机构可以实现液压系统的流量、压力以及功率等的控制，从而满足工业应用的需要。但是，分析中也发现液压系统的能耗较大，效率一般在30%左右。效率低使得设计液压系统的需求功率很大，真正做有用功的部分很少，而大部分功都损失变为热能。因为液压系统工作的最佳温度是40℃，这样损失的热能又得靠增加冷却器的散热面积来平衡，使得液压系统的设计投资很大，电动机、液压泵、油箱等的规格大大增加。这样不仅增加了设计成本，而且增加了运行成本。因此，提高液压系统的效率，也就是研究节能型液压系统，是液压研究领域中的重中之重。

In the previous contents, we have analysed the common mechanisms of hydraulic system. The flow, pressure and power of hydraulic system can be controlled by these-mechnisms to meet the need of industrial applications. However, in analysis, we also find that the energy consumption of hydraulic system is rather large. Its efficiency is generally about 30%. The low efficiency makes the requiring power of designing hydraulic

4.6 二次静压调节技术
(Secondary static pressure regulation technology)

system very large, but the power truly used for doing useful work is quite low, and most power are converted into heat energy, as the optimum working temperature of hydraulic system is 40 degrees Celsius. Such loss heat energy is balanced by increasing the radiating area of cooler so that the design investment of hydraulic system is a lot. The specifications of electric motor, hydraulic pump, oil tank and so on are greatly increased, thus, not only design cost but also operation cost is increased. Therefore, improving the efficiency of hydraulic system is the primary importance among priorities in the hydraulic research field, namely studying the energy-saving hydraulic system.

针对液压系统能耗的问题，德国汉堡国防工业大学的 H. W. Nikolaus 教授于 1977 年首先提出了二次调节传动的概念。二次调节传动系统是工作在恒压网络的压力偶联系统，能在四个象限内工作，可回收与重新利用系统的制动动能和重力势能。在系统中，液压马达能无损耗地从恒压网络获得能量，系统中可以同时并联多个负载，在各负载段可同时实现互不相关的控制规律。这就扩大了系统的工作区域，改善了系统的控制性能，降低了工作过程中的能耗和冷却费用，减少了设备的总投资。因为它既没有节流损失，又没有溢流损失，还回收系统的制动动能和重力势能，所以其节能效果比上述的回路系统都优越。

To the problem of energy consumption of hydraulic system, H. W. Nikolaus, the professor of Defence Industrial University of Hamburg in Germany, in 1977, firstly came up with the conception of secondary regulation transmission. The secondary regulation transmission system is a pressure coupling system that works in the constant pressure network and can operate in four quadrants, recycling and reusing the braking energy and gravitational potential energy of the system. In the system, the hydraulic motor can gain energy from the constant pressure network with no energy loss, and it can support multiple loads in parallel and realize irrelevant control laws in each load segment simultaneously. This way, it has enlarged the working space, improved the control performance of the system and reduced the energy consumption in the working process, cooling costs as well as the gross investment. Since that it has neither throttling loss nor overflow loss, and recycles the braking energy and gravitational potential energy of the system, so its energy-saving performance is superior to that of the system with above-mentioned circuits.

4.6.1 二次调节静液传动的概述 (Outline of secondary regulation hydrostatic transmission)

二次调节静液传动技术是 20 世纪 70 年代末从德国发展起来的一种液压传动技术。二次调节是对将液压能与机械能互相转换的液压元件进行调节，实现能量转换和传递的技术。如果把静液传动系统中机械能转化为液压能的元件（液压

第 4 章 电液比例容积控制
Chapter 4 Electro-hydraulic Proportional Volume Control

泵）称为一次元件或初级元件，则可把液压能和机械能相互转换的元件（液压马达/泵）称为二次元件或次级元件。在静液传动系统中，可以把液压能转换成机械能的液压元件是液压油缸和液压马达，液压油缸的工作面积是不可调节的，所以二次元件主要是指液压马达。同时，为了使二次调节静液传动系统能够实现能量回收，所需要的二次元件必须是可逆的静液传动元件。因此，对这类静液传动元件可称为液压马达/泵。但是，为了使许多不具备双向无级变量能力的液压马达和往复运动的液压缸也能在二次调节系统的恒压网络中运行，目前出现了一种"液压变压器"，它类似于电力变压器，用来匹配用户对系统压力和流量的不同需求，从而实现液压系统的功率匹配。

The secondary regulation hydrostatic transmission technology is a kind of hydraulic transmission technology developed from Germany in the late 1970s. The secondary regulation is a technology that regulates the hydraulic components used for converting the hydraulic energy and mechanical energy mutually and achieves the conversion and transmission of energy. If the component (hydraulic pump) that converts the mechanical energy into hydraulic energy in the hydrostatic transmission system is called first order component or primary component, then the component (hydraulic motor/pump) that converts the mechanical energy and hydraulic energy mutually is called secondary component or subordinated component. The component that can convert the hydraulic energy into mechanical energy in the hydrostatic system is hydraulic cylinder and hydraulic motor, but the working area of the hydraulic cylinder is unregulated, so the secondary component is mainly meant to hydraulic motor. At the same time, for making the secondary hydrostatic system recycle energy, the secondary component needed must be reversible hydrostatic transmission component. Thus, this kind of hydrostatic transmission component could be called as hydraulic motor/pump. However, in order to make many hydraulic motors without bidirectional stepless variable capacity and reciprocating motion hydraulic cylinder be able to operate in the constant pressure network of secondary regulation system, so far, a "hydraulic transformer" appears, it resembles the power transformer used for matching different requirements of system pressure and system flow for users' requirements so as to achieve the matching power of hydraulic system.

二次调节静液传动技术的实现建立在恒压系统基础之上，目前对二次调节传动技术的研究可以分为两大部分：

The achievement of secondary regulation hydrostatic transmission technology is based on the constant pressure system, so far, the research of secondary regulation transmission technology can be divided into two big parts:

一是利用二次调节传动技术的能量回收原理，一般用在被控对象频繁启停、

4.6 二次静压调节技术
(Secondary static pressure regulation technology)

重力势能变化大的工况，如摩天大楼的电梯、城市公交车和油口抽油机等，在国外有成功的应用实例。

Firstly, the application of energy recycling principle of the secondary regulation transmission technology is generally in the working conditions that the controlled object starts and stops frequently and gravitational potential energy changes greatly, such as the elevator in the skyscraper, the city bus, the pumping unit and so on, and there are also examples of successful application in foreign countries.

二是从控制的角度出发，利用一些复杂的算法改善液压泵/马达的动态特性。

Secondly, in terms of control, it utilizes some complex algorithms to improve the dynamic characteristic of hydraulic pump/motor.

二次调节对改善液压系统的传动效率非常有效，仿佛走出液压传动低效率的"普遍规律"。它与负载敏感回路的主要区别是不但实现了功率适应，而且还可以对工作机构的制动动能和重力势能进行回收和重新利用，同时多个执行元件可以并联在恒压网络上，并分别进行控制。

The secondary regulation is very effective for improving the transmission efficiency of the hydraulic system, as if it is out of the low efficiency which is regarded as "general law" of hydraulic transmission. The main difference from the load sensitive circuit is that it not only achieves the power adaption, but also can recycle and reuse the braking kinetic energy and the gravitational potential energy of the working mechanism, meanwhile, many execute components can be installed in parallel in the constant pressure network, which are controlled respectively.

基于能量回收与重新利用而提出的二次调节概念，对改善静液传动系统的效率非常有效。这种调节技术不但能实现功率适应，而且还可以对工作机构的制动动能和重力势能进行回收与重新利用，因而具有广阔的应用前景。具有位能变化的液压驱动卷扬起重机械，采用二次调节技术后可以回收位能。对于周期性液压车辆、机械，采用二次调节技术，不仅能储存其制动过程中的惯性能，而且可以在启动时释放回收的能量快速启动。同时，在恒压网络开式回路上可以连接多个互不相关的负载，在驱动负载的二次元件上直接控制其转角、转速、转矩或功率。二次调节静液传动系统在控制与功能上的特点，为解决静液传动技术中目前尚未解决的某些传动问题和替代有关传动技术，提供了有利的条件。

The secondary regulation conception is proposed based on energy recycling and reuse, so it is very effective for improving the efficiency of hydrostatic transmission system. The regulation technology can not only realize the power adaption, but also recycle and reuse the braking kinetic energy and the gravitational potential energy of the working mechanism, so it has a broad application prospect. The hoist and crane machine with changing potential energy by hydraulic drive can recycle the potential energy by using the secondary regulation technology. For the periodic hydraulic vehicles and machines, utilizing the secondary regulation technology can not only store the inertia energy produced in the braking process, but also release the energy recycled for the rapid start at the start-up stage. Meanwhile, it can connect many incoherent loads in the constant pressure opening circuit, and control directly its rotating angle, rotating speed, torque or power on the secondary components for driving the load. The characteristics of secondary regulation hydrostatic transmission system in control and function provides advantageous conditions for settling some unsolved transmission problems in the hydrostatic transmission technology or replacing relevant transmission technology.

二次调节静液传动技术由于提供了能量回收和重新利用的可能性，改善了液压系统的传动效率，在许多领域都有应用前景，特别是在能源日益紧张的今天具有重要的理论研究意义和实际应用价值。国外对此项技术的理论研究及应用日趋成熟，已将其成功应用于造船工业、钢铁工业、大型试验台、车辆传动等领域。但在国内目前仍只限于试验研究阶段，离实际应用尚有差距。为使此项技术早日在我国得到应用，在许多方面仍需进行细致的研究工作，尤其是在以下几方面：

As it offers the possibility of recycling and reusing the energy, the secondary regulation hydrostatic transmission technology improves the transmission efficiency of the hydraulic system, it has applicative prospects in many fields, especially nowadays, growing shortage of energy, so it has important theory researching significance and practical applicative value. The theory research and application of the technology abroad is maturing gradually, and the technology has been applied largely into the fields like shipbuilding industries, steel industries, large test desks, vehicle transmissions and so on. But at home, it is still only in the test researching phase and far from the practical application. In order to ensure the technology be applied earlier in our country, much intensive researching work is still needed in many aspects, especially the following aspects:

（1）二次元件本身的改进研究。二次元件作为二次调节系统的关键部件，其性能的优劣对系统整体性能的影响起决定作用。如何在提高元件性能和延长使用寿命的前提下降低成本，是一个有待解决的问题。

The improving study of secondary components themselves. The secondary component works as a key component, so the merit of its performance plays a decisive role in whole performance of the system. Under the premise of improving the performance of component and prolonging the service life, how to reduce the cost is an issue to be solved.

（2）对系统的能量回收与重新利用进行更细致的研究。给出能量回收相关条件，为实际应用奠定基础。

4.6 二次静压调节技术
(Secondary static pressure regulation technology)

To study more intensively on the energy recycling and reusing of the system. The corresponding conditions of energy recycling are given in order to lay a foundation for practical application.

(3) 控制方法的研究。对系统的位置、转速、转矩和功率进行复合控制研究，更好地改进其控制特性。

The research of control methods. Do research on the compound control of system position, rotating speed, torque and power, then improve its control characteristics better.

(4) 进一步加强与微电子技术和计算机技术的结合，拓宽二次调节系统的应用领域。

The combination with microelectronic technology and computer technology should be further strengthened, and the application field of secondary regulation system be broadened.

4.6.2 二次调节技术的发展 (The development of secondary regulation technology)

静液传动二次调节原理最早是由英国人 Pearson 和 Burrett 于 1962 年首次申请了专利。1977 年，德国人 Nikolaus 在德国注册了二次调节控制器的专利，并于 1980 年与 Kordak 在德国的 Mannesmann Rexroth 公司建立了第一个试验台。

The principle of hydrostatic transmission secondary regulation is first applied for a patent by the British Pearson and Burrett primarily in 1962. In 1977, the Germany Nikolaus registered the patent of secondary regulation controller in Germany, and constituted the first test-bed with Kordak in the Germany Mannesmann Rexroth company in 1980.

从 1980 年开始，联邦德国国防工业大学 (LHAS) 静液传动及控制实验室就开始二次调节直接转速控制的研究。同时，亚琛工业大学 (RWTH) 流体传动及控制研究所在著名教授 W. Backe 的领导下，在该领域也进行了卓有成效的研究，并开发出一些应用二次调节技术的新产品，而且得到了广泛应用。

Since 1980, the hydrostatic transmission and control laboratory of West Germany Defence Industrial University (LHAS) has started to do research on the secondary regulation direct rotating speed control. Meanwhile, under the lead of famous professor W. Backe, the institute of fluent transmission and control of Germany Aachen Industrial University (RWTH) also did fruitful research in this field and developed some new products applying the secondary regulation technology, which were widely used.

1980 年，W. Backe 和 H. Murrenhoff 教授进行了液压直接转速控制的二次调节静液传动系统的研究，他们用的二次元件的变量油缸是单出杆活塞缸。

In 1980, professor W. Backe and H. Murrenhoff carried out the research of secondary regulation hydrostatic transmission system of hydraulic direct rotating speed control, and the variable cylinder of secondary component that they used was single-rod piston cylinder.

1981~1987年，R. Kordark、W. Backe、H. Murrenhoff、W. Nikolaus 和 F. Metzner 等人先后提出了液压直接控制系统、液压先导控制调速系统和机液调速系统。但这些调速系统的控制性能不太理想，结构复杂，实现较困难。

Between 1981 and 1987, R. Kordark, W. Backe, H. Murrenhoff, W. Nikolaus, F. Metzner et al. brought forward the hydraulic direct control system, hydraulic pilot control speed governing system and mechanical-hydraulic speed governing system in sequence. But the control performance of the speed governing systems was not ideal enough, structurally complex and hard to be carried out.

1982~1987年，H. Murrenhoff、W. Backe 和 H. J. Hass 等人为提高系统的控制性能，对二次调节电液转速控制系统和电液转角控制系统进行了研究。转角控制是在二次调节转速控制系统的基础上增加一条马达输出轴的转角反馈回路。这种系统可以是单反馈控制回路，但其阻尼比较小，控制性能不太好。为提高系统的阻尼，改善系统的控制性能，引入二次元件变量油缸位移反馈，组成双返回路电液转速控制系统。

Between 1982 and 1987, for improving the control performance of system, H. Murrenhoff, W. Backe, H. J. Hass et al. carried out research on the secondary regulation electronic-hydraulic rotating speed control system and electronic-hydraulic rotating angle control system. The rotating angle control was based on the secondary regulation rotating speed control system adding with a rotating angle feedback circuit of a motor output axis. The system could be a single feedback control circuit, but its damping was very small, so the control performance was not very good. To improve the damping and the control performance of system, they constituted double reverse circuit electronic-hydraulic rotating speed control system by bringing in the shift feedback of secondary component variable cylinder.

1987年，F. Metzer 提出了数字-模拟混合转角控制系统，将经过电液力反馈转速控制的二次元件作为被控对象，控制算法采用数字 PID 控制，它能实现二次元件的转速、转角、转矩和功率控制。1988年，W. Holz 发表文章介绍此系统，并给出其应用的可能性。

In 1987, F. Metzer put forward the digital-simulation mixture rotating angle control system, regarding the rotating speed control secondary component of electronic-hydraulic-force feedback as the control object, whose algorithm was digital PID control, so it could realize the rotating speed, rotating angle, torque and power control of sec-

4.6 二次静压调节技术
(Secondary static pressure regulation technology)

ondary component. In 1988, Mr. W. Holz published an article to introduce the system and show the possibilities of its application.

1993 年，W. Backe 教授和 Ch. Koegl 提出基于二次调节技术的转矩伺服加载的思想。在该系统中，能量可以直接被恒压网络吸收，并研究了转速和转矩控制系统中两个参数的解耦问题。

In 1993, professor W. Backe and Ch. Koegl put forward the toque servo loading conception based on the secondary regulation technology. In the system, the energy could be directly sucked up by the constant pressure network. They also did research on the decoupling problems of two parameters in the system of rotating speed control and torque control respectively.

1994 年，R. Kordak 研究了具有高动态特性的电液转矩控制系统。

In 1994, R. Kordak researched the electronic-hydraulic torque control system with high dynamic characteristics.

1996 年，H. Murrenhoff 与 E. Weishaupt 综述了二次调节转速控制的最新发展，分析了油源压力与负载变化造成的工作点变化并最终对转速控制性能的影响。日本人 H. Nakazawa、S. Yokota 和 Y. Kita 则提出将二次元件-飞轮组合应用于汽车刹车装置，从而达到节能的目的。

In 1996, H. Murrenhoff and E. Weishaupt summed up the newest development of secondary regulation rotating speed control, and analyzed the effect of working point change caused by oil source pressure and load change on the rotating speed control performance. The Japanese H. Nakazawa, S. Yokota and Y. Kita put forward the secondary component-flywheel group applied into vehicle braking device so as to realize the goal of energy saving.

1999 年，H. Berg 与 M. Ivantysynova 提出 T-次调节转速和位置的新控制策略。根据二次调节静动态技术发展需要和适应过程控制，提出了一种具有强鲁棒性、高频宽的新控制结构。

In 1999, H. Berg and M. Ivantysynova put forward the new control strategy of T-times regulation rotating speed and position. According to the secondary regulation static-dynamic technology development requirements and adapted process control, they put forward a new control structure of strong robustness and large bandwidth.

4.6.3 二次调节静液传动的工作原理 (Operating principle of secondary regulation hydrostatic transmission)

二次调节系统与传统的液压控制系统不同，它是一种压力耦合系统，是对将液压能与机械能转换的液压元件所进行的调节（图 4-41）。这种调节的基本原理是，通过调节可逆式轴向柱塞元件（二次元件）的斜盘倾角来适应外负载的变化，类似于电力传动系统在恒压网络中传递能量。

The secondary regulation system is different from the traditional hydraulic control system for that it is a kind of pressure coupling system, and it regulates the hydraulic components that can realize the conversation between hydraulic energy and mechanical energy (Fig. 4-41). The basic principle of the regulation is to adapt the changing load by regulating the swash plate inclination of reversible axis plunger component (secondary component), it resembles that the electric power transmission system transfers energy in the constant pressure network.

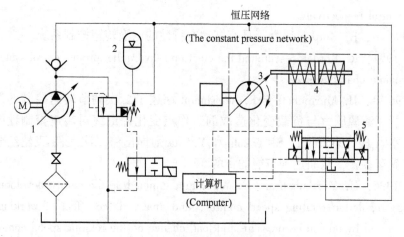

图 4-41　二次调节系统原理图

(Fig. 4-41　The schematic diagram of secondary regulation system)

1—恒压变量泵（The constant pressure variable pump）；2—蓄能器（Power accumulator）；
3—二次元件（Secondary component）；4—变量油缸（Variable cylinder）；
5—电液比例阀（Electric-hydraulic proportional valve）

其工作原理如下：当外负载增大时，将引起转速的下降。该变化由转速传感器检测送给控制器，再由控制器进行计算和判断，然后输出一个电信号给伺服阀，由伺服阀驱动变量缸使二次元件的斜盘倾角增大，从而使二次元件的排量增大，转矩增大，转速升高，以适应增大了的外负载。反之亦然。由于恒压油源部分的动态特性较好，所以在对二次调节静液传动系统进行分析与研究时，可以不考虑油源部分动态性能对系统输出的影响，并且可以认为恒压网络中的压力基本保持恒定不变。这样不仅能简化研究的复杂性，同时也能保证研究结果的准确性。其中二次元件不经过任何能引起能量损失的液压控制元件而直接与恒压网络相连，因而，可以无损失地从恒压网络中获取能量，从而达到节能的目的。

The operating principle is as follows: when the external load increases, the rotating

4.6 二次静压调节技术
(Secondary static pressure regulation technology)

speed will decrease, and the change is detected by rotating speed sensor and sent to the controller, then computed and judged by the controller, next it outputs a electrical signal to the servo valve, at last, the servo valve drives the variable cylinder to make the swash plate inclination increase so that the displacement of secondary component increases, torque increases, rotating speed increases so as to adapt to the load that has increased. Vice versa. Because of the good dynamic characteristic of constant pressure oil source, when analyzing and studying the secondary regulation hydrostatic transmission system, the effect of dynamic performance of oil source on the output can be ignored and the pressure in the constant pressure network is maintained basically unchanged. Thus it can not only simplify complexity of the system but also ensure the precision of research results. Among other things, the secondary component is connected to the constant pressure network without passing any hydraulic control component that can result in energy losses, so it can attain energy from the constant pressure network without losses and reach the goal of saving energy.

4.6.4 二次调节系统的转速控制 (Rotating speed control of secondary regulation system)

二次调节系统的转速控制就是在二次元件的输出端增加一个测速传感器，如测速电机或光电码盘，与控制器构成一个电反馈二次元件转速控制系统。实际工程中，为了能够改善系统性能，常将变量油缸的运动通过机械反馈和力反馈的形式加入到二次调节系统中。转速控制是二次调节静液传动系统中最基本的控制方案，位置、转矩和功率控制等其他方案都是在转速控制的基础上增加反馈通道完成的，而且转速必须得到监测或控制。

The rotating speed control of secondary regulation system is to add a speed sensor at the output port of secondary component, such as tachogenerator or photoelectric coded disk, which together with controller constitute an electrical feedback secondary component rotating speed control system. In the practical project, to improve the system performance, the motion of variable cylinder is generally added to the secondary regulation system by the form of mechanical feedback and force feedback. The rotating speed control is the most basic control scheme in the secondary regulation hydrostatic transmission system, and other control schemes like position, torque and power are all completed based on the rotating speed control by adding with feedback channels, and the rotating speed must be monitored and controlled.

图 4-42 为转速控制系统的原理图，n_s 为设定转速。当实测转速与设定转速不符时，它们的差值通过控制器驱动比例阀，比例阀的输出流量控制变量油缸，

而变量油缸的移动调节了二次元件的斜盘倾角，改变二次元件的排量，使其输出转矩与负载转矩相一致，即达到设定转速。例如，当负载转矩增大，输出转矩小于负载转矩时，二次元件的转速必然下降。这样实测转速小于设定转速，这两个转速之差通过控制器、比例阀，增大变量油缸的位移，也就增大了二次元件排量，进而增大了输出转矩，最终使驱动转矩和负载转矩相符，转速恢复到设定值，二次元件又在设定转速下工作。

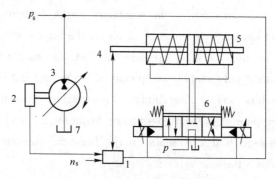

图 4-42 二次调节转速控制原理图

(Fig. 4-42 The schematic diagram of secondary regulation rotating speed control)
1—控制器（Controller）；2—转速测量仪（Rotating speed measuring apparatus）；
3—二次元件（Secondary component）；4—位移传感器（Shift sensor）；5—变量油缸
（Variable cylinder）；6—电液比例阀（Electric-hydraulic proportional valve）；7—油箱（Oil tank）

Fig. 4-42 is the schematic diagram of rotating speed control system, n is the set rotating speed. When the detected practical rotating speed doesn't match with the set one, their differential drives the proportional valve by controller, then the output flow of the proportional valve controls variable cylinder, and the motion of variable cylinder regulates the swash plate inclination of secondary component and changes the displacement of secondary component in order that the output torque is equal to the load torque, namely, it reaches the setting rotating speed. For example, when the load torque increases, the output torque is less than the load torque, the rotating speed of secondary component will inevitably decrease. Thus when the detected practical rotating speed is less than the setting rotating speed, the differential of the two rotating speeds passes through controller, the proportional valve increases the shift of variable cylinder namely increases the displacement of secondary component, and then increases the output torque and finally makes the drive torque match with the load torque, the rotating speed comes back to the setting value, and the secondary component will work at the setting rotating speed.

通常情况下，二次调节系统的黏性阻尼较小，并且不是真正需要调速的负

4.6 二次静压调节技术
(Secondary static pressure regulation technology)

载，因此可以把它看做是二次元件的内部损耗。忽略二次元件的弹性负载，则二次元件的力矩平衡方程为：

Under normal circumstances, the viscous damping of secondary regulation system is small, and it is the load without really needing speed regulation, so it could be regarded as internal loss of secondary component. Ignoring the elastic load of secondary component, the torque balance equations of secondary component are

$$M = J\omega + M_2 \tag{4-39}$$

$$M = Vp_L - B\omega \tag{4-40}$$

式中　M——二次元件输出转矩（The output torque of secondary component）；

　　　J——二次元件和负载的总转动惯量（The total moment of inertia of secondary component and load）；

　　　ω——二次元件输出转速（The output rotating speed of secondary component）；

　　　M_2——负载转矩（Load torque）；

　　　V——二次元件排量（The displacement of secondary component）；

　　　p_L——负载压差（The load differential pressure）；

　　　B——二次元件和负载的黏性阻尼系数（The viscous damping coefficient of secondary component and load）。

假定系统处于平衡状态，如果负载扭矩 M_2 变大，由式可知，转速 ω 的变化率为负，ω 下降，测速泵转速 ω_{st} 下降，变量缸右移，使得二次元件排量 V 增大，二次元件输出转矩 M 上升，输出转速 ω 回升，直到回到原平衡值为止。即

$$M_2 \uparrow \rightarrow \omega \downarrow \rightarrow \omega_{st} \downarrow \rightarrow V \uparrow \rightarrow M \uparrow \rightarrow \omega \uparrow$$

Assuming that the system is in equilibrium, if the load torque M_2 increases, it can be seen from the equation, the change rate of rotating speed ω is minus, ω decreases, the rotating speed of speed pump ω_{st} decreases, the variable cylinder moves right, the displacement V of secondary component increases, the output torque M of secondary component increases, the output rotating speed ω increases, until it comes back to the original balance value. Namely,

$$M_2 \uparrow \rightarrow \omega \downarrow \rightarrow \omega_{st} \downarrow \rightarrow V \uparrow \rightarrow M \uparrow \rightarrow \omega \uparrow$$

4.6.5 二次调节系统的转矩控制 (Torque control of secondary regulation system)

在二次调节转速控制系统的基础上增加一个马达输出轴的转矩传感器，即构成转矩反馈回路，即一个电反馈二次元件的转矩控制系统。二次调节转矩控制系统的基本工作原理是：在恒压网络中，通过调节二次元件斜盘倾角来改变二次元

件排量，以适应负载（工作机构）转矩的变化，从而使负载按设定的规律变化。对于工作在恒压网络的二次调节系统，其系统工作压力近似为常量，只需通过调节二次元件的斜盘倾角改变其排量，即可实现调节转矩的目的。

On the basis of secondary regulation rotating speed control system, adding with a torque sensor of motor output axis, it constitutes a torque feedback circuit, namely a torque control system of secondary component electrical feedback. The basic working principle of secondary regulation torque control system is: in the constant pressure network, it changes the displacement of secondary component by regulating the swash plate inclination of secondary component, so as to adapt changing torque of the load (operation mechanism) and make the load change according to the setting law. For the secondary regulation system working in the constant pressure network, the system working pressure is approximately a constant, so it changes the displacement by only regulating the swash plate inclination of secondary component so that reaching the goal of regulating the torque.

二次元件的输出转矩表达式如下（The output torque expression of secondary component is as follows）：

$$M_2 = \frac{p_0 V_2}{2\pi} \tag{4-41}$$

$$M_2 = p_0 V_{2\max} \frac{a_2}{a_{2\max}} = p_0 V_{2\max} \frac{y}{y_{\max}} \tag{4-42}$$

二次元件的力矩平衡方程如下（The torque equilibrium equation of secondary component is as follows）：

$$p_0 V_2 = J \frac{d^2\varphi}{dt^2} + R_H \frac{d\varphi}{dt} + M \tag{4-43}$$

于是得（Accordingly, we can get）：

$$\frac{d^2\varphi}{dt^2} = \frac{1}{J}\left(\frac{p_0 V_{2\max}}{2\pi} \frac{a_2}{a_{2\max}} - M_R - M\right) \tag{4-44}$$

式中　M_2——二次元件输出转矩（The output torque of secondary component）；

　　　p_0——恒压网络工作压力（The working pressure of constant pressure network）；

　　　V_2——二次元件的排量（The displacement of secondary component）；

　　　$V_{2\max}$——二次元件最大排量（The maximum displacement of secondary component）；

　　　a_2——二次元件斜盘倾角（The swash plate inclination of secondary component）；

　　　$a_{2\max}$——二次元件最大斜盘倾角（The maximum swash plate inclination of secondary component）；

　　　φ——二次元件输出转角（The output rotating angle of secondary component）；

4.6 二次静压调节技术
(Secondary static pressure regulation technology)

J——二次元件和负载转动惯量之和（The result moment of inertia of secondary component and load）；

y——液压缸活塞位移（The shift of hydraulic cylinder piston）；

y_{max}——液压缸活塞最大位移（The maximum shift of hydraulic cylinder piston）；

M_R——摩擦转矩（The friction torque）；

M——负载转矩（The load torque）；

R_H——二次元件阻尼系数（The damping coefficient of secondary component）。

可见，改变斜盘倾角的大小和方向，就可以线性地改变输出转矩的大小和方向。若负载转矩一定，通过改变二次元件的斜盘倾角使二次元件输出转矩大小和方向改变，即可实现对负载加速或减速、正转或反转的控制。转矩控制系统的原理如图 4-43 所示。

It is clear that changing the size and direction of swash plate inclination is able to change the size and direction of the output torque linearly. If the load torque is certain, the control of accelerating or decelerating and positive rotation or negative rotation of the load could be realized by changing the swash plate inclination angle of secondary component so as to change the size and direction of secondary component output torque. The principle of torque control system is showed as Fig. 4-43.

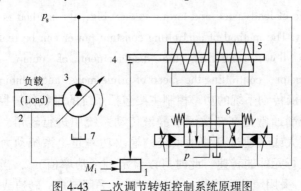

图 4-43 二次调节转矩控制系统原理图

(Fig. 4-43 The schematic diagram of secondary regulation torque control system)
1—控制器（Controller）；2—转矩转速仪（Torque tachometer）；3—二次元件（Secondary component）；
4—位移传感器（Shift sensor）；5—变量油缸（Variable cylinder）；
6—电液比例阀（Electronic-hydraulic proportional valve）；7—油箱（Oil tank）

4.6.6 二次调节系统的功率控制 (Power control of secondary regulation system)

液压装置传递的功率往往较大，所以在设计液压系统时需要根据输出端的负载情况考虑经济性、效率和安全性等问题。液压系统的功率控制可以分为省功率控制、过载保护控制和恒功率控制。液压系统恒功率控制主要有两种情况，一种是必须按一定功率驱动负载的情况，另一种是经常在原动机（发动机或电动机）

最大输出功率状态下使用的情况。前者的负载驱动条件有恒功率限制,后者的主要目的是提高工效。此外,还有一种需把包括原动机与传动装置在内的驱动源的功率效率经常保持在最大值的控制方式,当负载速度和力矩的变化范围较大时,这种方式在提高效率方面是有利的。恒功率控制的方法可以通过控制旁通流量、控制泵排量、控制马达排量、控制原动机速度以及把它们组合起来的方法来实现。

The transferring power of hydraulic devices is generally very large, so the economic, efficiency and security ought to be taken into account according to the load conditions of output port while designing the hydraulic system. The power control of hydraulic system can be divided into saving power control, overload protection control and constant power control. There are two main situations in the constant power control of hydraulic system, one is that the load must be driven in a definite power, another is that it is usually used in the situation that prime mover (engine or electric motor) is in the maximum output power. The load driving conditions of the former is constant power control, and the main purpose of the latter is to improve the working efficiency. Besides, there is a control method that keeping the power and efficiency of drive source including prime mover and transmission mechanism usually remaining the maximum, and when the variation range of load speed and torque is very big, the method is good for improving the efficiency. The method of controlling constant power can be realized by controlling the bypass flow, controlling the displacement of pump, controlling the displacement of motor, controlling the speed of prime mover and combining all of them.

二次调节静液传动系统的功率控制主要包括系统的过载保护控制和恒功率控制。过载保护控制通常是在二次调节静液传动系统中进行转角、转速或转矩控制。当负载力过大或转速过高使得系统的输出功率超出系统动力源的额定功率时,需要降低执行元件的转速,此时可以设定一个功率值 p_n,当系统的输出功率值小于 p_n 时,采用恒功率控制;大于 p_n 时,采用转角、转速或转矩控制。

The power control of secondary regulation hydrostatic transmission system mainly includes the systematic overload protection control and constant power control. The overload protection control is usually in the secondary regulation hydrostatic transmission system, when it carries out rotating angle, rotating speed and torque control, and the load force is too big or the rotating speed is too high to make the systematic output power exceed the rated power of systematic power source, then it will need to reduce the rotating speed of actuator, at this time, we can set a power value p_n, and it will adopt constant power control when the output power is smaller than p_n and adopt rotating angle, rotating speed or torque control when the value is bigger than p_n.

4.6 二次静压调节技术
(Secondary static pressure regulation technology)

二次调节恒功率控制是指二次元件通过自身的闭环反馈控制来实现输出功率的控制。由于二次调节系统属于马达排量控制方式,根据

The secondary regulation constant power control means that the secondary component realizes the controlling of output power by itself closed-loop feedback control. As the secondary regulation system belongs to the motor displacement controlling method, according to

$$P_i = p_s q_2 \tag{4-45}$$

式中　P_i——二次元件输入功率(The input power of secondary component);
　　　p_s——系统压力(The system pressure);
　　　q_2——二次元件输入流量(The input flow of secondary component)。

和(And)

$$P_0 = M_2 \omega_2 \tag{4-46}$$

式中　P_0——二次元件输出功率(The output power of secondary component);
　　　M_2——二次元件输出转矩(The output torque of secondary component);
　　　ω_2——二次元件转速(The rotating speed of secondary component)。

可知实现系统的恒功率控制方法(We can know the control method of realizing the system constant power):

(1) 通过检测二次元件的输入流量并反馈到控制器,与实际要求值相比较,用这个差值来控制二次元件的排量,使输出功率与期望值相符,如图 4-44 所示。

By detecting the input flow of secondary component and giving feedback to the controller, comparing it with practical requiring value, using the differential value to control

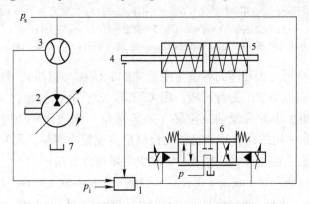

图 4-44　二次调节流量检测恒功率控制系统原理图
(Fig. 4-44　The schematic diagram of secondary regulation flow detection constant power control system)
1—控制器(Controller); 2—二次元件(Secondary component); 3—流量计(Flowmeter);
4—位移传感器(Shift sensor); 5—变量油缸(Variable cylinder);
6—电液比例阀(Electronic-hydraulic proportional valve); 7—油箱(Oil tank)

the displacement of secondary component, the output power is controlled so as to coincides with the expectation value, as is showed in the Fig. 4-44.

（2）通过检测二次元件的输出转矩与转速，然后用两者的乘积与实际要求值进行比较，用来调节二次元件的排量。当使转速处于某一合理范围（不为零、不超速）时，此方式可实现较精确的功率控制。如图 4-45 所示。

By detecting the output torque and rotating speed of secondary component, then comparing the product of two quantities with practical requiring value, so as to regulate the displacement of secondary component, when the rotating speed is in a reasonable range (no zero or no overspeed), the method can realize quite precise power control, as is shown in the Fig. 4-45.

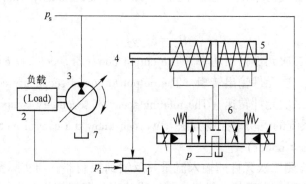

图 4-45　二次调节转矩、转速检测恒功率控制系统原理图

(Fig. 4-45　The schematic diagram of secondary regulation torque and rotating speed detection constant power control system)

1—控制器（Controller）；2—扭矩转速仪（Torque tachometer）；3—二次元件（Secondary component）；
4—位移传感器（Shift sensor）；5—变量油缸（Variable cylinder）；
6—电液比例阀（Electronic-hydraulic proportional valve）；7—油箱（Oil tank）

（3）通过检测二次元件的转速与变量油缸的位移（扭矩），然后用两者的乘积（功率）与实际要求值进行比较，用来调节二次元件的排量，如图 4-46 所示。在用能反映转矩大小的变量油缸位移（斜盘倾角）来表示转矩时，因摩擦转矩的存在，所以有一定的误差。此途径的检测量实现较为容易，同时，这些检测量也可以用在其他控制方式中，故试验研究中常采用此方法。

By detecting the rotating speed of secondary component and the shift (torque) of variable cylinder, then comparing the product (power) of the both with the practical requiring value, and it is used for regulating the displacement of secondary component, as is showed in the Fig. 4-46. While using the shift (inclination of swash plate) of variable cylinder to reflect the size of torque, as the existence of friction, there will be a definite error. The detecting amounts of the method are easily attained, at the same

4.6 二次静压调节技术
(Secondary static pressure regulation technology)

time, these detecting amounts can be used in other control modes, so the method is usually used in the experimental investigations.

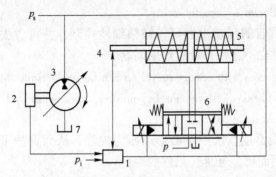

图 4-46 二次调节转速、摆角检测恒功率控制系统原理图

(Fig. 4-46 The schematic diagram of secondary regulation rotating speed and pivot angle detection constant power control system)

1—控制器 (Controller); 2—转速传感器 (Rotating speed sensor); 3—二次元件 (Secondary component); 4—位移传感器 (Shift sensor); 5—变量油缸 (Variable cylinder); 6—电液比例阀 (Electronic-hydraulic proportional valve); 7—油箱 (Oil tank)

4.6.7 二次调节静液传动系统的特点 (Characteristics of secondary regulation hydrostatic transmission system)

二次调节静液传动系统的特点主要表现为 (The characteristics of secondary regulation hydrostatic transmission system are mainly performed):

(1) 用户本身可以作为能量的输出者 (回馈惯性能、重力势能), 向系统提供能量并在必要时储存能量, 从而降低了整个系统的能量消耗。

The user can be the output port of energy (feedback inertia energy and gravitational potential energy), supply energy for the system and store energy if necessary, and reduce the energy consumption of the total system.

(2) 它是压力耦联系统, 系统中的压力基本保持不变, 恒压油源的工作压力直接与二次元件相连。因此, 在系统中没有原理性的节流损失, 提高了系统效率。

It is pressure coupling system, and the pressure in the system is basically maintained unchanged, and the working pressure of constant pressure oil source is connected with secondary component directly. Therefore, there is no fundamental throttle loss in the system, so it improves the efficiency of system.

(3) 通过加入液压蓄能器, 可以使用户的加速功率提高到数倍于一次元件的输出功率, 因此可装用较小的一次元件, 节约了成本。

It can increase the accelerating power of users to many times the output power of primary component by adding a hydraulic accumulator, so it can install a smaller primary component in order to reduce the cost.

（4）带能量回收的制动过程可以被精细地控制，并可省去昂贵的防抱死延迟制动系统，同时也不存在系统过热的问题。

The braking process with energy recycling can be finely controlled, and the expensive anti-lock delay brake system can be omitted, meanwhile, it doesn't exist the problem of system overheat.

（5）液压蓄能器使系统中不会形成压力尖峰，从而保护了液压元件不受尖峰压力的冲击，因而可显著延长元件的寿命。

The hydraulic accumulator can stop the system from forming the pressure peak, so it can protect the hydraulic component from the shock of the peak pressure, and prolong the service life of component dramatically.

（6）二次元件工作于恒压网络，可以并联多个互不相关的负载，实现互不相关的控制规律，而液压泵站只需按负载的平均功率之和进行设计安装。

The secondary components works in the constant pressure network, which can parallelly connect many irrelevant loads, and realizes irrelevant controlling laws, but the hydraulic power unit only needs to carry out design and installment according to the sum of the average power of the loads.

（7）由于负荷均匀，泵组的工况变化有限，噪声状况得到改善。

As the load is homogenous, and the working condition changes of pump package are limited, so the noise circumstance is improved.

4.6.8 二次调节技术的主要应用（Main applications of secondary regulation technology）

由于二次调节静液传动系统具有许多优点，使它在很多领域得到广泛的应用。国外已将其成功应用于造船工业、钢铁工业、大型试验台、车辆传动等领域。第一套配备有二次调节闭环控制的产品是无人驾驶集装箱转运车CT40，它建在鹿特丹的欧洲联运码头（ECT）。

For there are so many advantages of secondary regulation hydrostatic transmission system, it has been widely applied in many fields. It has been successfully applied into fields like shipbuilding industries, steel industries, large test rig and vehicle transmission and so on abroad. The first product equipped with secondary regulation closed-loop control is pilotless container transfer car CT40, which is built at the Europe combined transport (ECT) dock of Rotterdam.

4.6 二次静压调节技术
(Secondary static pressure regulation technology)

德国的海上浮油及化学品清污船——科那西山特号，其液压传动设备配置有二次调节反馈控制系统。该系统可以使预选的撇沫泵和传输泵设备的转速保持恒定，并使之不受因传输介质黏度的变化而引起的外加转矩的影响。

The hydraulic transmission equipments of Germany maritime spill oil and chemicals clean-up steamer are equipped with secondary regulation feedback controlling system. The system can keep the rotating speed of the preselected skimming pump and transmission pump devices as constant, and prevent it from the effect of external torque caused by the change of transmission medium viscosity.

德累斯顿工业大学建立的通用试验台，应用了二次调节反馈控制的特点。可以进行能量回收并具有高反馈控制精度，满足实际中的严格要求。

A general test rig established by Dresden University of Technology adopted the characteristics of secondary regulation feedback control. It could recycle energy and has high feedback controlling precision to meet the strict requirements in practice.

用于汽车撞击试验的试验台，由于采用了二次调节技术，不仅能够使汽车按照规定的曲线加速，而且由于在高压管道上没有节流点而达到最好的效率，整个设备的液压油不需要进行冷却。市内公交汽车在频繁启动过程中，大量的动能被白白浪费掉。力士乐公司为老式公交汽车配备的驱动装置应用了二次调节技术，当汽车刹车时，二次元件进入泵工况，受负载拖动，向蓄能器回馈能量；而当汽车启动时，二次元件进入马达工况，蓄能器和恒压变量泵一起向马达输出能量，从而加速启动过程。显然，这种驱动方式可以降低汽车的成本，节约燃料，更重要的是减轻城市污染。这对环境严重污染的现今社会具有重大的现实意义。

Because the test desk used for the automobile crashing experiments adopted the secondary regulation technology, it can not only make the autos accelerate as the specified curve, but also achieve the best efficiency as there is no throttle spot on the high pressure pipes and the hydraulic fluid of the whole facilities need not cooling. A lot of kinetic energy is wasted in the frequent starting process of city buses. The Rexroth enterprise equipped the traditional city buses with a drive set applying the secondary regulation technology. When the autos brake, the secondary components turn into the working conditions of pump, driven by the load, and the energy is fed back to the accumulator. When the autos start, the secondary components turn into working conditions of motor, and the accumulator and the constant pressure variable pump supply energy to the motor together and accelerate the start process. Evidently, the driving method can reduce the autos cost, save fuel and more primarily relief the pollution of city. It has impressive realistic significance in the today society that the environment is badly polluted.

除了汽车领域和试验台应用，力士乐公司还将二次调节技术应用在石油行业。通常抽油机是采用一套齿轮传动系统，具有相同的提升和下降速度。这种系统的效率特别是在开采高黏度石油时很低。而采用二次调节技术的开发设备具有高填充率和高循环频率，并且可以回收活塞杆下降时的势能。

Except for the applications of automobile field and test desk, the Rexroth enterprise also adopted the secondary regulation technology into the petroleum industries. In generally, the adopted oil pumping unit is a gear transmission system, and has the same lifting speed and descent speed. The systematic efficiency is quite low especially when exploiting the high viscosity petroleum. The exploiting devices with secondary regulation technology have high filling rate and high circulation frequency, and can recycle the potential energy generated by descending of the piston rod.

总结二次调节系统的应用，主要体现在以下几个方面：

Summing up the applications of secondary regulation system, it is mainly reflected in the following aspects:

（1）回收液压驱动卷扬机械的势能；

Recycling the potential energy of hoisting machines driven by hydraulic;

（2）回收液压驱动摆动机械的惯性能；

Recycling the inertia energy of swing machines driven by hydraulic;

（3）群控作业机械和试验装置的综合节能。

The comprehensive energy-saving of group operating machinery and test devices.

下面介绍几个关于二次调节技术的例子。

The followings are some examples about secondary regulation technology.

A　公交车二次调节技术的应用（The secondary regulation technology application of city bus）

如图 4-47 所示，二次调节技术应用于公交车，当车辆开始制动状态时，通过控制器控制使二次元件的斜盘倾角发生变化，转变为泵工况，二次元件作为泵工况开始向系统输入能量，直到车辆完全停止，完成能量的回收。当车辆再次处于起步状态时，制动时储存在液压蓄能器中的能量释放出来，与一次元件共同提供起步动能。

As shown in the Fig. 4-47 is that the secondary regulation technology is applied into buses, when the autos begin the braking condition, the changing of swash plate inclination of secondary component is controlled by controller and turns it into the working condition of pump, and the secondary component as the pump working condition starts to supply energy to the system, and it will not stop recycling energy until the autos stop completely. When the autos are in starting condition again, the energy stored in the hydraulic energy accumulator in braking condition is released, supplying starting kinetic

4.6 二次静压调节技术
(Secondary static pressure regulation technology)

energy together with primary components.

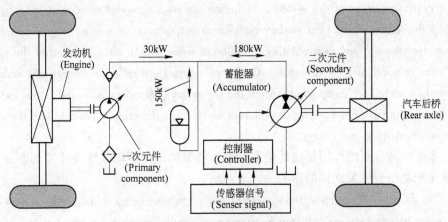

图 4-47 公交车二次调节系统简图

(Fig. 4-47 The secondary regulation system sketch of buses)

系统的控制原理如图 4-48 所示，系统的控制器可以接受系统的反馈量，包括二次元件的转速、二次元件输出轴的转矩信号、系统压力和使用者的操作意图等相关传感器信号，综合处理后对系统进行控制。能量的回收与利用主要是通过二次元件实现，当二次元件由马达转为泵工况时，换向阀动作，蓄能器储存能量。

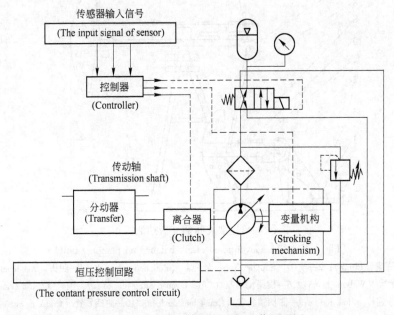

图 4-48 公交车二次调节系统工作原理图

(Fig. 4-48 The working schematic diagram of bus secondary regulation system)

第4章 电液比例容积控制
Chapter 4 Electro-hydraulic Proportional Volume Control

The control principle of system is shown as Fig. 4-48, and the system controller can receive feedbacks of the system, including the rotating speed of secondary components, the torque signal of secondary component output axis, the sensor signal like the systematic pressure and manipulating intention of users and so on, and control the system after comprehensively processed. The recycling and utilization of energy are mainly realized by secondary component, when the secondary component is turned into working condition of pump from working condition of motor, the direction valve acts and the energy accumulator stores energy.

系统中离合器的作用是在能量回收时,将发动机与传动系统间动力切断,消除减速制动过程中发动机的影响。

The function of clutch in the system is that when the energy is recycled, it cuts off the power between engine and transmission system and eliminates the effect of engine in the deceleration braking process.

B 抽油机二次调节技术的应用 (The secondary regulation technology application of oil pumping unit)

如图4-49所示,当抽油机工作时,驴头悬点上作用的负载是变化的。工作分为两个冲程,抽油机上冲程时,驴头悬点需提起抽油杆柱和液柱,在抽油机未

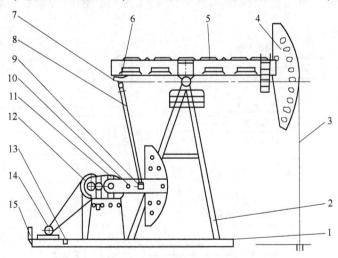

图 4-49 常规曲柄平衡抽油机

(Fig. 4-49 The conventional crank balance oil pumping unit)

1—底座 (Plinth); 2—支架 (Support); 3—悬绳器 (Polished rod eye); 4—驴头 (Ass-head);
5—游梁 (Walking beam); 6—横梁轴承座 (Cross beam bearing block); 7—横梁 (Cross beam);
8—连杆 (Link rod); 9—曲柄销装置 (Crank bolt device); 10—曲柄装置 (Crank device);
11—减速器 (Speed reducer); 12—刹车保险装置 (Brake protection device); 13—刹车装置 (Braking device);
14—电动机 (Electric motor); 15—配电箱 (Power distribution cubicle)

4.6 二次静压调节技术
(Secondary static pressure regulation technology)

进行平衡的条件下，电动机就要付出很大的能量，这时电动机处于电动状态。在下冲程时，抽油机杆柱对电动机做功，使电动机处于发电机的运行状态。抽油机未进行平衡时，上、下冲程的负载极度不均匀，这样将严重地影响抽油机的四连杆机构、减速箱、电动机的效率和寿命，恶化抽油杆的工作条件，增加它的断裂几率。为消除这些缺点，一般在抽油机的游梁尾部或曲柄上或两处都加上了平衡重。这样一来，在悬点下冲程时，要把平衡重从低处抬到高处，增加平衡重的位能。为了抬高平衡配重，除了依靠抽油杆柱下落所释放的位能外，还要电动机付出部分能量。在上冲程时，平衡重由高处下落，把下冲程时储存的位能释放出来，帮助电动机提升抽油杆和液柱，减少了电动机在上冲程时所需给出的能量。

As is shown in the Fig. 4-49, when the oil pumping unit is working, the load that acts on the ass-head polished rod is changing. The task is divided into two strokes, when the oil pumping unit is in the upward stroke, the ass-head polished rod ought to lift the sucker rod and the liquid column, before the oil pumping unit is in the balance condition, the electric motor has to pay out much energy and it is in the electric condition at this time. When the oil pumping unit is in the downward stroke, work is done on the electric motor by the sucker rod, which makes electric motor be in the condition of the generator. Before the oil pumping unit under balance condition, the loads of upward stroke and downward stroke are extremely uneven, which will badly effect the efficiency and life of four link mechanism, reduction gear box and electric motor, and deteriorate the working condition of sucker rod and increase its fracture probability. In order to eliminate these faults, a counterweight should be added to the tail or the crank or even the both of the walking beam generally. In this way, when the polished rod is in the downward stroke, it will lift the counterweight from the low position to the high position to increase its potential energy. To lift the counterweight, besides depending on the potential energy released when the sucker rod falls, the electric motor also needs to pay out part energy. When it is in the upward stroke, the counterweight falls from the high position, releasing the energy stored in the downward stroke, helping electric motor lift sucker rod and liquid column, and reducing the energy that the electric motor ought to pay out in the upwards stroke.

从以上常规的机械式抽油机的工作原理可知，该抽油机有如下缺点：

We can see from the above common mechanical oil pumping unit working principle, the oil pumping unit has the following disadvantages：

(1) 负载在下降过程中产生的重力势能完全浪费了。

The gravitational potential energy produced in the falling process of the load is totally wasted.

(2) 抽油机在上升和下降的过程中不能实现无级调速。

The oil pumping unit can not achieve stepless speed regulation in the upward and downward process.

(3) 当在井下发生卡泵现象时，不能对井上设备进行很好的保护。

When the pump can't work in the down-hole, we can't protect the equipments on the ground well too.

为了解决这些问题，科技人员研制了一种运用二次调节技术的液压抽油机。图 4-50 即为一种基于二次调节技术的液压抽油机工作原理图。

In order to solve these problems, the scientific and technical personnels developed a hydraulic oil pumping unit applying the secondary regulation technology. The Fig. 4-50 is a working principle diagram of hydraulic oil pumping unit based on the secondary regulation technology.

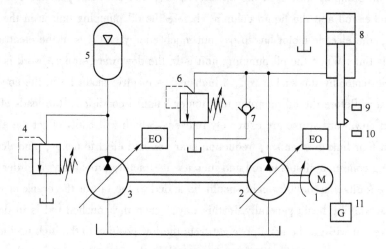

图 4-50 二次调节液压抽油机工作原理图

(Fig. 4-50 The working schematic diagram of secondary regulation hydraulic oil pumping unit)
1—电动机（Electrical motor）；2, 3—液压泵/马达（Hydraulic pump/motor）；4, 6—溢流阀（Relief valve）；5—液压蓄能器（Hydraulic energy accumulator）；7—单向阀（Check valve）；8—液压缸（Hydraulic cylinder）；9, 10—行程开关（Travel switch）；11—负载（Load）

二次调节液压抽油机的液压系统原理如图 4-50 所示。该系统中的两个二次元件（系统中的二次元件即为液压泵/马达）与一个电动机刚性连接，根据二次元件中斜盘倾角的不同，可工作于液压马达和液压泵两种状态。

The hydraulic system principle of secondary regulation hydraulic oil pumping unit is shown as Fig. 4-50. The two hydraulic secondary components in the system are rigidly connected with an electric motor, and the secondary component in the system is hydraulic pump or motor, according to the difference of the swashplate angle in the secondary

4.6 二次静压调节技术
(Secondary static pressure regulation technology)

component, it can work in two conditions as hydraulic motor and hydraulic pump.

二次调节抽油机工作过程如下：

The working process of secondary regulation oil pumping unit is as follows:

开机启动时，控制器发出指令调节二次元件 2 的斜盘倾角为零，使其输出流量为零。使二次元件 3 工作于液压泵工况，给蓄能器充压。

When the engine starts, in order to make the output flow become zero, the controller sends out the order to regulate the swash plate dip angle of secondary component 2 at zero, it make the secondary component 3 operate in the hydraulic pump condition to charge pressure to accumulator.

在液压缸上行过程中，电动机带动同轴的二次元件 2 和 3 工作，二次元件 3 在辅助能源蓄能器作用下工作于液压马达工况，控制器调节二次元件 2 工作于液压泵工况。此时二次元件 2 的驱动力来自于电动机和蓄能器驱动下的二次元件 3，其运动速度可通过预调电位计调节二次元件的排量来实现。

When the hydraulic cylinder moves upwards, the electric motor drives the coaxial secondary component 2 and 3, then the secondary component 3 operates in the hydraulic motor condition under the function of auxiliary power accumulator, and the controller regulates the secondary component 2 operating in the hydraulic pump condition. At this time, the driving force of secondary component 2 is from the secondary component 3 driven by the electric motor and power accumulator, its movement velocity can be regulated by the pre-setting potentiometer to regulate the displacement of the secondary component.

当液压缸碰到行程开关 9 时，二次元件 2 过零点，由液压泵工况转换成液压马达工况。液压缸在重力势能作用下向下运动，使液压缸中的油液向二次元件输出，此时液压缸相当于液压泵。

When the hydraulic cylinder hits the stroke switch 9, the secondary component 2 surpasses zero point, and it converts into hydraulic motor condition from hydraulic pump condition. The hydraulic cylinder moves downwards at the function of gravitational potential energy, causing the oil in the hydraulic cylinder to be outputted into the secondary component, meanwhile, the hydraulic cylinder corresponds to the hydraulic pump.

同时，二次元件 3 在控制器指令作用下，斜盘过零点，变成液压泵工况。其驱动力来自电动机和工作于液压马达工况的二次元件 2。二次元件 3 输出的液压油进入蓄能器储存起来，在上冲程中释放。

At the same time, the secondary component 3 receives the instruction of controller, the swash plate passes the zero point, and turns into hydraulic pump condition. The driving force is from the electric motor and the secondary component 2 working in the hydraulic motor condition. The hydraulic oil outputted by secondary component 3 enters the power accumulator and gets stored, which will be released in the upward stroke.

储存在蓄能器中的能量在下一个提升负载周期时释放,带动工作于液压马达工况的二次元件3,与电动机一起带动二次元件2工作,为液压缸提供所需能量,实现回收能量的再利用。

The energy stored in the power accumulator will be released in the next load lifting cycle, and drive the secondary component 3 which is working in the hydraulic motor condition. Then it drives the secondary component 2 together with the electric motor to supply the energy needed for the hydraulic cylinder, which finally achieves the reuse of the restoring energy.

第5章 电液比例控制基本回路
Chapter 5 Basic Electro-Hydraulic Proportional Control Circuits

在液压系统中，由若干个液压元件按照一定的规律组合构成的，能完成某一特定基本功能的液压回路，称为液压基本回路。如果在一个液压基本回路中含有电液比例元件，那么该回路就可以称为电液比例控制基本回路。如果在一个液压系统中含有电液比例控制基本回路，则该系统就可以称为电液比例控制系统。

In the hydraulic system, basic hydraulic circuit is a hydraulic circuit structure which is made up of a number of hydraulic components by a certain law and can complete a particular function. If a basic hydraulic circuit contains electro-hydraulic proportional components, it can be called the basic electro-hydraulic proportional control circuit. If a hydraulic system contains electro-hydraulic proportional control circuits, it can be called the electro-hydraulic proportional control system.

在液压基本回路中，为完成同一个工作目的，可以由不同的元件组合去完成；即使是选择的元件相同，由于其组合方式不同，其构成的回路的性能也可能大不相同。因此，了解和熟悉各种比例控制基本回路，是正确使用和设计电液比例控制系统的基本条件。

Basic hydraulic circuits can be composed of many different components in different ways to finish the same work. Even choosing the same components, due to different combinations of components, the circuits may differ quite far in performance from each other. Thus, knowing various kinds of basic proportional control circuits well is the basic requirement to use and design electro-hydraulic proportional system properly.

5.1 电液比例压力控制回路 (Electro-hydraulic proportional pressure control circuit)

压力控制回路是任何一个液压系统都必不可少的基本回路。它的基本功能有两个：在正常工况下，压力控制回路向系统提供具有合适压力的油液，满足液压执行器对油液压力和流量的需求，从而输出合适的力或力矩；在异常工况下，为系统提供压力保护，即系统卸荷或使油液在安全压力下溢流。

第 5 章 电液比例控制基本回路
Chapter 5　Basic Electro-Hydraulic Proportional Control Circuits

Pressure control circuit is the essential circuit for any hydraulic system. It has two basic functions. Under normal working conditions, it provides the appropriate pressure oil to the system in order to meet the pressure and flow demands of the hydraulic actuator, thereby to output appropriate force or torque; Under abnormal working conditions, it provides pressure protection for the system, namely, system unloading and overflowing under the safe pressure.

电液比例压力控制回路可以实现系统的无级调压。换句话说，几乎可以使系统压力跟踪任意形状的压力-时间（行程）曲线，使升压、降压过程平稳且迅速。电液比例技术在压力控制回路中的应用，既提高了系统的性能，又使系统大大简化，但是其电气控制技术较复杂，成本也较高。

Electro-hydraulic proportional pressure control circuit can realize stepless pressure regulation of the system. In other words, it can almost achieve any shape of the pressure-time (stroke) curves and make the process of pressure arising and pressure reducing smooth and rapid. The electro-hydraulic proportional technology applicated in pressure control loop, not only improves the system performance but also simplifies the system greatly, while its electrical control technology is complicated and expensive.

5.1.1　比例溢流调压回路 (Proportional relief pressure-regulating circuit)

在比例调压回路中，最常见的是采用比例溢流阀来进行调压。

In proportional pressure-regulating circuit, the most common way is to realize the pressure adjustment by proportional relief valve.

比例溢流调压回路是将比例溢流阀并联到泵的出口，通过改变比例溢流阀的控制电信号，可以在比例溢流阀的调节范围内任意设定系统压力。

In the proportional relief pressure regulating circuit, the proportional relief valve is parallelly connected to the pump outlet, and through changing the input signal of the proportional relief valve, the system pressure can be randomly set in its regulating range.

溢流调压的原理是在控制容腔内的压力升高时排走多余的流量，避免流量在控制容腔内的继续积累而使压力过高。

The principle of pressure relief regulating is to drain away excess flow when the pressure of the control chamber rises, avoiding the flow accumulated in the control chamber to raise the pressure too high.

5.1.1.1　采用直动式比例溢流阀的调压回路 (Proportional pressure regulating circuit using direct proportional relief valve)

图 5-1 为采用直动式比例溢流阀的调压回路。为了保证安全，比例溢流阀调

5.1 电液比例压力控制回路
(Electro-hydraulic proportional pressure control circuit)

压回路通常都要加入限压的安全阀。

Fig. 5-1 shows a pressure regulating circuit using direct proportional relief valves. In order to ensure the safety, a safety valve is usually put in the proportional relief pressure regulating circuit.

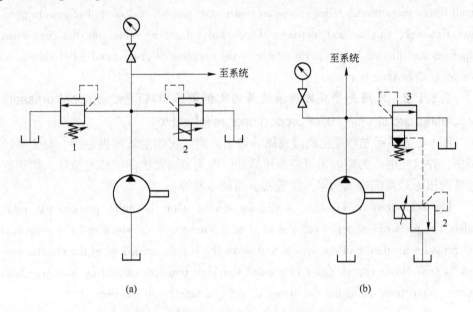

图 5-1 采用直动式比例溢流阀的调压回路
(Fig. 5-1 Pressure regulating circuit using direct proportional relief valve)
(a) 直接使用直动式比例溢流阀 (Directly using the direct proportional relief valve);
(b) 直动式比例溢流阀做远程调压 (Using direct relief valve for remote pressure regulating)
1—直动式溢流阀 (Direct relief valve); 2—直动式比例溢流阀 (Direct proportional relief valve);
3—先导式溢流阀 (Pilot-operated relief valve)

图 5-1a 适用于流量小的情况,比例溢流阀并联在液压泵的出口构成调压回路,传统直动式溢流阀做安全阀,在比例溢流阀失调时保护系统。

Fig. 5-1a is applicable to the case of small flow. The proportional relief valve is parallelly connected with the outlet of hydraulic pump, composing pressure regulating circuit. The traditional direct relief valve works as safety valve to protect the system when the proportional relief valve doesn't work.

图 5-1b 适用于大流量的情况,直动式比例溢流阀与先导式溢流阀的遥控口连接,进行远程调压。先导式溢流阀做主阀,负责系统调压及保护系统安全。在该方案中,虽然只要一个小型的直动式比例溢流阀就可以对系统压力进行比例控制,但是由于遥控需要连接管道,控制容积增加,而且受普通先导式溢流阀性能限制,控制性能较差。

Fig. 5-1b is applicable to the case of large flow. The direct proportional relief valve is connected with the remote control port of the traditional pilot-operated relief valve, which used for remote pressure regulating. The pilot-operated relief valve is used as the main valve to regulate the system pressure and keep the system safe. In this scheme, a small direct proportional relief valve can realize the proportional control of system pressure. However, this method increases the control volume by increasing the connecting pipelines and limited by the performance of the common pilot-operated relief valve, its control performance is poor.

5.1.1.2 采用先导式比例溢流阀的比例调压回路(Proportional pressure regulating circuit using pilot proportional relief valve)

图 5-2 为采用先导式比例溢流阀的调压回路。比例溢流阀做主阀，负责系统调压，并且在电流为零时可以使系统卸荷。但为预防过大的故障电流输入而引起过高的压力对系统造成伤害，设置安全阀是必要的。

Fig. 5-2 shows a pressure regulating circuit with the pilot proportional relief valve. The proportional relief valve is used as the main valve, which to be in charge of the pressure regulating of the system and make the system unload when the electric current is zero. However, in order to prevent the high pressure caused by overlarge fault current input from damaging the system, setting a safety valve is necessary.

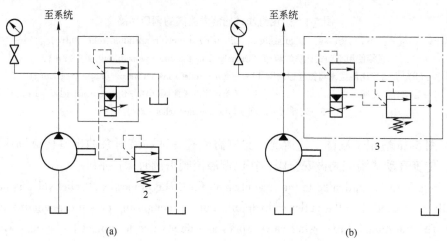

图 5-2 采用先导式比例溢流阀的调压回路

(Fig. 5-2 Pressure regulating circuit using pilot proportional relief valve)

(a) 使用不带限压阀的先导式比例溢流阀 (Using the pilot proportional relief valve without pressure limiting valve);
(b) 使用带限压阀的先导式比例溢流阀 (Using the pilot proportional relief valve with pressure limiting valve)
1—先导式比例溢流阀 (Pilot proportional relief valve); 2—直动式溢流阀 (Direct relief valve);
3—带限压阀的比例溢流阀 (Proportional relief valve with pressure limiting valve)

5.1 电液比例压力控制回路
(Electro-hydraulic proportional pressure control circuit)

直接使用先导式比例溢流阀与图 5-1b 所示的方案相比，由于先导式比例溢流阀的先导阀经过了专门的设计，主阀结构虽然与普通的先导式溢流阀基本相同，但其性能参数在设计中也做了相应的优化，因此控制性能与可靠性方面都更好一些。所以，该方案适用于流量较大，同时控制性能要求较高的场合。

Comparing the program by directly using pilot proportional relief valves with that shown in Fig. 5-1b, the pilot valve of the proportional relief valves is specially designed and manufactured. Although the main valve's structure is basically the same as that of common pilot-operated relief valve, its performance parameters have been optimized and its control performance and reliability are better. Therefore, this program is suitable for the case with large flow and high control performance demand.

5.1.2 比例减压回路 (Proportional pressure reducing circuit)

比例减压回路用于在单泵供油的液压系统中，某个支路需要的工作压力低于主系统压力值，或者某个支路需要稳定的可调低压力值的情况。

The proportional pressure reducing circuit is used in the single-pump hydraulic oil supply system, and it should be used when a branch's working pressure is lower than that of system or the branch requires an adjustable and stable low pressure.

在比例减压回路中，通过改变比例压力阀的输入信号，可以在某一条支路上获得任意降低了的系统压力。

In the proportional pressure reducing circuit, an arbitrary reduced system pressure can be obtained at a branch by changing the input signal of the proportional pressure valve.

5.1.2.1 采用直动式比例溢流阀的比例减压回路(Proportional pressure reducing circuit using direct proportional relief valve)

在图 5-3 中，直动式电液比例压力阀与传统先导式减压阀的遥控口相连接，遥控减压阀的设定压力。同样，由于控制容腔增大，控制性能较差。

In Fig. 5-3, the direct-acting electro-hydraulic proportional pressure valve is connected to the remote control port of traditional pilot-operated reducing valve to control the setting pressure of the pressure reducing valve. And because of the increase in control cavity, its control performance is poor.

5.1.2.2 采用两通比例减压阀的比例减压回路(Proportional pressure reducing circuit using two-way proportional pressure reducing valve)

图 5-4 为采用比例减压阀的基本回路。通过改变比例减压阀的控制电流，可以在缸伸出时获得低于系统压力的多种压力值。但是，由于两通减压阀的结构，在液压缸回程时，只能通过单向阀快速回油，不能控制回油的压力。

Fig. 5-4 shows a basic circuit with proportional pressure reducing valve. Through changing the control current of the proportional pressure reducing valve, multiple pres-

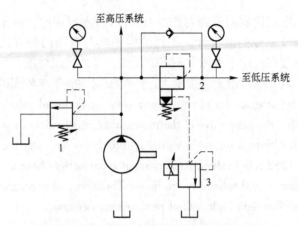

图 5-3 采用直动式比例溢流阀的比例减压回路

(Fig. 5-3 Proportional pressure reducing circuit using direct proportional relief valve)

1—直动式溢流阀（Direct-acting relief valve）；2—先导式减压阀（Pilot-operated reducing valve）；
3—直动式比例溢流阀（Direct proportional relief valve）

sure values below the system pressure can be obtained when the cylinder is extending out. However, because of the structure of two-way pressure reducing valve, when the hydraulic cylinder is in return stroke, it can only get the oil returned rapidly through the check valve but can't control the pressure of return oil.

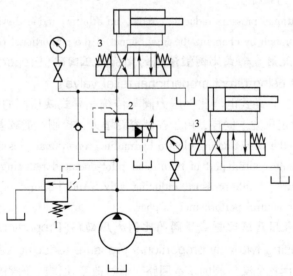

图 5-4 采用两通比例减压阀的比例减压回路

(Fig. 5-4 Proportional pressure reducing circuit using two-way proportional pressure reducing valve)

1—溢流阀（Pressure relief valve）；2—先导式比例减压阀（Pilot proportional pressure reducing valve）；
3—电磁换向阀（Solenoid directional control valve）

5.1 电液比例压力控制回路
(Electro-hydraulic proportional pressure control circuit)

5.1.2.3 使用三通比例减压阀的比例减压回路(Proportional pressure reducing circuit using three-way proportional pressure reducing valve)

图 5-5 为采用三通比例减压阀的基本回路,三通减压阀有一个通道是直接与油箱连接的。图中液压缸负载为重力负载,通过改变三通比例减压阀的控制电流,在活塞双向运动时保持恒压控制;也可以靠改变液压缸无杆腔的压力来控制液压缸的上升和下降。

Fig. 5-5 shows a basic circuit with three way proportional pressure reducing valve. There is a channel directly connected with the tank in three-way proportional pressure reducing valve. The load of hydraulic cylinder in the figure is gravity load. It maintains constant pressure control of the two-direction piston movement by changing the control current of three-way proportional pressure reducing valve, and it can also control the hydraulic cylinder to rise and fall by changing the pressure of non-rod chamber of the hydraulic cylinder.

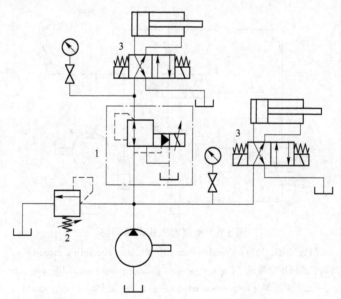

图 5-5 使用三通比例减压阀的比例减压回路
(Fig. 5-5 Proportional pressure reducing circuit using three-way proportional pressure reducing valve)
1—三通比例减压阀 (Three way proportional pressure reducing valve); 2—溢流阀 (Relief valve);
3—电磁换向阀 (Solenoid directional control valve)

5.1.3 比例容积式调压回路 (Proportional volumetric pressure regulating circuit)

比例容积式调压回路是指采用比例压力调节变量泵对回路进行压力控制的回路。

Proportional volumetric pressure regulating circuit refers to the circuit which is used for pressure control using proportional pressure control variable displacement pump.

与溢流调压的原理不同,容积调压是在控制腔的压力升高时,利用压力升高的信息作为流量多余的反馈,并以此来控制变量泵使输出流量减少,从而控制系统压力。

Different from the pressure relief regulating principle, the process of volumetric pressure regulating is that when the pressure is rising, it will use the information of the increasing pressure as the feedback of the excess flow to control the variable pump and decrease the output flow, so as to control the system pressure.

如图 5-6 所示,在液压回路中使用比例调节的恒压变量泵,可以构成比例压力调节回路。在负载压力未达到设定压力时,泵以最大流量供油,这时供油压力追随负载变化。在工作压力达到设定压力时,供油压力与负载无关,流量适应负载要求。系统的设定压力即变量泵的截流压力,它由进入比例溢流阀的控制电流的大小来设定。

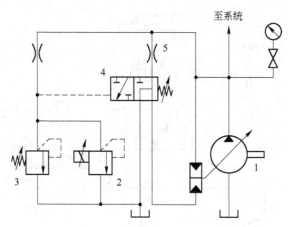

图 5-6 比例容积式调压回路

(Fig. 5-6 Proportional volumetric pressure regulating circuit)

1—由比例溢流阀调节的恒压变量泵(Constant pressure variable pump regulated by proportional relief valve);
2—比例溢流阀(Proportional relief valve); 3—安全阀(Safety valve);
4—液动换向阀(Hydraulically operated direction control valve); 5—阻尼孔(Damping hole)

As shown in Fig. 5-6, using the proportional constant pressure variable pump in the hydraulic circuit can constitute a proportional pressure control loop. When the load pressure does not meet the setting pressure, oil is supplied at the maximum flowrate by the pump and the oil pressure changes with the load. When the working pressure reaches the setting pressure, the oil pressure has nothing to do with the load and the flow meets the load requirement. So the system's setting pressure is the variable pump's closure pressure which is set by the control current in the proportional relief valve.

5.1 电液比例压力控制回路
(Electro-hydraulic proportional pressure control circuit)

5.1.4 比例调压回路的应用 (Applications of proportional pressure regulating circuit)

电液比例调压回路的主要应用有：

There are several main applications of the electro-hydraulic proportional pressure regulating circuit：

(1) 多级压力控制 (Multi-stage pressure control)；

(2) 系统卸荷 (System unloading)；

(3) 力控制系统 (Force control system)。

比例调压的一个具体应用为钻深孔时的心轴夹紧压力控制回路。

A specific application of proportional pressure regulating is mandrel clamping pressure control circuit for deep-hole drilling.

钻深孔时，需对心轴的夹紧力进行自动控制，以避免损坏昂贵的刀具，其控制回路如图 5-7 所示。控制信号由 CNC（计算机数控）系统产生，比例阀上装有内装压力传感器和闭环控制用的电子器件。

When the deep-hole is being drilled, the mandrel clamping force is required to be controlled automatically to avoid damaging the expensive tools, whose control circuit is shown in Fig. 5-7. The control signal is generated by the CNC (computer numerical control) system, and the proportional valve is installed with built-in pressure sensor and electronic devices for closed-loop control.

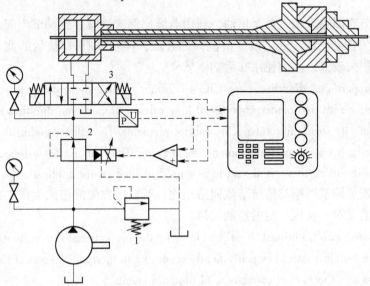

图 5-7 心轴夹紧压力控制回路

(Fig. 5-7 Mandrel clamping pressure control circuit)

1—溢流阀 (Relief valve)；2—比例减压阀 (Proportional pressure reducing valve)；
3—电磁换向阀 (Solenoid directional control valve)

5.2 电液比例速度控制回路（Electro-hydraulic proportional speed control circuit）

要对液压执行器实行速度控制即调速，可以采用两种方法，这就是改变进入执行器的流量，或改变执行器的排量。对液压缸来说，还可以改变它的有效工作面积。

There are two ways to control the hydraulic actuator's speed: changing the actuator's displacement or changing the actuator's flow. Changing the cylinder's effective working area is also a way to control the cylinder's speed.

电液比例速度调节，根据上述原理有三种常用的方式：比例节流调速、比例容积调速以及比例容积节流调速。使用电液比例速度调节回路，可以实现执行器跟踪任意形状的速度－时间（行程）曲线。

According to the above principles, there are three common ways to achieve the electro-hydraulic proportional speed regulation: proportional throttling speed control, proportional volumetric speed control and proportional volumetric throttling speed control. Using electro-hydraulic proportional speed regulating circuit, the actuator can achieve tracking speed-time (stroke) curve of arbitrary shape.

5.2.1 比例节流调速回路（Proportional throttle regulating circuit）

比例节流回路采用定量泵供油，利用电液比例流量阀（比例节流阀、比例调速阀）作为控制元件。通过改变节流口的开度，实现进/出口流量的调节。但由于其节流损失较大，不宜使用在大功率场合。

The proportional throttling circuit uses constant displacement pump to supply oil, and uses the electro-hydraulic proportional flow valve (proportional throttle valve, proportional velocity regulating valve) as control elements. Through changing the opening of the orifice, it can regulate the import/export flow. But, at the same time, because of the large throttling loss, it should not be used in high power applications.

比例流量调节回路与传统节流回路相比，可以很方便地按照生产工艺及设备负载特性的要求，实现一定的控制规律。

Compared with traditional throttling circuits, the proportional flow regulating circuit can achieve certain control law conveniently according to the requirements of the production process and the load characteristic of the equipments.

5.2.1.1 使用比例流量阀的比例节流回路（Proportional throttle regulating circuit using proportional flow valve）

如图 5-8 所示，根据在回路中的位置比例流量阀又分为三种形式：进口节

5.2 电液比例速度控制回路
(Electro-hydraulic proportional speed control circuit)

流、出口节流、旁路节流。由于这三种节流调速回路在速度负载特性及功率特性上各有自己的特点,在使用时按照承受负值负载的能力、运动平稳性及启动性能的要求选用。

As Fig. 5-8 shows, according to the position of proportional flow valve in the loop, the proportional throttle regulating loop can be divided into three types: the meter-in circuit, the meter-out circuit and the bypass throttling circuit. Because the three throttling speed control circuits all have their own characteristics in the speed-load characteristic and power characteristic. When used in practice, they are selected according to the requirements such as the capacity to bear the negative load, motion smoothness and the startup performance.

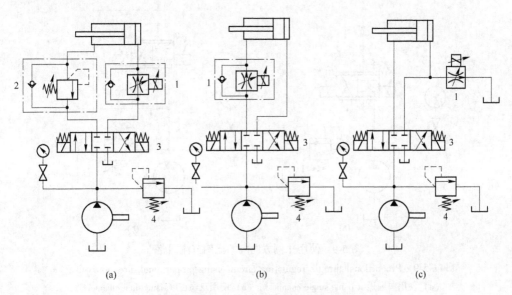

图 5-8 使用比例流量阀的比例节流回路

(Fig. 5-8 Proportional throttle regulating circuit using proportional flow valve)
(a) 进口节流回路 (Meter-in circuit); (b) 出口节流回路 (Meter-out circuit);
(c) 旁路节流回路 (Bypass throttling circuit)
1—比例流量阀 (Proportional flow valve); 2—单向溢流阀 (Check relief valve);
3—电磁换向阀 (Solenoid directional control valve); 4—溢流阀 (Relief valve)

在图 5-8a 中,可以使用一个单向溢流阀,在节流时形成一定的背压,从而在一定程度上提高响应速度,预防爬行等不良后果。

In Fig. 5-8a, it can form a certain back pressure when throttling by using a check relief valve, thereby it can improve the response speed to a certain extent and prevent crawling and other adverse consequences.

5.2.1.2 使用比例方向阀的比例节流回路(Proportional throttle circuit using proportional direction valve)

如图 5-9 所示,比例方向阀可以用来构成比例节流回路。两位四通比例方向阀常用作比例节流阀使用,图 5-9a 只利用其中一个通道,图 5-9b 同时使用两个通道,可以使流通能力加倍。

As Fig. 5-9 shows, proportional direction valve can be used to form proportional throttle regulating circuit. Two-position four-way proportional direction valve is always used as proportional throttle valve. In Fig. 5-9a, only one channel is used. While in Fig. 5-9b, two channels are used, which can double the flow capacity.

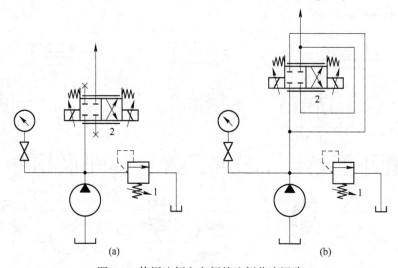

图 5-9 使用比例方向阀的比例节流回路

(Fig. 5-9 Proportional throttle regulating circuit using proportional direction valve)
(a) 使用单通道 (Using single channel); (b) 使用双通道 (Using dual channels)
1—溢流阀 (Relief valve); 2—比例方向阀 (Proportional directional valve)

5.2.2 比例容积调速回路 (Proportional volumetric speed control circuit)

比例容积调速回路采用比例排量调节变量泵与定量执行器,或定量泵与比例排量调节马达等的组合来实现流量调节,通过改变泵或马达的排量实现调速。

The proportional volumetric speed control circuit uses the combination of proportional variable displacement pump and constant displacement actuator, or constant displacement pump and proportional displacement motor to realize flow regulation, and to achieve speed regulation by changing the pump or motor displacement.

容积调速回路的优点是效率高,适用于大功率系统;但在相应速度及控制精度上都不如节流调速。但其不存在节流损失,故应该设置安全阀。

5.2 电液比例速度控制回路
(Electro-hydraulic proportional speed control circuit)

The volumetric speed control circuit has the advantages of high efficiency and applicability for large power systems, but the corresponding speed and control accuracy are inferior to those of throttle governing. There is no throttle pressure loss, so we should set safety valve.

采用比例排量变量泵可以用于容积调速回路, 其应用基本回路如图 5-10 所示。它是通过改变比例排量泵中减压阀的控制电流改变泵的排量来改变进入液压执行器的流量, 从而达到调速的目的。

The proportional displacement variable pump can be used to volumetric speed control loop, and the application of its basic circuit is shown in Fig. 5-10. It achieves the purpose of speed control by changing the control current of pressure reducing valve in proportional displacement pump to change the pump displacement which can change the flow into the hydraulic actuator.

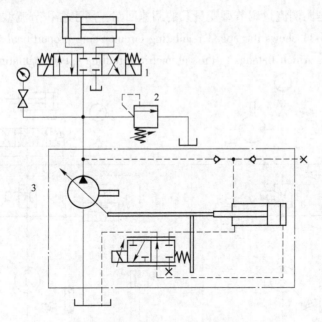

图 5-10 使用比例变量泵的比例容积式流量调节回路

(Fig. 5-10 Proportional volumetric flow regulating circuit using proportional variable pump)

1—电磁换向阀 (Solenoid directional control valve); 2—溢流阀 (Relief valve);
3—比例排量调节变量泵 (Proportional displacement variable pump)

5.2.3 比例容积节流式流量调节回路 (Proportional volumetric throttling flow regulating circuit)

为了能够同时发挥节流调速和容积调速回路的优点, 通常将容积调速与节流调速结合起来，构成比例容积节流式流量调节回路。

In order to simultaneously obtain the advantages of throttling governing and volumetric speed control, the two circuits are often combined together to constitute a proportional volumetric throttling flow regulating circuit.

比例流量调节变量泵内部有负载压力补偿，输出流量与负载无关，具有很高的稳流精度，可以方便地用电信号控制系统各工况所需流量，并同时做到泵的出口压力与负载压力相适应。

The proportional flow control variable pump has internal load pressure compensation. Its output flow is independent of the load and has high precision of steady flow, so it can be conveniently controlled by the electric signal to follow the required flow of the system in all working conditions, and at the same time the pump pressure suits the load pressure.

图 5-11 是比例流量调节型变量泵的调速回路，属于容积节流型的调速回路。

The Fig. 5-11 shows the speed regulating circuit using proportional flow regulating variable pump, which belongs to the volumetric throttling speed regulating circuit.

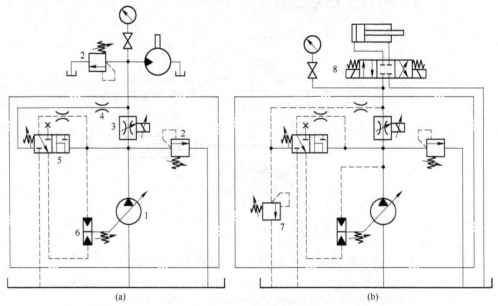

图 5-11 比例容积节流式流量调节回路

(Fig. 5-11 The proportional volumetric throttling flow regulating circuit)

(a) 不带压力控制（Without pressure control）；(b) 带有压力调节（With pressure regulation）

1—变量泵（Variable pump）；2—溢流阀（Relief valve）；3—比例调速阀（Proportional speed regulating valve）；
4—阻尼孔（Damping hole）；5—减压阀（Pressure reducing valve）；6—变量机构（Variable mechanism）；
7—手动压力调节阀（Hand operated pressure regulating valve）；8—电磁换向阀（Solenoid directional control valve）

5.2 电液比例速度控制回路
(Electro-hydraulic proportional speed control circuit)

图 5-11a 为不带压力控制的比例流量调节,由于该泵不会回到零流量处,系统必须设置足够大的溢流阀,使在不需要流量时能以合理的压力排走所有的流量。

Fig. 5-11a shows the proportional flow regulation without pressure control. A large enough relief valve must be set to drain away all the flow at a reasonable pressure when there needs no flow, because the pump will not return to the zero position.

图 5-11b 是一种带有压力调节的比例流量调节。通过手动压力调节阀 7 可以调定泵的截流压力。当压力达到调定值时,泵便自动减少排出流量,维持输出压力近似不变,直至截流。但有时为了避免变量活塞的频繁移动,上述的溢流阀仍是必要的。

Fig. 5-11b shows the proportional flow regulation with pressure control. We can set the pump's closure pressure by adjusting the valve 7 manually. When the pressure reaches the setting value, the pump will reduce the output flow automatically and maintain the output pressure constantly until closure. But sometimes the above-mentioned relief valve is still necessary to avoid the variable piston's frequent movement.

5.2.4 电液比例速度控制回路的应用 (Applications of electro-hydraulic proportional speed control circuit)

5.2.4.1 使用比例流量阀控制多执行器(Using proportional flow valve to control multi-actuator)

图 5-12 为使用一个比例节流阀对多个执行器进行多速控制的回路。在该回路中,一般每次只对其中的一个执行器进行控制。同时,该回路也可以改成进口、出口或者旁路节流,连接方法类似。

Fig. 5-12 shows a circuit that uses one proportional throttle valve to control multiple actuators' speed. In the circuit, it controls only an actuator at one time. Meanwhile, the circuit can also be changed to meter-in, meter-out and bypass throttling, and the connection methods are similar.

在图 5-12a 中,只通过溢流阀来保证节流阀的入口压力稳定,从而减少压力波动对流量的影响;在图 5-12b 中,利用溢流阀进行压力补偿,保证流量阀进出口压力差恒定,从而能对流量进行精确地控制。如果采用比例调速阀,由于自带定差减压阀进行压力补偿,可以省略掉图中的溢流阀。

In Fig. 5-12a, it only uses a relief valve to ensure the inlet pressure of throttle valve stable, so as to reduce the impact of pressure fluctuations on the flow. In Fig. 5-12b, it uses relief valve for pressure compensation to ensure the pressure difference of flow valve's inlet and outlet constant, thus it can accurately control the flow. It can

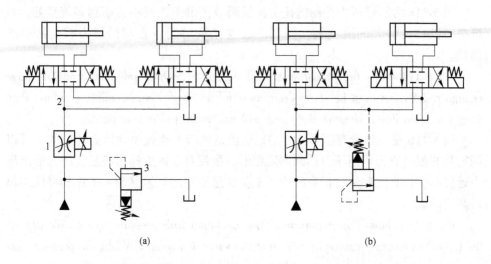

图 5-12 使用比例流量阀控制多执行器的调速回路

(Fig. 5-12 Speed regulating circuit using proportional flow valve to control multi-actuator)
(a) 流量阀进口压力控制(Inlet pressure control by flow valve);(b) 带压力补偿(With pressure compensation)
1—比例流量阀(Proportional flow valve);2—电磁换向阀(Solenoid directional control valve);
3—溢流阀(Relief valve)

leave out the relief valve in the figure if it uses the proportional speed regulating valve because it has self-contained constant difference pressure-reducing valve to realize pressure compensation.

5.2.4.2 使用比例调速阀的双向同步回路(Two-way synchronizing circuit using proportional speed regulating valve)

图 5-13 为使用比例调速阀的比例同步回路。这种回路的特点是双向调速、双向同步,上升行程为进口节流,下降行程为回油节流。回油节流有助于防止因自重下滑时的超速运行。

Fig. 5-13 shows a synchronizing circuit using proportional speed regulating valve. This circuit is characterized by two-way speed control and bidirectional synchronization. The up stroke uses meter-in control and the down stroke uses meter-out control, which is conducive to preventing the overspeed operation due to dead weight when it is slipping.

回路中液控单向阀用于平衡负载的自重。另外四个单向阀为一组,构成桥式整流回路,使正反向行程通过调速阀的流量方向一致。

The pilot operated check valve in the loop is used to balance the dead weight of the load. The other four check valves form a bridge rectifier circuit as a group, which makes the flow direction in both strokes through the speed regulating valve coincident.

5.2 电液比例速度控制回路
(Electro-hydraulic proportional speed control circuit)

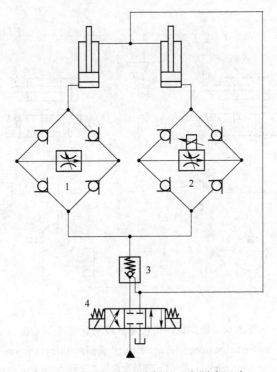

图 5-13 使用比例调速阀的双向同步回路

(Fig. 5-13 Two-way synchronizing circuit using proportional speed regulating valve)

1—调速阀（Speed regulating valve）；2—比例调速阀（Proportional speed regulating valve）；
3—液控单向阀（Hydraulic operated check valve）；4—方向阀（Directional control valve）

5.2.4.3 使用比例排量变量泵的双向同步回路（Two-way synchronizing circuit using proportional displacement variable pump）

图 5-14 为使用比例排量泵的容积调速式比例同步回路。该回路也是一种具有双向调速、双向同步功能的回路。速度控制采用电气遥控设定，位置互相跟随。

Fig. 5-14 shows a proportional synchronizing circuit using proportional displacement pump. It is a circuit that has functions of two-way speed control and bidirectional synchronization. The speed control uses electrical remote settings, which makes their positions follow each other.

由于是容积控制，没有节流损失，适用于大功率系统和高速的同步系统。同时，由于两个执行器的供油系统完全独立，更适合于两个液压缸相距很远但又要求同步精度很高的系统中。

Because of volumetric control, it has no throttling loss so that it is suitable for large power system and high speed synchronous system. At the same time, because the two

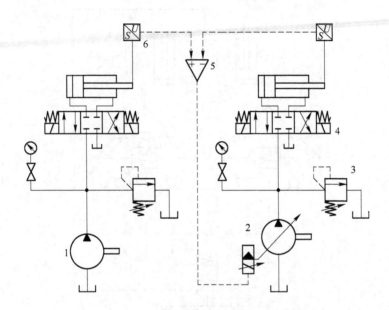

图 5-14 使用比例排量变量泵的双向同步回路

(Fig. 5-14 Two-way synchronizing circuit using proportional displacement variable pump)

1—定量泵（Fixed displacement pump）；2—比例排量变量泵（Proportional displacement variable pump）；
3—安全阀（Safety valve）；4—电磁换向阀（Solenoid directional control valve）；
5—放大器（Amplifier）；6—位移传感器（Displacement sensor）

actuators' oil supply systems are completely independent, it is more suitable for the system whose two hydraulic cylinders are far apart but it also requires good synchronization precision.

5.2.4.4 纺织机拉力控制回路(Tension control circuit of textile machine)

如图 5-15 所示，在纺织机上，拉力控制是由调整制动力的比例控制系统实现的。

As Fig. 5-15 shows, in the textile machine, the control of tension is accomplished by a proportional control system adjusting the braking force.

比例阀上装有集成电子器件，在闭环控制中，由传感器送来的反馈信号可控制制动力。与传统的压力系统不同的是，这个系统具有从零压开始精确调整的特点。

The proportional valve is equipped with integrated electronic devices, and the feedback signal sent by the sensor can control the braking force in the closed-loop control. Different from the traditional pressure system, the system has the characteristic that it can adjust pressure accurately from zero-pressure.

5.3 电液比例方向及速度控制回路
（Electro-hydraulic proportional direction and speed control circuit）

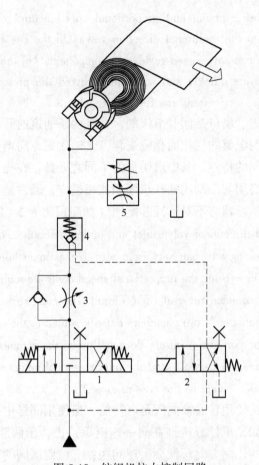

图 5-15 纺织机拉力控制回路
(Fig. 5-15 The tension control circuit of textile machine)
1，2—电磁换向阀（Solenoid directional control valve）；
3—单向节流阀（Check throttle valve）；4—液控单向阀（Hydraulic operated check valve）；
5—比例流量阀（Proportional flow valve）

5.3 电液比例方向及速度控制回路（Electro-hydraulic proportional direction and speed control circuit）

与传统的方向阀不同，比例方向阀兼有方向控制与流量控制双重功能，可以实现液压系统的换向以及速度的比例控制。使用比例方向阀一方面可以节省调速元件，另一方面能迅速准确地实现工作循环，避免压力尖峰及满足切换性能的要求，延长元件的使用寿命。

Proportional direction valve has dual function of direction control and flow control, and it can realize the reversing and proportional speed control of hydraulic system, which is different from the traditional direction valve. On the one hand, using proportional direction valve can save speed regulating components. On the other hand, it can realize the working cycle quickly and accurately, avoid the pressure peak, meet the performance changing, and extend the life of components.

比例方向阀的进、出口是同时节流的，并且两条通道的开口面积可以在从零到最大之间随着控制电流的变化而相应变化，这给比例方向调速回路带来很多与普通换向阀回路不同的特性。其中最明显的不同之处是：在比例控制中，执行器两工作腔的面积比必须与控制阀的开口面积比相适应，通常是对称执行器与对称阀芯配用，不对称执行器与不对称阀芯配用（如面积比为 2：1）。

The proportional directional valve inlet and outlet is simultaneously throttling, and the two channels' opening area can vary from zero to maximum following the change of the control current. These bring the proportional speed control circuit a lot of characteristics distinct from the common reversal valve circuit. The most striking difference is that the area ratio of the actuator's two chambers must be suited to the open area ratio of the control valve in the proportional control. We usually use the symmetric actuator in conjunction with the symmetrical spool, and the unsymmetrical actuator in conjunction with the unsymmetrical spool (such as the area ratio is 2：1).

在有些情况下，使用普通换向阀是可行的，但使用同样中位机能的比例方向阀就会出问题。例如，单杆液压缸在同一速度下，进、出两腔的流量互不相同，如果使用对称阀芯的比例方向阀就会出现问题；三位四通阀的中位机能，即无信号状态下自然位置上各油口的连通情况，它们对回路的性能也有十分重要的影响，相同中位机能的比例方向阀由于结构的不同，在使用到同一回路，在比例阀中位时有可能出现问题。

In some cases, it's feasible to use the common directional control valve, but it will get into trouble when using the proportional directional valve with the same median function. For example, the flows in the two chambers of single-rod cylinders are different under the same speed, so there will be problems if we use proportional direction valve with symmetrical spool. The median function of three-position four-way valve, namely the oil ports' connection situation at the natural state without any signal, is very important to the performance of the circuit. For proportional direction valves whose structure are different with the same median function, when they are used in the same circuit, there may be problems when the proportional valves are at the middle position.

5.3 电液比例方向及速度控制回路
(Electro-hydraulic proportional direction and speed control circuit)

5.3.1 对称执行器比例方向控制回路 (Proportional directional control circuit of the symmetric actuator)

对称执行器包括液压马达，面积相等的双出杆液压缸，以及面积比接近1：1的单杆液压缸。这类执行器可由对称开口的中闭型（O型）和加压型（P型）以及泄放型（Y型）的比例方向阀进行控制。

The symmetric actuator includes the hydraulic motor, the two-rod cylinder of the same area and the single rod cylinder whose area ratio is close to 1 : 1. All these actuators can be controlled by the O-proportional directional valve, the P-proportional directional valve and the Y-proportional directional valve.

5.3.1.1 封闭（O）型比例方向阀换向回路 (O-proportional directional valve reversing circuit)

图 5-16 是使用封闭型比例方向阀的基本换向回路。这种回路的特点是阀处于中位时，执行器的进出油口全封闭，活塞或马达被锁定。

Fig. 5-16 shows a basic directional control circuit using O-proportional directional valve. The characteristics of this circuit are that when the valve are in the middle position, the actuator's inlet and outlet are all closed while the piston or the motor is locked.

当惯性负载较大、阀芯转换较快时，会产生一些不良的现象，例如引起某腔的压力过分升高，或另一腔的压力过分降低，出现抽空或空穴。两种现象都会使运动不稳定。因此，使用这种回路时应注意运动减速时的高压保护，并采取防止空穴产生的措施。

When the inertial load is heavy and the spool converts quickly, there will produce some undesirable phenomena, such as causing an excessive increase in the pressure of a certain chamber, or an excessive decrease in the pressure of the other chamber to creat an evacuation or a cavitation. The two phenomena will both cause an unstable motion. Therefore, when this circuit is used, we should pay attention to the high pressure protection when slowing down and take actions to prevent the cavitation.

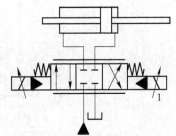

图 5-16 使用封闭型比例方向阀的基本换向回路

(Fig. 5-16 Basic directional control circuit using O-proportional directional valve)

1—电液比例换向阀（Electro-hydraulic proportional directional valve）

当负载突然减小或负载因节流而获得一个减速度，都会使出油侧节流口压差 Δp_2 大大增加，这样可使 $\Delta p_1 < \Delta p_2$，如图 5-17 所示，因而液压的左腔将产生抽空

或空穴，使系统的运动速度不受控制。可见在大惯性负载时采取适当措施是必要的。

When the load decreases suddenly or the load obtains a deceleration because of throttling, the differential pressure of the throttle in the outlet side Δp_2 will increase greatly. It makes $\Delta p_1 < \Delta p_2$, as is shown in Fig. 5-17, thus the left chamber will get an evacuation or a cavitation and the system's speed will get out of control. So it is necessary to take some appropriate measures in the case of heavy inertia load.

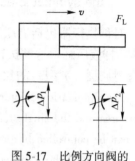

图 5-17 比例方向阀的基本换向回路原理图

(Fig. 5-17 The principle diagram of basic reversing circuit using proportional direction valve)

图 5-18 是一种较好的使用封闭型比例方向阀的方向速度控制回路。执行器可以是液压马达或对称液压缸。

Fig. 5-18 is a better direction and speed control circuit using O-proportional directional valve. The actuators can be hydraulic motors or symmetric hydraulic cylinders.

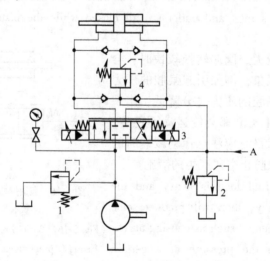

图 5-18 使用封闭型比例方向阀的方向速度控制回路

(Fig. 5-18 Direction and speed control circuit using O-proportional directional valve)

1, 4—溢流阀 (Pressure relief valve); 2—背压阀 (Back pressure valve);
3—电液比例换向阀 (Electro-hydraulic proportional directional valve)

溢流阀 4 用于吸收压力冲击，其调整压力应大于最高工作压力。两个补油单向阀用于出现真空时补油，它的开启压力应在 0.05MPa 左右。

The pressure relief valve 4 is used to absorb the pressure impact and its adjusting

5.3 电液比例方向及速度控制回路
(Electro-hydraulic proportional direction and speed control circuit)

pressure should be higher than the maximum operating pressure. The two recharging oil check valves are used for recharging oil when there is a vacuum, and its opening pressure should be around 0.05MPa.

如果这个油马达回路只是整个液压系统的一部分,那么其他部分的回油可与补油单向阀的进油口相连,并加上调整压力为 0.3MPa 左右的背压阀。这样可使防真空保护更为理想。

If the oil motor circuit is only a part of the hydraulic system, the return oil from other parts can be connected with the inlet of the oil recharging check valve. And it will make the anti-vacuum protection more desirable with a back pressure valve whose adjusting pressure is set at 0.3MPa or so.

5.3.1.2 加压(P)型比例方向阀换向回路(P-proportional directional valve reversing circuit)

图 5-19a 为 P 型比例方向阀的基本回路。当阀在中位时,A、B 油口与 P 油口几乎是关闭的,只允许小流量通过对两腔加压,而 T 油口是完全关闭的。

Fig. 5-19a shows a basic P-proportional directional valve reversing circuit. When the P-proportional directional valve is in the middle, the A, B and P oil ports are almost closed, just allowing small flow through to compress the two chambers. However, the T oil port is completely closed.

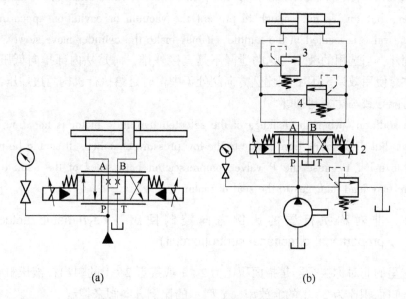

(a)　　　　　　　　　　　(b)

图 5-19 加压型比例方向阀换向回路

(Fig. 5-19　P-proportional directional valve reversing circuit)

1,3,4—溢流阀(Pressure relief valve);2—电液比例换向阀(Electro-hydraulic proportional directional valve)

这种回路的优点是中位时能提供小量的油流，补偿执行器的泄漏，可减少空穴出现对机器的损坏。对双杆液压缸和马达，这一小流量足以补偿泄漏，以及小惯性下防止真空出现。但是对于大惯量系统，上述的流量是不够的，可以如图5-19b所示在执行器两端跨接两个限压溢流阀。但是这种跨接式溢流阀只适用于对称执行器。

The advantages of this circuit are providing a small amount of oil when it is in the middle, compensating for leakage of the actuator and reducing the damage to the machine when cavitation occurs. To the two-rod cylinder and the motor, this small flow is enough to compensate for the leakage and prevent the vacuum appearing under small inertia. But to the large inertia system, the above-mentioned flow is not enough. In Fig. 5-19b, two limiting pressure relief valves can be crossed over the two sides of the actuators, but the crossover relief valve is only applicable to the symmetric actuator.

对于差动缸，这种方式产生的流量与需要补充的流量不相等。当液压缸外伸行程时，有杆腔的流量可能会经跨接溢流阀向无杆腔泄油，但却不足以防止真空或空穴出现，而且当阀位于中位时，还有可能使液压缸产生缓慢的移动。

To the differential cylinder, the flow generated in this way is not equal to the flow which needs to compensate to the system. When the cylinder is at the extending stroke, the flow of the rod chamber may cross the pressure relief valve and spill to the non-rod chamber, but this is not enough to prevent the vacuum or cavitation appearing. And when the valve is located in the middle, it may make the cylinder move slowly.

此外，当选用的液压马达的泄漏不是直接外排，而是从内部排向低压腔时，就应注意使用效果。因P型比例方向阀处于中位时是给执行机构两边加压，此压力有可能导致马达密封损坏。

In addition, when the leakage of the selected hydraulic motor is not draining out directly, but discharging from inside to the low pressure chamber, it should be paid attention to using. Because the P valve compresses on both sides of the actuator, this pressure may lead damages to the seal of motor.

5.3.2 非对称执行器的比例方向控制回路 (Asymmetric actuator's proportional directional control circuit)

这里的非对称执行器是指面积比为2∶1或接近2∶1的单出杆液压缸。它主要由开口面积比为2∶1的泄放型（Y型）的比例方向阀来控制。

The asymmetric actuator refers to the single-rod cylinder whose area ratio is 2∶1 or close to 2∶1. It is mainly controlled by the Y proportional directional valve whose open area ratio is 2∶1.

5.3 电液比例方向及速度控制回路
(Electro-hydraulic proportional direction and speed control circuit)

图 5-20a 为泄放型（Y 型）的比例方向阀基本应用回路。Y 型阀处在中位（自然位置）时，供油口封闭，而两工作油口通过节流小孔与油箱相连。由此，阀处在中位时，不会在两腔建立起高压。

Fig. 5-20a shows a basic application circuit using Y proportional directional valve. When the Y type valve is at the middle position (the natural position), the oil supply port is closed and the two working ports are connected with the oil tank by the orifices. As a result, when the valve is at the middle position, it will not set up a high pressure between the two chambers.

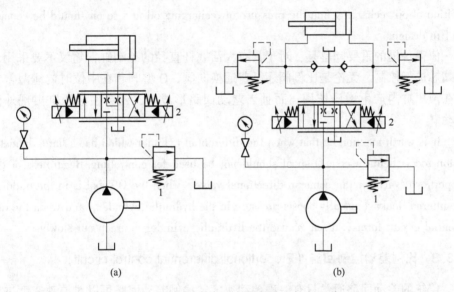

(a)　　　　　　　　　(b)

图 5-20　非对称执行器的比例方向控制回路

(Fig. 5-20　The proportional directional control circuit of the asymmetric actuator)

1—溢流阀 (Pressure relief valve); 2—电液比例换向阀 (Electro-hydraulic proportional directional valve)

通常，普通的 Y 型方向阀在中位时，其控制的液压缸是可以浮动的。但对 Y 型比例方向阀却无此功能。因为它处于中位时，连通两工作腔的开口很小，不足以通过较大的流量。同样，它也不能从油箱吸油，以防止空穴产生。为了防止真空状态出现和惯性引起的压力峰值，需要加上适当的回路。

Usually, when the common Y directional valve is at the middle position, its controlling cylinder can be floating. But the Y proportional directional valve doesn't have this function because when it is at the natural position, the opening area between the two working chambers is too small to allow a large flow through. And it can't absorb oil from the tank to prevent cavitation appearing. In order to prevent the vacuum and the pressure peak caused by the inertia, it's necessary to add the appropriate circuit.

图 5-20b 是适用于避免真空状态和惯性压力冲击的典型回路。两单向阀用于

真空时补油，它们的开启压力应该很低。两个溢流阀把工作腔与油箱相连，用于压力保护。为了更有效地保护系统，设计时应考虑单向阀的开启压力尽量低些、补油管的尺寸大小、补油点连接的地方以及补油的压力等问题。

Fig. 5-20b is a typical circuit which is used for avoiding the vacuum and the inertial pressure shock. The two cheek valves are used to recharge oil when there exists a vacuum, their opening pressure should be very low. The two pressure relief valves connect the working chamber with the tank for pressure protection. In order to protect the system more effectively, a low opening pressure of the check valve, the pipeline's size, the position of oil recharging and the pressure of recharging oil and so on should be considered in design.

值得一提的重要问题是，对于具有大活塞杆直径的差动缸，建议不要采用O型阀芯进行控制。无论是比例阀还是普通换向阀，O型阀芯在中位时内泄漏常常会在液压缸内产生增压作用；而且当差动回路形成时，还会使液压缸出现外伸蠕动。

It is worth mentioning that when the differential cylinder which has a large diameter piston rod is being used, O-spool should not be used for controlling. Regardless of the proportional valve or the common directional valve, when the O-spool is in the middle, the internal leakage usually boosts pressure in the hydraulic cylinder, and when the differential circuit forms, it may cause the hydraulic cylinder to creep out slowly.

5.3.3 比例差动控制回路 (Proportional differential control circuit)

传统的差动回路通常只有一种差动速度。比例差动回路可以对差动速度进行无级调节。有几种方法可以实现差动控制，所使用的比例阀芯的形式通常是Y型和YX_3型。

The traditional differential circuit usually has only one differential speed. While the proportional differential circuit can steplessly control the differential speed. There are several ways to achieve the differential control. The proportional valve spools usually used are the Y type and the YX_3 type.

由于比例阀的阀芯工作位置是连续的，很容易制造出专门适合于实现差动控制的阀芯，使差动回路获得简化。

Because the proportional valve spool's working position is continuous, it is easy to make the spool which is specifically for the differential control, so it can simplify the differential circuit.

图 5-21a 是一种典型的差动回路，它是利用Y型阀芯实现的。左电磁铁通电时液压缸差动向前伸出，右电磁铁通电时返回。通常普通方向阀的差动速度不可

5.3 电液比例方向及速度控制回路
(Electro-hydraulic proportional direction and speed control circuit)

调,但是从图中可以看出,使用比例方向阀后,在两个方向上速度都连续可调,这点与普通方向阀有所不同。

The Fig. 5-21a is a typical differential circuit which is using the Y-type spool. When the left electromagnet is energized, it is the differential forward controlling. When the right electromagnet is energized, it is to return. Usually, the common directional valve's differential speed is nonadjustable, but it can be seen from the figure that the speed in both directions can be adjusted continuously which is different from the common directional valve.

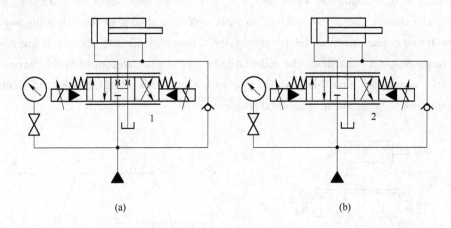

图 5-21 基本的差动回路

(Fig. 5-21 Basic differential circuit)

(a) 使用 Y 型阀芯的差动回路 (Differential circuit using Y-valve);
(b) 使用 YX_3 型阀芯的差动回路 (Differential circuit using YX_3-valve)

1—Y 型比例方向阀 (Y proportional directional valve); 2—YX_3 型比例方向阀 (YX_3 proportional directional valve)

差动速度的调节是控制从 P 到 A 的开口面积变化来实现的。由于在 B 管处装入单向阀,使阀芯中位时不具 Y 型阀的特点。为此,可把一个节流小孔与单向阀并联。

The differential speed is adjusted by changing the opening area from P to A. Because a check valve is set in the pipe B, its spool doesn't have the Y-valve's characteristics when it is in the middle. So an orifice can be parallel to the check valve.

图 5-21b 是利用 YX_3 型阀芯来实现差动的,事实上 YX_3 型阀芯是专门用于实现差动回路的。显然,这种差动回路想要获得最大推力,可在有杆腔出口处加一个二位三通电磁阀,以改变该处的油路使其通油箱。

Fig. 5-21b uses YX_3 type spool to realize differential motion. In fact, the YX_3 spool is specially designed for the realization of differential circuit. Obviously, if the maximum thrust is needed from the differential circuit, it can be achieved by adding a two-position three-way electromagnetic valve at the exit of the rod chamber, in order to link the oil line to the tank.

图 5-22a 所示为特殊的阀芯来实现比例差动回路。这是一个四位阀,要获得这种阀是容易的,只要通过加工阀芯便可以实现。该控制回路可以对外伸运动实现无级调速,其最大速度由差动回路确定,因而加大了调速范围。差动连接平滑地过渡到最大推力连接,使回路大为简化,因此这种比例阀有推广使用价值。

Fig. 5-22a shows that a special spool can be used to realize proportional differential motion. It is a four-position valve and it is easy to get this valve by processing the spool. The control circuit can achieve stepless speed regulation in the extending movement. Its maximum speed is determined by the differential circuit; thereby it increases the range of speed regulation. The differential connection is transformed to the maximum thrust's connection smoothly, which simplifies the circuit greatly. So this differential proportional valve is worth being used widely.

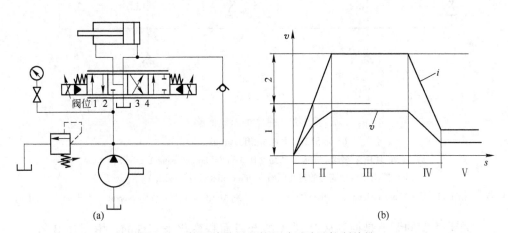

图 5-22 使用特殊阀芯的差动回路及控制特性

(Fig. 5-22 Differential circuit with special spool and its control characteristic)
(a) 差动回路 (Differential circuit); (b) 控制特性 (Control characteristic)
1—全力模式 (Full mode); 2—差动模式 (Differential mode);
Ⅰ—全力加速段 (Full acceleration stage); Ⅱ—差动加速段 (Differential acceleration stage);
Ⅲ—差动快速段 (Differential fast stage); Ⅳ—减速过渡段 (Deceleration transition stage);
Ⅴ—全力工作段 (Full working stage)

这种回路的外伸工作过程是:当无输入控制信号时,阀位 2 是自然中位,活塞不动;当从放大器来的控制信号处于较低水平时,阀的工作位置逐渐过渡到 3 的位置,这是全力工作模式,液压缸提供最大的加速力,使活塞尽快加速;当达到全流速度后,如果继续增大控制电流,则阀位由 3 过渡到 4,这时是差动工作

5.3 电液比例方向及速度控制回路
(Electro-hydraulic proportional direction and speed control circuit)

模式,由于 B 到 T 的油路被关闭,使油液通过单向阀与 P 汇合,形成最高流量,活塞这时的速度为最大,并且与信号成比例可调。

The circuit's extending process is: when there is no control signal input, the valve position 2 is in the middle and the piston doesn't move. When the control signal from the amplifier is low, the valve's working position gradually turns to 3, which is the full mode where the hydraulic cylinder provides the maximum accelerative force in order to make the piston accelerate as soon as possible. If the control current continues to be increased to the full-flow speed, the valve's position will transform from 3 to 4, which is the differential mode. Because the oil-way from B to T is closed, it makes the oil join the P through the check valve forming the maximum flow rate. The speed of the piston is the highest at this time and it can be adjusted proportionally following the signal.

在行程末端,控制信号恢复到较低水平,活塞又工作在全力模式,完成需要的工作循环。图 5-22b 所示为四位专用差动阀在工作过程中的控制电流与行程的关系曲线。从图中曲线可以看出,两种工作模式可以平滑转换。其突出的优点是:在工作压力及流量不变的情况下,启动时可获得最大加速力;在空行程中可获得可调的差动快速,且在工作行程中又可获得最大推力。

At the end of stroke, the control signal comes back to a lower level, and the piston works in the full mode again to complete the required cycle. The Fig. 5-22b shows the relation curve between the control current and the stroke when the 4-position special differential valve is at work. It can be seen from the curve in the figure that the two operational modes can convert smoothly. The prominent advantage is: when the working pressure and flow remain unchanged, it can achieve the maximum acceleration when it starts; It can acquire an adjustable differential speed in the idle stroke and the maximum thrust in the working stroke.

5.3.4 其他比例方向阀控制的实用回路 (Other practical control circuits using proportional directional valve)

5.3.4.1 重力平衡回路(Gravity balancing circuit)

在垂直运动的重物或超越负载的场合下,运动部件的运动速度会超过供油能力所能达到的速度,这时液压缸供油腔可能出现真空。运动部件超速失控,容易发生意外,因此液压系统中要设置平衡回路。

When the weight moves vertically or the weight is beyond the load, the moving component's speed will exceed the speed that the oil deliverability can reach and the cylinder's oil chamber may have a vacuum. It is necessary to set a balancing circuit in the hydraulic system because when the moving component's speed is out of control, it is possible to get an accident.

平衡回路是在超速方向上设置一个适当的阻力。对立式液压缸来说，就是在下行的方向上设置平衡阻力，使之产生足够的背压，以便与自重或与运动方向同向的负载相平衡。

The balancing circuit sets an appropriate resistance in the speeding direction. For the hydraulic vertical cylinder, the circuit will set a balance resistance in the downward process to produce an adequate back pressure for the balance of the dead weight or the load of the same direction of movement.

用比例方向阀控制垂直上升、下降速度的典型回路如图 5-23 所示，平衡元件采用单向溢流阀，溢流阀的调整压力应稍大于运动部件自重在液压缸下腔所形成的压力，单向阀用于通过上行时的工作流量。

The Fig. 5-23 shows the typical circuit where the proportional directional valve controls vertically up and down speed. The relief valve is used as the balancing component. The pressure relief valve's adjusted pressure should be slightly larger than the pressure in the cylinder's lower chamber caused by the moving component's weight. The check valve is used for passing the work flow in the up stroke.

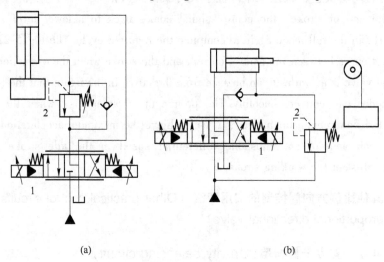

(a)　　　　　　　　　　(b)

图 5-23　重力平衡回路

(Fig. 5-23　Gravity balancing circuit)

(a) 重力平衡回路 (Gravity balancing circuit);
(b) 带重力平衡的差动控制回路 (Differential control circuit with gravity balancing)
1—电液换向阀 (Electro-hydraulic reversing valve); 2—溢流阀 (Pressure relief valve)

这种回路下行时，由于背压的存在，运动比较平稳。平衡用的溢流阀应采用球式锥形阀芯的直动式溢流阀，以减小或消除因泄漏而造成的缓慢下降。

5.3 电液比例方向及速度控制回路
(Electro-hydraulic proportional direction and speed control circuit)

When the circuit is in the down stroke, the movement is relatively stable because of the back pressure. The direct-acting relief valve with ball and cone spool should be used as the relief valve for balance to reduce or eliminate the slow fall caused by the leakage.

5.3.4.2 比例同步控制回路(Proportional synchronous control circuit)

对于带有多个执行器同时驱动同一负载运动的液压系统,由于每个液压缸的制造质量、摩擦力、泄漏、负载及结构变形上的差异,如果不采用适当的同步措施,各缸的行程会不同步,导致产品质量下降或设备不能正常工作。同步回路是为了克服上述的影响,通过改变进入其中一些或全部液压执行器的流量来实现同步。通常以其中一个执行器的位置作参考,改变进入其他执行器的流量来达到位置跟随而同步。可见同步回路本质上是个位置控制回路,因为控制变量的信息来自对位置误差的检测。

For the hydraulic system with multiple actuators and driving the same load movement, because of the diversity in every hydraulic cylinder's manufacturing quality, friction, leakage, load and structural deformation, if appropriate synchronous measures are not taken, the stroke of every cylinder will not be synchronized, leading to the decline in product quality or abnormal operation. In order to overcome these effects, the synchronous circuit achieves synchronization by changing the flow entering into some or all of the hydraulic actuators. Usually one of the actuators' position is taken as reference, while the flow into the other actuators is changed to achieve the synchronization of the position. It can be seen that the synchronization circuit is a position control loop in essence, because the information of control variables comes from the detection of position error.

对比例同步回路,位置误差的检测都是利用位置传感器来进行的,因而位置同步精度高,容易实现双向同步。根据采用的比例元件可分为比例方向阀同步回路、比例调速阀同步回路和比例变量泵同步回路。前两者属于节流控制式,后者属于容积控制式。

For the proportional synchronous circuit, the position error is detected by position sensor, so the position synchronizing precision is high and it can easily realize the bidirectional synchronization. According to different proportional components, the proportional synchronizing circuits can be divided into three kinds: proportional direction valve synchronizing circuit, proportional speed control valve synchronizing circuit and proportional variable pump synchronizing circuit. The first two belong to throttle control and the latter belongs to volume control.

A 进油同步回路 (Oil supply synchronous circuit)

图 5-24 所示为比例阀控制进油的同步回路。液压缸的上行速度可以通过调

速阀2来调节，比例方向阀根据位置传感器的反馈信号，连续地控制阀口开度，输出一个与手调节流阀相应的流量。

Fig. 5-24 shows a synchronous circuit controlled by proportional valve. The upward speed of the hydraulic cylinder is regulated by throttle valve 2. According to the feedback signal of position sensor, the proportional directional control valve continuously control the opening of the valve port, thus outputting a corresponding flow to a hand-operated throttle valve.

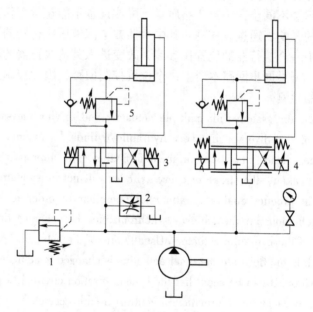

图 5-24　进油同步回路

(Fig. 5-24　Oil supply synchronous circuit)

1—溢流阀 (Pressure relief valve); 2—调速阀 (Speed regulating valve);
3—电磁换向阀 (Solenoid directional control valve); 4— 比例方向阀 (Proportional directional valve);
5—背压阀 (Back pressure valve)

当出现位置偏差时，比例放大器求得一控制信号，调整比例阀的开口，使其朝减小偏差的方向变化，直至偏差消失。因此，这是一种位置闭环控制系统，控制精度主要取决于位置传感器的检测精度与比例阀的响应特性，理论上该回路没有积累误差。上述回路要求比例阀有较大的通流能力。采用比例阀放油同步系统可以使选用较小通径的比例阀，从而降低成本。

When there is position deviation, the proportional amplifier will obtain a control signal, and adjust the opening of the proportional valve to reduce the deviation, until the deviation disappears. Therefore, this is a closed-loop position control system, and its control precision primarily depends on the detection accuracy of the position sensor

5.3 电液比例方向及速度控制回路
(Electro-hydraulic proportional direction and speed control circuit)

and the response characteristics of the proportional valve. In theory there is no accumulated error in the loop. The above-mentioned circuit requires the proportional valve to have a large flow capacity. The synchronous system of oil drainage with proportional valve can use a proportional valve whose diameter is quite small to reduce the cost.

B 比例阀放油同步回路 (Oil drainage synchronous circuit with proportional valve)

图 5-25 为使用比例方向阀的放油同步回路。该回路由两个完全相同的定量泵分别向液压缸单独供油,比例阀 3 通过控制两个液控换向阀来控制液压缸的速度。

Fig. 5-25 shows an oil drainage synchronous circuit using proportional direction valve. The circuit uses two identical constant displacement pumps to supply oil to the hydraulic cylinder respectively. The proportional valve 3 controls the speed of the hydraulic cylinder by controlling two pilot operated directional control valves.

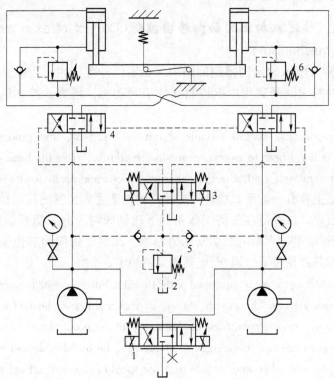

图 5-25 比例阀放油同步回路

(Fig. 5-25 Oil drainage synchronous circuit with proportional valve)

1, 3—比例方向阀 (Proportional direction valve); 2—溢流阀 (Relief valve);
4—液控换向阀 (Pilot operated directional control valve); 5—单向阀 (Check valve);
6—单向溢流阀 (Check relief valve)

如果两个油缸的位置出现不同步，则横梁会倾斜，传感器检测到以后，控制比例阀 1 的比例电磁铁使运动较快一侧的定量泵通过比例阀 1 排走部分流量，使其控制的液压缸速度慢下来，达到位置同步。由于比例阀 1 通过的流量只是纠偏用的小流量，可以选用较小的通径。

If the positions of two hydraulic cylinders are not synchronized, the beam will tilt. When sensors 1 detecting this, it controls the solenoid proportional valve 1 to let the constant displacement pump at the faster side discharge partly flow though proportional valve 1, and then slows down the controlled hydraulic cylinder until realize position synchronization. Since the flow through the proportional valve 1 is only used for small corrective flow, so a small size one can be chosen.

5.3.5 比例方向速度控制回路的应用（Applications of proportional direction speed control circuit）

5.3.5.1 传送机构上的轴向移动控制（Control of axial movement on transmitting mechanism）

在如图 5-26 所示的比例控制回路中，在弱信号范围内，为提高分辨率致使特性呈非线性的比例阀利用中枢控制插件板的功能，精确地控制机构的速度和位置。

In the proportional control circuits shown in Fig. 5-26, the proportional valve, whose feature is non-linear so as to improve the resolution, uses the functions of central control board to precisely control the mechanism's speed and position within weak signal.

在液压缸上装有一数字式线性传感器以输出速度和位置的反馈信号。储能系统保持油压恒定，以保证在任何工作条件下比例阀两端有一适量的压降 Δp。紧急状态下，通/断电磁阀断电而切断压力油源，此时比例阀阀芯位于零遮盖的中间位置。工作条件恢复后，通/断电磁阀又可通电。

The hydraulic cylinder is equipped with a digital linear sensor to output speed and position feedback signals. The energy storage system maintains constant oil pressure to ensure that in any working conditions the both ends of the proportional valve have an appropriate drop pressure Δp. In emergency situations, the on/off solenoid valve turns off to cut off the pressure oil source. At this time the spool of proportional valve is located in the middle of the zero lap. When the work conditions recover, the on/off solenoid valve can get energized again.

5.3.5.2 炼钢设备上的塞孔柱控制（Control of the jack pillar of steel equipment）

在钢厂中，特殊设计制造的坚固电液装置可以保证其可靠性和优良性能。本

5.3 电液比例方向及速度控制回路
(Electro-hydraulic proportional direction and speed control circuit)

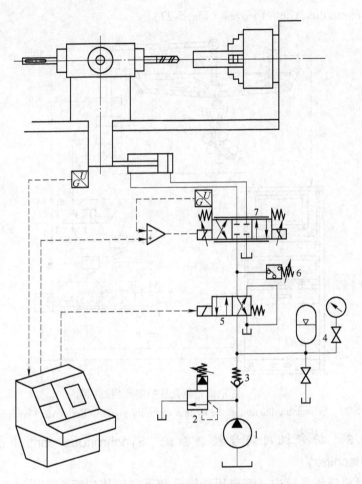

图 5-26 传送机构上的轴向移动控制回路

(Fig. 5-26 Control loop of axial movement on transmitting mechanism)

1—液压泵（Hydraulic pump）；2—溢流阀（Pressure relief valve）；3—单向阀（Check valve）；
4—蓄能器（Accumulator）；5—换向阀（Directional control valve）；6—压力继电器（Pressure relay）；
7—比例方向阀（Proportional direction valve）

例显示调整钢水液位的专用电液系统，包括一个装有集成电子器件的比例阀，一个手动应急装置，一个配有电位器式传感器的液压伺服缸以及一个安装有控制阀的油路板（图 5-27）。

The application is used in the steel factory. The firm electro-hydraulic device which is specially designed and manufactured can ensure its reliability and excellent performance. This example shows the electro-hydraulic system specialized for adjustment of molten steel level, including a proportional valve with integrated electronic devices, a manual emergency device, a hydraulic servo cylinder with potentiometer sensor and a manifold for installing control valves (Fig. 5-27).

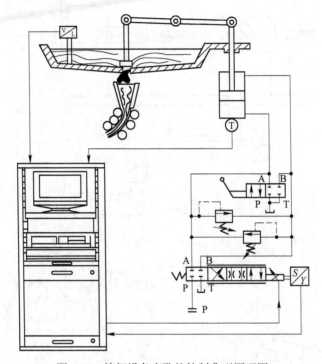

图 5-27 炼钢设备塞孔柱控制典型原理图

(Fig. 5-27 Schematic diagram of the system controlling jack pillar of steel equipment)

5.3.5.3 折弯机用同步控制系统(Synchronous control system in bending machine)

在折弯机液压系统中，完成压梁提升和下降动作的两个液压缸必须同步运动，并应具有非常高的位置精度（图 5-28）。

In the bending machine system, the two hydraulic cylinders must move synchronously and should have very high position accuracy during the lifting and declining of the rolled beam (Fig. 5-28).

利用两只闭环控制的比例阀可以成功地实现上述控制。闭环控制系统利用装在压梁上的位置传感器得到控制信号。

The above-mentioned control can be successfully achieved by using two proportional valves of closed-loop control. Closed-loop control system uses the position sensor on the pressure beam to get the control signals.

5.3.5.4 控制高空作业车的自动调平(Automatic leveling circuit of high-altitude operation vehicle)

此典型应用实例说明，可用一套专用电子器件控制的比例阀来实现平台的自

5.3 电液比例方向及速度控制回路
(Electro-hydraulic proportional direction and speed control circuit)

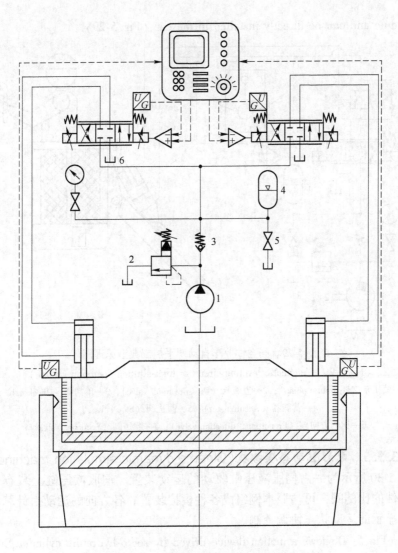

图 5-28 折弯机同步控制典型原理图

(Fig. 5-28 Typical bending machine synchronous control principle diagram)

1—液压泵（Hydraulic pump）；2—溢流阀（Pressure relief valve）；3—单向阀（Check valve）；
4—蓄能器（Accumulator）；5—截止阀（Stop valve）；6—比例方向阀（Proportional direction valve）

动调平。这套电子器件包括角位移传感器以及以闭环方式工作的电子放大器。整个电液系统是一结构紧凑的模块，可直接装在车上使用（图 5-29）。

This typical application indicates that a proportional valve controlled by a set of specialized electronic device can be used to control the automatic leveling of the platform. The electronic devices include an angular displacement sensor, as well as an electronic amplifier working in closed loop way. The whole electro-hydraulic system is a compact module and can be directly installed in the car (Fig. 5-29).

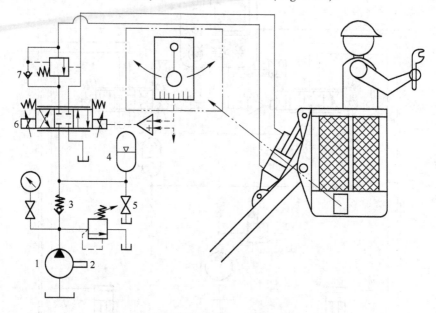

图 5-29 高空作业车自动调平控制典型原理图

(Fig. 5-29 Automatic leveling circuit of high-altitude operation vehicle)

1—液压泵(Hydraulic pump); 2—溢流阀(Pressure relief valve); 3—单向阀(Check valve);
4—蓄能器(Accumulator); 5—截止阀(Stop valve);
6—比例方向阀(Proportional direction valve); 7—顺序阀(Sequence valve)

5.3.5.5 电影院动态模拟机(Cinema dynamic simulation machine)

图 5-30 所示为一由伺服液压缸驱动的运动装置。伺服液压缸上装有带集成电子器件的比例阀,可仿照本例组成各种模拟装置。各方向的运动由计算机根据屏幕上显示的运动状态协调控制。

The Fig. 5-30 shows a motion device drived by servo-hydraulic cylinder. Servo-hydraulic cylinder is equipped with proportional valve with integrated electronic devices, which can compose various simulation devices following this case. Movement in all directions is coordinately controlled by the computer according to the state of motion displayed on the screen.

5.3.5.6 三通比例插装阀在注塑机上的应用(Applications of three-way proportional cartridge valve on injection molding machine)

在注塑机中,插装式比例阀实现三个控制功能,即注射速度控制、挤压成型和锁模。电控通过集成有电子器件和位置传感器的比例阀来实现,其位置传感器在主阀芯上形成闭环,实现精确控制并具有高动态性能(图 5-31)。

In the injection molding machine, proportional cartridge valve achieves three

5.3 电液比例方向及速度控制回路
(Electro-hydraulic proportional direction and speed control circuit)

control functions, namely, injection speed controlling, extrusion molding and model locking. Electronic control is realized by the proportional valve with integrated electronic devices and position sensors, and the position sensor forms a closed-loop on the main spool to achieve precise control and have a high dynamic performance (Fig. 5-31).

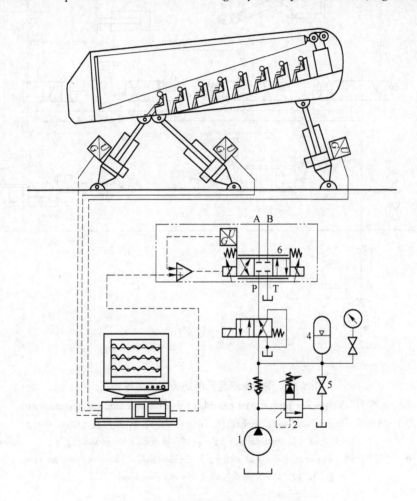

图 5-30 电影院动态控制典型原理图

(Fig. 5-30 Typical schematic diagram of cinema dynamic control)

1—液压泵 (Hydraulic pump); 2—溢流阀 (Relief valve); 3—单向阀 (Check valve);
4—蓄能器 (Accumulator); 5—截止阀 (Accumulator stop valve); 6—比例方向阀 (Proportional direction valve)

5.3.5.7 阀门的定位(Valve positioning)

用电液系统可以很容易地实现阀门的遥控。由变化范围为 4~20mA 的给定输入电流确定初始位置,比例阀可在闭环控制中根据现场监视器的输出信号进行可靠的控制。远方阀门可由集成电子放大器(即计算机)产生的 4~20mA 的信号

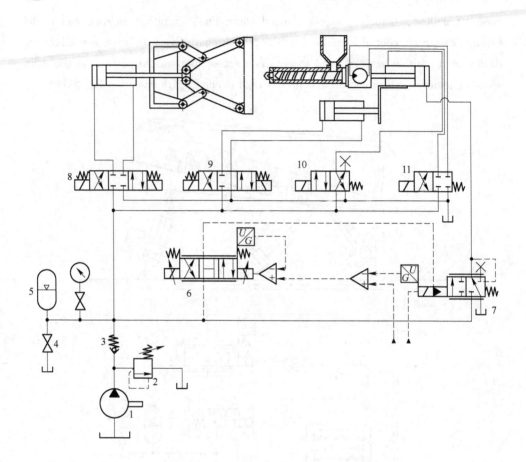

图 5-31　注塑机三通比例插装阀典型原理图
(Fig. 5-31　Three-way proportional cartridge valve on injection molding machine)
1—液压泵 (Hydraulic pump); 2—溢流阀 (Relief valve); 3—单向阀 (Check valve);
4—截止阀 (Accumulator stop valve); 5—蓄能器 (Accumulator);
6—比例方向阀 (Proportional direction valve); 7—三通插装阀 (Three-way cartridge valve);
8, 9, 10, 11—电磁换向阀 (Solenoid directional valve)

直接控制。由一通/断电磁阀操纵的单向阀提供安全闭锁（图 5-32）。

It's easy to achieve the remote control of the valve with the electro-hydraulic system. The initial position is determined by the input current whose variation range is 4 ~ 20mA, and the proportional valve in the closed-loop can be reliably controlled according to the output signals of the on-site monitors. Remote valve can be controlled directly by integrated electronic amplifier, namely the signal ranging 4 ~ 20mA generated by computer. A check valve which is controlled by an on/off solenoid valve provides a safe latch (Fig. 5-32) .

5.4 比例复合回路 (Proportional compound circuit)

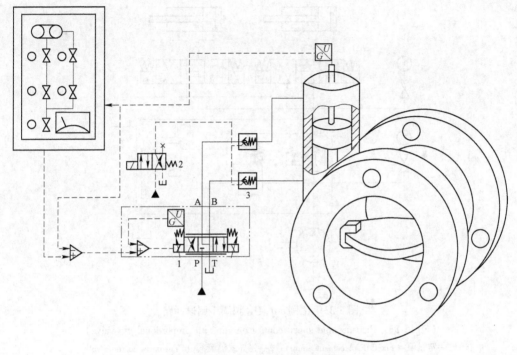

图 5-32 阀门定位控制典型原理图

(Fig. 5-32 Valve positioning hydraulic system)

1—比例方向阀 (Proportional direction valve); 2—电磁换向阀 (Solenoid directional valve);
3—液控单向阀 (Pilot-controlled check valve)

5.4 比例复合回路 (Proportional compound circuit)

5.4.1 比例压力-流量复合阀调压调速回路 (Pressure and speed control circuit with proportional pressure-flow compound valve)

使用 p-q 复合阀可以使系统变得简单,并且控制性能也可以达到要求。p-q 复合阀利用定差溢流阀做压力补偿,使泵的输出压力适应负载压力,从而使泵供油时没有过剩的压力 (图 5-33)。

Using p-q compound valve can make the system simple, and can make the control performance meet the requirements. p-q compound valve uses a constant difference relief valve for pressure compensation, making the output pressure of the pump adapt to the load pressure, so that there is no excess pressure when the pump supplies oil (Fig. 5-33).

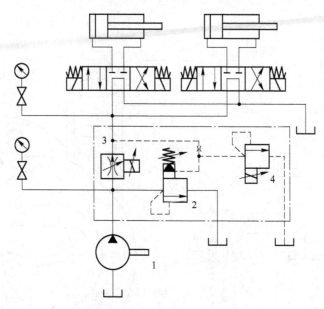

图 5-33　比例 p-q 复合阀调压调速回路

(Fig. 5-33　Pressure and speed control circuit with proportional p-q valve)

1—定量泵（Fixed displacement pump）；2—先导式溢流阀（Pilot-operated relief valve）；
3—比例流量阀（Proportional flow valve）；4—比例溢流阀（Proportional relief valve）

所需的流量控制由比例流量阀 3 进行控制；主溢流的先导式溢流阀 2 按系统最高压力来调定，从而保证系统的安全；在各个工况下，系统的压力可以由比例溢流阀 4 来进行调整。

The desired flow is controlled by proportional flow valve 3, and the pilot-operated relief valve 2 of main relief valve is adjusted, according to the maximum pressure of system, to ensure the security of the system. In every condition, the pressure of the system can be adjusted by proportional relief valve 4.

5.4.2　比例压力/流量调节型变量泵回路（Proportional pressure/flow regulating variable pump circuit）

在压力/流量调节型变量泵系统中，压力由比例压力阀 1 控制，输出流量由比例流量阀 2 控制，从而使变量泵压力和流量都能适应负载的变化（图 5-34）。

In proportional pressure/flow regulating variable pump system, the pressure is controlled by the proportional pressure valve 1, and the output flow is controlled by proportional flow valve 2. So both the pressure and flow of the variable pump can be adaptive to the change of load (Fig. 5-34).

该变量泵系统用于工作循环复杂、工况变化频繁、动静特性要求较高的

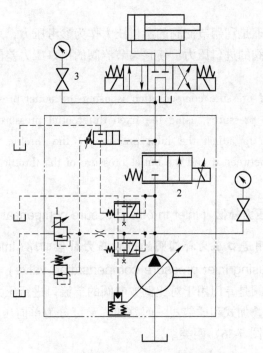

图 5-34 比例 p/q 变量泵调压调速回路

(Fig. 5-34　Pressure and speed control circuit with proportional p/q variable volume pump)

1—比例压力阀（Proportional pressure valve）; 2—比例流量阀（Proportional flow valve）;
3—电磁换向阀（Solenoid directional control valve）

场合。

The variable pump system is used in the condition that working cycle is complex, working condition changes frequently, and the requirements of static and dynamic characteristics are relatively high.

5.5　应用于比例节流的压力补偿回路（Pressure compensating circuit using proportional throttle）

为了提高使用比例节流元件（包括比例节流阀、比例方向阀）流量的控制精度，需要在控制节流口面积的同时对节流口两端的压差进行控制，保证控制流量尽可能不受负载或者供油压力变化的影响。

In order to improve the flow control precision of using proportional throttle elements (including proportional throttle valve and proportional direction valve), the throttle area and the throttle pressure difference of both ends need to be controlled simultaneously, so as to ensure the control flow not to be influenced by the load or the pressure change of supplied oil.

压力补偿的原理是利用节流阀的出口压力作为参考压力，采用定差减压阀或溢流阀来调节节流阀的进口压力，使它与节流阀的出口压力差值稳定在一个恒定的数值上。

The principle of pressure compensation is using the outlet pressure of the throttle valve as a reference pressure, using the fixed differential pressure reducing valve or pressure relief valve to adjust the inlet pressure of the throttle valve, making the pressure difference between it and the outlet pressure of the throttle valve stabilize at a constant value.

5.5.1 进口节流压力补偿 (Inlet throttle pressure compensation)

5.5.1.1 使用进口压力补偿阀的进口压力补偿回路(Inlet pressure compensation circuit using inlet pressure compensation valve)

进口压力补偿阀是专门用于对比例方向阀的节流口进行压力补偿的元件。进口压力补偿阀分为叠加式和插装式，而叠加式又可分为单向压力补偿（图5-35）和双向压力补偿（图5-36）两类。

Inlet pressure compensation valve is specialized for pressure compensation to the orifice of the proportional direction valve. Inlet pressure compensation valve is divided into sandwich valve and cartridge valve, and the sandwich one can also be divided into two types: unidirectional pressure compensation valve (Fig. 5-35) and bidirectional pressure compensation valve (Fig. 5-36).

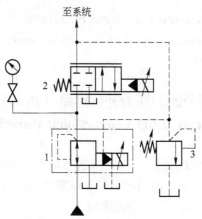

图 5-35　单向压力补偿阀应用回路

(Fig. 5-35　Application circuit of check pressure compensation valve)

1—三通比例减压阀 (There-way proportional pressure reducing valve);
2—比例换向阀 (Proportional directional control valve); 3—直动式溢流阀 (Direct-acting relief valve)

5.5 应用于比例节流的压力补偿回路
(Pressure compensating circuit using proportional throttle)

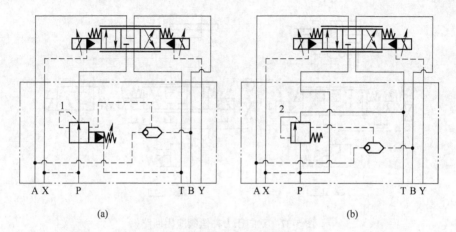

图 5-36 双向压力补偿阀应用回路

(Fig. 5-36 Application circuit of bidirectional pressure compensation valve)

(a) 两通压力补偿 (Two-way pressure compensation valve);

(b) 三通压力补偿 (There-way pressure compensation valve)

1—定差减压阀 (Constant difference pressure-reducing valve); 2—定差溢流阀 (Constant difference relief valve)

单向压力补偿阀的补偿元件是三通定差减压阀；两通双向压力补偿阀的补偿元件是定差减压阀；三通压力补偿阀的补偿元件是定差溢流阀。

The compensation element of unidirectional pressure compensation valve is three-way fixed difference pressure reducing valve; the compensation element of two-way bidirectional pressure compensation valve is fixed difference pressure reducing valve; the compensation element of three-way pressure compensation valve is constant difference relief valve.

5.5.1.2 使用普通减压阀的进口压力补偿回路 (Inlet pressure compensation circuit using ordinary pressure reducing valve)

该回路存在梭阀能否正确选择反馈压力的问题，所以只能使用在速度变化慢，运动部件质量不大，并以摩擦负载为主的场合（图5-37）。

The loop has a problem that whether the shuttle valve can correctly select the feedback pressure or not, so it can only be used in situations where the speed change is slow, moving parts are small in mass or friction load is principal (Fig. 5-37).

以电磁换向阀代替梭阀选择反馈压力，从而避免了不正确压力反馈的问题。

The electromagnetic reversing valve is used instead of the shuttle valve to select the feedback pressure, thus avoiding problems of generating incorrect pressure feedback.

为了使梭阀只感应正确的负载压力，并且防止减速制动时出现的高压，在回路中设置了压力保护回路（图5-38）。

In order to make the shuttle valve only induct correct load pressure and prevent cre-

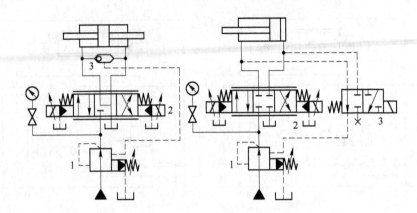

图 5-37 双向压力补偿阀应用回路

(Fig. 5-37 Application circuit of bidirectional pressure compensation valve)

1—减压阀 (Pressure reducing valve); 2—比例方向阀 (Proportional direction valve);
3—梭阀/电磁换向阀 (Shuttle valve /Solenoid directional control valve)

ating high pressure when retarding brake, a pressure protection circuit is installed in the circuit (Fig. 5-38).

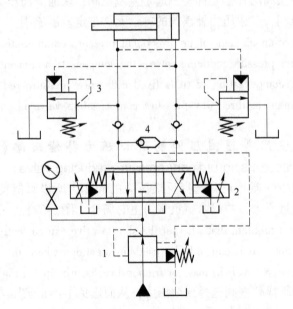

图 5-38 带压力保护的双向压力补偿阀应用回路

(Fig. 5-38 Application circuit using bidirectional pressure compensation valve with pressure protection)

1—减压阀 (Pressure reducing valve); 2—比例换向阀 (Proportional directional valve);
3—溢流安全阀 (Relief safety valve); 4—梭阀 (Shuttle valve)

5.5 应用于比例节流的压力补偿回路
(Pressure compensating circuit using proportional throttle)

5.5.2 出口节流压力补偿 (Outlet throttle pressure compensation)

出口压力补偿回路（图 5-39）可以利用减压阀来设计，也可以采用出口压力补偿器。

Outlet pressure compensation circuit (Fig. 5-39) can use the pressure reducing valve as well as the outlet pressure compensator to design.

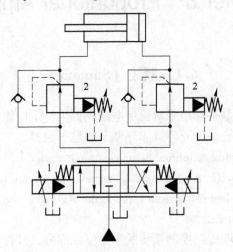

图 5-39 双向出口节流压力补偿回路

(Fig. 5-39 Bidirectional outlet throttle pressure compensation circuit)
1—比例换向阀 (Proportional reversing valve); 2—减压阀 (Pressure reducing valve)

通过比例方向阀的压差可以由减压阀来调整，从而可在较低的压差下获得较高的流量控制精度。如果需要在两个方向上进行精确的调速，在油孔 A 侧串联一只相同的减压阀即可。

The differential pressure through the proportional direction valve can be adjusted by pressure reducing valve, so that a higher flow control accuracy can be obtained under lower differential pressure. If precise speed control is needed in both directions, a same pressure reducing valve needs to be cascaded at the A side of oil hole.

第 6 章 比例放大器

Chapter 6　Proportional Amplifier

6.1　概述（Summary）

比例控制放大器是一种用来对比例电磁铁提供特定性能电流，并对电液比例阀或电液比例控制系统进行开环或闭环调节的电子装置。

A proportional control amplifier is an electronic device which provide proportional solenoid valve with specific performance of current, and make open loop or close loop adjustment for an electro-hydraulic proportional valve or an electro-hydraulic proportional control system.

比例放大器是电液比例控制系统的重要组成部分，其性能及可靠性对整个系统起着十分重要的作用。最初的比例控制放大器设计为模拟恒压式，控制性能较差，且多为开环控制。而后发展为恒流式。由于恒流式能够抑制负载阻抗热特性的影响，并且恒流式带铁芯感性负载动态性能优于恒压式，恒压式逐渐被恒流式所取代。如今现代比例控制放大器几乎无一例外地采用恒流式结构。

A proportional amplifier is an important part of an electro-hydraulic proportional control system whose performance and reliability plays a very important role in the whole system. Original proportional control amplifiers are designed as analog constant pressure type. Their control performance is relatively poor, and most of them are open loop control systems. Then the constant flow type is developed. The constant flow type can inhibit effects of the heat load impedance characteristic, and the constant flow type proportional control amplifier with iron core perceptual load has better dynamic performance than the one of constant pressure type. The constant pressure type is gradually replaced by the constant flow type. Now most modern proportional control amplifiers, almost without exception, use the constant flow structure.

6.1.1　比例放大器的基本技术要求（Fundamental requirements of the proportional amplifier）

为了使比例系统获得预期的功能，必须对比例阀进行正确的控制。这一任务是由比例放大器来完成的。概括起来，对比例放大器的基本要求是：

In order to obtain expected functions of the proportional system, we must control the proportional valve properly. This task is accomplished by a proportional amplifier. To sum up, there are some fundamental requirements of a proportional amplifier:

(1) 实现从电压信号到电流信号的转换，并提高与输入电压成比例且功率足够的控制电流。

Realizing the transformation from a voltage signal to a current signal, and improving the control electric current which is proportional to the input voltage and have enough power.

(2) 具有足够的功率来驱动比例电磁铁中衔铁上的负载，其输入阻抗大，输出阻抗与比例电磁铁线圈阻抗相匹配。

Enough power to drive loads of a proportional solenoid armature. Big input impedance and output impedance must be matched with coil impedance of the proportional solenoid.

(3) 输入端有各种标准信号的接口，易于实现与不同信号发生装置的连接，采用标准电源。信号的波形、幅值和频率符合电液比例阀的静态和动态性能要求。

Having an input terminal with all kinds of standard signals interface, so that, it can be easy to connect to different signal generators. It should use standard power. The waveform, amplitude and frequency of signal should satisfy static and dynamic performance requirements of an electro-hydraulic proportional valve.

(4) 易于检测控制信号和反馈信号，具有基本的故障诊断和元件保护功能。

Testing controlling signals and feedback signals easily and possessing basic fault diagnosis and components protection functions.

(5) 能产生正确有效的控制信号。例如，为了减小比例元件零位死区的影响，放大器应具有幅值可调的初始电流功能；为减小滞环的影响，放大器的输出电流应含有一定频率和幅值的颤振电流分量；为减小系统过渡过程的冲击，对阶跃输入信号能自动生成速率可调的斜坡信号。

Producing a correct and effective control signal. For example, in order to reduce effects from the proportional elements' zero dead-zone, the amplifier should have function of making initial current with adjustable amplitude. To reduce influence of hysteresis, the amplifier's output current must contain flutter component with certain frequency and amplitude. In order to reduce the impact of the system transition process, for a step input signal, it can automatically produce ramp signal with adjustable rate.

综上所述，比例放大器是一个能够对弱电的控制信号进行整形、运算和功率放大的电子控制装置。为了简化应用技术，目前已发展出适应不同场合和具有完善功能的标准电控器单元，可以适应不同的比例元件或控制要求。利用这些单元，可以方便地组成电气控制系统。一般情况下，仅需按技术要求选用标准单元，并加上必要的简单外围电路，便可组成一个完整的电气控制系统。

To sum up, the proportional amplifier is an electronic control device which can realize the shaping, operation and power magnification of a weak electric control signal. In order to simplify the application technology, at present we have developed some electric controlling units which have perfect functions to fit different situations. The electric controlling units can also adapt to different proportional components or controlled request. Generally, you only need to use the standard unit and necessary peripheral circuits according to technical requirements to form a complete electrical control system easily.

6.1.2 比例放大器的分类 (Classification of the proportional amplifier)

比例放大器有以下多种分类方法 (There are many ways to classify proportional amplifiers)：

（1）按放大器输出控制电流的通路数，可将比例放大器分为单通路和双通路两种类型。

According to the number of controlled electric current routes amplifiers can be divided into two types: single-channel and binary-channel.

单通路比例放大器用于控制带单个电磁铁的比例元件，如比例压力阀或比例流量阀，以及单电磁铁驱动的比例方向阀等。双通路比例放大器用于控制三位比例方向阀、压力-流量复合控制阀、压力-流量复合控制泵等带有两个比例电磁铁的比例元件。需要注意的是，双通路比例放大器工作时，只有其中一个比例电磁铁起实质性的控制作用。当控制三位比例方向阀时，比例放大器根据信号的极性选择一个起作用的比例电磁铁；当对压力-流量复合控制阀、压力-流量复合控制泵进行压力控制时，压力阀电磁铁起作用，流量阀电磁铁的控制信号是一个定值。

A single-channel proportional amplifier is used to control proportional elements with individual solenoid, such as proportional pressure valves, proportional flow valves and proportional direction valves that are controlled by single proportional solenoid. A binary-channel proportional amplifier is used to control three-position proportional direction valves, pressure-flow control valves, pressure-flow control pumps and any other proportional components which have two proportional solenoids. It is important to note that when binary-channel proportional amplifiers work, only one of the proportional solenoid works: three position proportional direction valves are controlled, according to polarity of the signal, the proportional amplifier chose one acting proportional solenoid; when the pressure of pressure-flow control valves and pressure-flow control pumps are controlled, a pressure valve solenoid works, and control signal of the solenoid in the flow valve is a fixed value.

（2）按放大器内是否带反馈通路，可将比例放大器分为开环控制和闭环控制两种类型。

According to whether the amplifier has feedback channels or not, it can be divided into two types: open-loop proportional control and closed-loop control.

开环控制比例放大器没有测量电路、反馈单元和反馈通路 PID 调节器，通常带有颤振信号发生器。闭环放大器用来控制带电反馈比例阀和比例泵，设置有测量放大电路、反馈比较环节和信号调节器。

An open-loop control proportional amplifier doesn't have a measuring circuit, a feedback unit and a feedback channel PID controller, but has a flutter signal generator in general. A closed-loop amplifier is used to control charged feedback proportional valves and proportional pumps. A closed-loop amplifier is equipped with a measuring amplification circuit, a feedback comparing link and a signal conditioner.

（3）按放大器内运算信号的类型可将比例放大器分为数字式和模拟式。

According to the type of signals in operation, amplifier can be divided into the digital type and the analog type.

模拟式比例放大器按连续信号的方式工作，加在比例电磁铁线圈两端的信号为连续直流电压，功耗较大。数字式比例放大器又分为数字信号放大器和开关式放大器。开关式放大器的功放管工作在截止或饱和区，即开关状态，加在比例电磁铁线圈两端的信号为脉冲电压，功耗小。开关式比例放大器以 PWM 式为主。数字信号放大器内部采用数字芯片完成信号运算。

An analog proportional amplifier works with continuous signal. The signal which is added on both ends of the proportional solenoid coil is continuous direct voltage. It has more power dissipation. Digital proportional amplifiers are divided into digital signal amplifiers and switch type amplifiers. The power amplification tubes of switch type amplifiers work in the cut-off or saturation zone, namely switch state. The signal which is added on both ends of the proportional solenoid coil is pulsed voltage. It has less power dissipation. Switch type proportional amplifiers are mainly PWM type. A digital signal amplifier adopts digital chips to accomplish the signal operation.

此外，根据所控制比例电磁铁的类型分类，还有单向和双向比例放大器。单向放大器用来控制单向比例电磁铁，双向放大器用来控制双向比例电磁铁，两者采用不同的放大电路。

Moreover, according to the type of controlled proportional solenoid, there are also unidirectional and bidirectional proportional amplifiers. The unidirectional proportional amplifier is used to control the unidirectional proportional solenoid and the bidirectional amplifier is used to control the bidirectional solenoid, different amplifier circuits are adopted between the two.

6.1.3　比例放大器的主要电路（Main circuits of the proportional amplifier）

比例控制放大器一般由比例放大器、电源电路、控制信号输入编程电路、逻辑控制电路、功率放大级和检测放大电路等部分组成（图6-1）。

A proportional control amplifier is generally composed of a proportional amplifier, a power supply circuit, a programming circuit for inputting the control signal, a logic control circuit, a power amplifier stage and a detection amplifier circuit, and so on (Fig. 6-1).

6.1.3.1　比例放大器(Proportional amplifier)

比例放大器又称为电控器（图6-1中点划线内的部分）。它本身包含多种有特定功能的基本电路，是电气控制系统的核心部分。信号的处理及放大是在比例放大器内完成的。它与外围电路一起构成比例阀的电气控制系统。通常它有一个或两个输出通道，把经过放大的功率信号送到比例电磁铁。双输出通道的用于控制比例方向阀，单输出通道的用于控制其他比例元件。比例放大器通常用于力控制型比例电磁铁或行程控制型电磁铁。用于控制后者的都带有位置检测反馈通道，如图6-1所示，该反馈通道专门用于阀芯的位置反馈。

A proportional amplifier is also called an electronic controller. It is shown in the dot dash line of figure 6-1. It is the core of an electrical control system and contains a variety of specific basic functional circuits. Signal processing and amplification are completed within a proportional amplifier. The proportional amplifier together with its peripheral circuits makes up an electrical control system for proportional valves. Usually a proportional amplifier has one or two output channels which are used for delivering a magnified power signal to the proportional solenoid. The proportional amplifier with double output channels is used to control proportional directional valves. Those which have single channel are used to control other proportional elements. Proportional amplifiers are generally used for power control type proportional solenoid or stroke control type solenoid. The proportional amplifier which is used for stroke control type solenoid has a feedback channel for position detection. As shown in Fig. 6-1, this feedback channel dedicated to valve core position feedback.

6.1 概述（Summary）

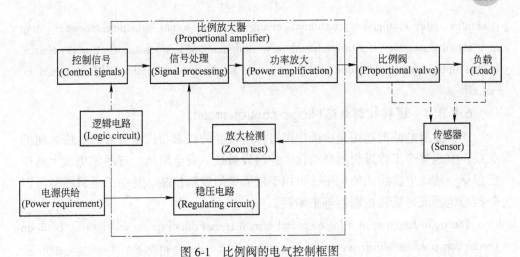

图 6-1　比例阀的电气控制框图
(Fig. 6-1　Electrical control diagram of the proportional valve)

6.1.3.2　电源电路(Power circuit)

比例控制放大器电源电路的主要作用是：从标准电源中获得并分离出比例控制放大器正常工作所需的各种直流稳定电源，并且在电网电压、负载电流及环境温度允许范围内变化时，保证输出直流电压的稳定性。同时，还兼具电源电压极性反接、过流、短路自我保护自恢复的非熔断式保护功能，以保证比例控制放大器的工作可靠性。

The main functions of the power circuit of a proportional control amplifier are getting and separating all kinds of DC stable power supply, which is needed for the proportional amplifier working regularly, from the standard power, and ensuring the output voltage stability when electrified wire netting voltage, load current and ambient temperature changed in permitted scope. At the same time, the power circuit has self-protection function of resisting and recovering from polarity reverse connected of the power supply voltage, overflowing and shout out, so that it can ensure functional reliability of the proportional control amplifier.

6.1.3.3　控制信号输入编程电路(Control signal input programming circuit)

控制信号输入编程电路用于产生给定值，作为输入控制信号的参考值。通常采用电位器组或信号发生器来产生阶跃变化或连续变化的控制信号。如果要求比例阀跟踪输入信号连续变化，这时应切除比例放大器内的斜坡信号发生器。

A control signal input programming circuit is used to generate a given value as a reference value of the input control signal. A control signal input circuit usually generates a step changed or continuous changed control signal by a potentiometer group or a signal generator. If a proportional valve is demanded to track the continuous change of the input signal, the slope signal generator in the proportional amplifier should be cut off.

6.1.3.4 逻辑控制电路(Logic control circuit)

逻辑控制电路的功用是协调生产过程对性能的要求与控制信号的生成之间的关系，实现整个工作过程的自动化，并对设备提供安全措施。有些比例放大器中已包括一些继电器组成的电路，可用于较简单的逻辑控制。复杂的逻辑控制可以外接继电器或可编程控制器等来实现。

The main function of a logic control circuit is coordinating the relationship between production process requirements on performance and the generation of control signals, realizing the automation of whole work process, at the same time, supplying security measures for equipments. Some proportional amplifiers have included some circuits composed by relays and can be used for relatively simple logic control. Complex logic control can be achieved by connecting external relays or programmable controllers.

6.1.3.5 功率放大级(Power amplifier stage)

功率放大级是比例控制放大器的核心单元。它的作用是对各种控制信号进行综合和放大，向电-机械转换装置提供足够大的驱动电流。对于比例电磁铁，通常所需的驱动电流为 0.8~1.2 A。

A power amplifier stage is a core unit of a proportional control amplifier. Its functions are synthesizing and amplifying all kinds of control signals, and supplying an enough drive current for an electricity-mechanical transmission. It usually requires a drive current about 0.8 ~ 1.2A for a proportional solenoid.

6.1.3.6 检测放大电路(Detection amplifier circuit)

闭环控制中，传感器检测实际输出值并构成反馈。构成闭环有两种形式，一种称为阀内闭环，常见的是阀芯位移的反馈；另一种称为外闭环，是对系统实际控制量的反馈。后者是闭环控制系统，前者对控制量来说仍然是开环的。

An actual output value is detected by sensors in the closed-loop control and the value is sent to the input end as a feedback. There are two forms of the closed-loop. One is called "closed-loop of inner valve" like spool displacement feedback, the other is called "outside closed-loop" which is the feedback of actual controlled variables of systems. The latter is a closed-loop control system, and the former is still an open loop to the control variable.

6.1.4 运算放大器简介 (Introductions of operational amplifier)

运算放大器（常简称为"运放"）是具有很高放大倍数的电路单元，在实际电路中，通常结合反馈网络共同组成某种功能模块。由于早期应用于模拟计算机中，用以实现数学运算，故得名"运算放大器"，此名称一直延续至今。运放是一个从功能的角度命名的电路单元，可以由分立的器件实现，也可以实现在半导体芯片当中。随着半导体技术的发展，如今绝大部分的运放是以单片的形式存在。运放的种类繁多，广泛应用于几乎所有的行业当中。

An operational amplifier (often referred to as "op amp") is a circuit element with high magnification. In the actual circuit it is usually associated with a feedback network to compose some functional modules. It is used for arithmetical operation in early analog computers, so named "op amp", which has been in use until nowadays. An op amp is a circuit element which is named from the view of functional point, which can be achieved with separate devices, and also can be achieved with semiconductor chips. With the development of semiconductor technology, today, most of op amps exist in the form of single chip. There are many types of the op amp, and it's widely used in almost all occupations.

采用集成电路工艺制作的运算放大器，除保持了原有的很高增益和输入阻抗之外，还具有精巧、廉价和可灵活使用等优点，因而在有源滤波器、开关电容电路、数-模和模-数转换器、直流信号放大、波形的产生和变换，以及信号处理等方面，得到广泛的应用。

An operational amplifier which is made by using integrated circuit technology not only maintains the original high gain and input impedance, but also has many advantages of exquisite, low-cost and flexible, so it is widely used in active filters, switched capacitor circuits, Digital-analog and analog-digital converters, DC signal amplifications, wave generation and transformation, signal processing and so on.

直流放大电路在工业技术领域中，特别是在一些测量仪器和自动化控制系统中，应用非常广泛。如在一些自动控制系统中，首先要把被控制的非电量（如温度、转速、压力、流量和照度等）用传感器转换为电信号，再与给定量进行比较，得到一个微弱的偏差信号。因为这个微弱的偏差信号的幅度和功率均不足以推动显示仪表或者执行机构，所以需要把这个偏差信号放大到需要的程度，再去推动执行机构或送到仪表中进行显示，从而达到自动控制和测量的目的。因为被放大的信号多数是变化比较缓慢的直流信号，分析交流信号放大的放大器由于存在电容器这样的元件，不能有效地耦合这样的信号，所以也就不能实现对这类信号的放大。能够有效地放大缓慢变化直流信号的最常用器件就是运算放大器。运

第6章 比例放大器
Chapter 6 Proportional Amplifier

算放大器最早被发明作为模拟信号的运算（实现加、减、乘、除、比例、微分和积分等）单元，是模拟电子计算机的基本组成部件，由真空电子管组成。目前所用的运算放大器，是把多个晶体管组成的直接耦合的具有高放大倍数的电路，集成在一块微小的硅片上。

A DC amplifier circuit is widely used in the field of industrial technology, especially in measuring instruments and automatic control systems. For example, in some automatic control systems, non electrical quantity should be converted (such as temperature, speed, pressure, flow, illumination, etc.) to an electrical signal firstly, and then compared with a given quantity to get a weak deviation signal. The amplitude and the power of the weak deviation signal are not sufficient to drive the meter and the actuator, therefore, this deviation signal needs to be amplified to a required degree, and then to drive the actuator or to be displayed in the meter, so as to achieve the goals of automatic control and measurement. Because most of the amplified signal is the DC signal which is slowly changed, the amplifier which is used for analysis of AC signal amplification has capacitors, so it can't couple the signal effectively and amplify this kind of signal. An operational amplifier of the most commonly used can enlarge the slow varying DC signal effectively. The earliest operational amplifier was invented as an analog signal computing unit (Realize add, subtract, multiply, divide, ratio differential and ratio integration, etc.), which is a basic component of analog electronic computers. It is composed by vacuum tubes. Operational amplifiers we used are high magnification circuits which are composed of many transistors that directly coupled in a tiny silicon chip.

第一块集成运放电路是美国仙童（Fairchild）公司发明的 μA741，在20世纪60年代后期广泛流行。直到今天，μA741仍然是各大学电子工程系中讲解运放原理的典型案例。

The first integrated operational amplifier circuit is the μA741 which is invented by United States Fairchild. It is widely prevalent in the late 1960s. Even today μA741 is still a typical case to teach the principle of operational amplifier in department of electronic engineering in every university.

6.2 电源电路（Power circuit）

电源电路包括电源供给电路、滤波电路和稳压电路等。图6-2所示为电源电路组成部分。通常在比例放大电路的外部电路将220V交流电降压整流成24V直流电，然后接入比例放大器。滤波和稳压电路通常在比例放大电路的内部。电源电路的功能是为整个电气控制系统提供能源。

6.2 电源电路 (Power circuit)

A power circuit includes a power supply circuit, a filtering and voltage-stabilizing circuit, etc. Fig 6-2 shows the components of a power circuit. The 220V AC power is usually changed to 24 V DC power by step-down and rectifying at the external circuit of the proportional amplifier and sent to a proportional amplifier. The filtering and voltage-stabilizing circuit is usually included in the inner of the proportional amplifier. The function of a power circuit is supplying energy sources for the whole electrical control system.

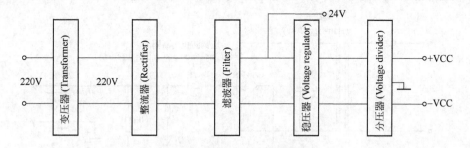

图 6-2 电源电路组成部分

(Fig. 6-2　Components of a power circuit)

电源电路包括两部分，就是在外的电源供给电路和在内的稳压电路。比例放大器外的供给电路可以采用 24V 的直流电源，但更常用交流 220V 电源，经整流、滤波后转换成 24V 的电源。在内的稳压电路的主要作用是分离出比例放大器正常工作中所需的各种直流稳压电源。对电源电路的要求是在外界电压变化、温度变化及负载变化的允许范围内，保证输出的直流电压稳定不变。并要求电源电路具有过载保护功能，在各种情况下仍能正常工作。

A power circuit includes two parts, namely a power supply circuit external and a voltage-stabilizing circuit internal. 24V dc power could be adopted as source of the supply circuit outside a proportional amplifier, but it is more common to use 220V ac power through rectification, filtering and then converted into 24V dc power to instead of it. The main function of the voltage-stabilizing circuit is to isolate all kinds of dc voltage-stabilized sources which are needed to make a proportional amplifier working well. A requirement of the power circuit is to ensure the output voltage DC constant when external voltage, temperature and load changed in permitted scope. Something which should not be ignored is that the power supply circuit must have overload protection function so that it can work regularly in all circumstances.

电源供给装置将交流电网电压（220V、50Hz）变成未经滤波的全波整流电压（24V），再输给比例控制放大器插板。比例控制放大器对输入电压进行滤波、稳压，并通过选择新的参考点，得到以新参考点为基准零点，能满足比例控制放大器中运算放大器等器件正常工作所需的稳定电压，如图 6-3 所示。

The power supply device change AC voltage (220V、50Hz) to full wave commutating voltage (24V) without filter, and then send the full wave commutating voltage to a proportional control amplifier. The proportional control amplifier will filter and stabilize the input voltage, then it will chose a new reference point as a datum mark, and obtain the stabilized voltage which can meet the requirement of the operational amplifier to work normally in the proportional control amplifier, as shown in Fig. 6-3.

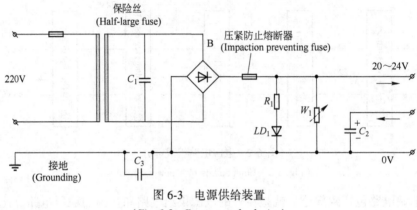

图 6-3 电源供给装置
(Fig. 6-3 Power supply device)

6.2.1 整流电路 (Rectification circuit)

利用二极管的单向导电性，将交变电压变换成单向脉动电压的电路，称为整流电路。在比例电控技术中广泛使用的是单相桥式整流电路。单相桥式整流电路及波形如图 6-4 所示。

A rectification circuit uses a unidirectional electrical conductivity of diode to transform alternating voltage into unidirectional pulse voltage. A single-phase bridge rectifier circuit is widely used in proportional electronic control technology. A single-phase bridge rectifier circuit and its waveforms are shown as Fig. 6-4.

6.2.2 滤波电路 (Filter circuit)

整流后输出的电压是脉动直流电压。其中交流分量较大，还不适宜于电子放大器的正常工作。因此需要有滤波电路把脉动直流电压变成平滑的电压。这种滤波电路实质上是一个低通滤波电路。

Rectified output voltage is pulsating DC voltage. It is not suitable for regular work of electronic amplifiers because of its large alternating current. Therefore we need a filter circuit to transform pulsating DC voltage into smooth voltage. This filter circuit is actually a LPF circuit.

6.2 电源电路 (Power circuit)

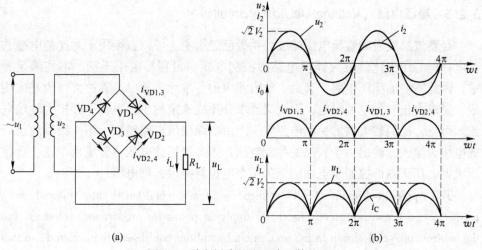

图 6-4 单向桥式整流电路及其波形

(Fig. 6-4 Single-phase bridge rectifier circuit and its waveform)
(a) 单向桥式整流电路 (Single-phase bridge rectifier circuit);
(b) 整流电路输出波形 (Rectifier circuit output waveform)

比例放大器中较常使用电容滤波电路，它是在整流电路的输出端安放一个较大的电容器与负载并联，利用电容的充放电作用，在整流电路导通时，电容充电；当整流电路截止时，电容对负载放电，使负载电压趋向平滑，如图 6-5 所示。

A capacitance filter circuit is commonly used in a proportional amplifier. It has a capacitor which is installed at the output terminal of rectification circuit and parallel-connected with load. The functions of capacitor, charging and discharging, are used when the rectifier circuit gets on and cuts off. When the rectifier circuit gets on, the capacitor will be charging. When the rectifier circuit cuts off, the capacitor will be discharging, so as to smooth load voltage, as shown in Fig. 6-5.

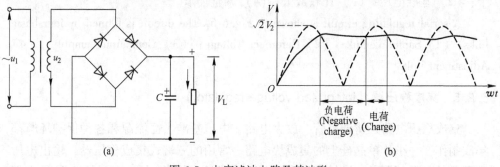

图 6-5 电容滤波电路及其波形

(Fig. 6-5 Capacitance filter circuit and its waveform)
(a) 电路 (Circuit); (b) 波形 (Waveform)

6.2.3 稳压电路 (Voltage stabilizing circuit)

经整流滤波后的直流电压的波纹系数已经很小，可以直接用于对比例电磁铁供电；但若用于比例放大器内电路的控制电压，很容易由于干扰，不能满足要求，因此需增加稳压电路。又由于放大器内的各单元多数以运算放大器为基础构成，要求由正、负电源供电，因此又要求分压成稳定的正负电源，并扩大输出电流。可见，比例放大器的稳压电路实质上由二级稳压电路组成：第一级以滤波后的电压为输入，输出一个较高的稳定电压；第二级把此电压再分成两个正、负稳压电源，用于内部控制供电和向外部指令电位器提供标准电压。

The ripple coefficient of DC voltage is already very small after filtered by a rectifier, and it can be directly used for supplying power for proportional solenoid, but the control voltage of circuit in the proportional amplifier can be easily interfered, so that it can't meet the requirements. Therefore, it is required to add a voltage stabilizing circuit. The majority of each unit in the amplifier is based on the operational amplifier, and it is required by the positive and negative power. So it needs to be divided into two stable powers which are positive and negative, and enlarge the output current. It is shown that the voltage stabilizing circuit of a proportional amplifier is composed of two stages: the first stage takes filtered voltage as an input, and outputs higher stabilized voltage; the second stage divides this voltage into positive and negative voltage-stabilized sources to supply power for internal control circuit and supply standard voltage for external command potentiometer.

6.2.4 串联型稳压电路 (Serial regulating circuit)

串联型稳压电路如图6-6所示，整个电路由四个基本环节组成：(1) 取样环节；(2) 基准电压；(3) 比较放大；(4) 调整环节。

A serial regulating circuit is shown in Fig. 6-6. The circuit is formed by four basic links: (1) Sampling link; (2) Reference voltage; (3) Comparator amplify; (4) Adjustment link.

6.2.5 集成稳压器 (Integrated voltage regulator)

集成稳压电路是将调整管、放大电路、电压基准、恒流源和各种保护环节都集成制作在一小片单晶硅上的集成块电路。常用的三端式集成稳压器，输出电压有可调的和不可调的，有输出正电压的和输出负电压的（图6-7，图6-8）。它们的管脚排列及接线如图6-9所示。将它们组合起来，可以形成双电源稳压电路。

An integrated voltage-stabilizing circuit is a circuit where an adjustment tube, an

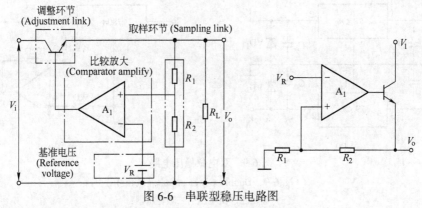

图 6-6 串联型稳压电路图

(Fig. 6-6 Serial regulating circuit)

amplification circuit, a voltage reference, a constant current source and various protection links are integrated in a small piece of silicon. The three terminals integrated voltage regulator is commonly used. Its output voltage can be adjustable or not adjustable, positive or negative (Fig. 6-7, Fig. 6-8). Its pin configuration and wiring are shown in Fig. 6-9. They are combined to form a double power supply voltage circuit.

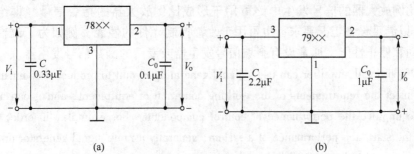

图 6-7 输出电压不可调节的三端集成稳压器接线图

(Fig. 6-7 Wiring diagram of three-terminal integrated voltage regulator whose output voltage can't be adjusted)

(a) 输出正电压 (Output positive voltage); (b) 输出负电压 (Output negative voltage)

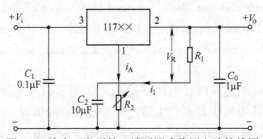

图 6-8 输出正电压的三端可调式稳压电路接线图

(Fig. 6-8 Wiring diagram of three-terminal voltage adjustable regulator which is used to output positive voltage)

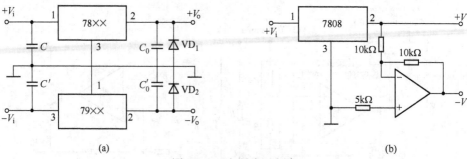

图 6-9 双电源稳压电路

(Fig. 6-9 Double power supply circuit)

(a) 输出正、负电压的三端稳压器；

(Three-terminal regulator which is used to output positive and negative voltage);

(b) 输出负电压的三端稳压器

(Three-terminal regulator which is used to output negative voltage)

6.3 控制信号发生电路（Control signal generation circuit）

比例放大器的信号发生电路可用于形成比例放大器的指令信号（即给定信号），以满足设备工况要求；也可用于改善控制元件的性能，例如为了改善系统的启动和停止性能，通常设有斜坡信号发生器或专门的函数信号发生器。

Proportional amplifier can use a signal generating circuit to produce a command signal to meet the requirements of the working condition of equipment, and it can also be used to improve the performance of control components. For example, in order to improve the start-stop performance of a system, generally a ramp signal generator or a special function signal generator is installed in the system.

本质上，参考电压作为输入控制信号提供给比例放大器。对于开环系统，它就是实际的控制输入；对于闭环系统，它还要与反馈信号比较，得出的偏差值才是真正的控制信号。

Essentially, a reference voltage is provided to a proportional amplifier as an input control signal. For an open-loop system, it is an actual control input. But for a closed-loop system, it needs to be compared with a feedback signal, and the obtained deviation is the actual control signal.

在控制连续变化的过程时，有时需要利用周期函数发生器来产生控制信号，有时要利用非周期函数发生器来产生特定的非线性控制信号。

In the process of continuous change control, sometimes a periodic function generator is needed to generate control signal, or an aperiodic function generator is used to produce a specific nonlinear control signal.

6.3 控制信号发生电路 (Control signal generation circuit)

为了改善控制性能，适应不同的对象和工况，比例放大器中还设有多种信号处理和发生电路。例如为了改善迟滞特性，通常设有颤振信号发生器等。下面介绍常用的信号发生与处理电路。

In order to improve control performance to adapt to different objects and working conditions, the proportional amplifier also has a variety of signals occurring and processing circuits. For example, in order to improve the hysteresis characteristics, generally a flutter signal generator is installed. Some commonly used signal occurrence and processing circuits will be introduced in this chapter.

6.3.1 周期信号发生器 (Periodic signal generator)

6.3.1.1 斜坡信号发生器 (Slope signal generator)

斜坡信号发生器用于控制信号的上升变化速度和下降变化速度，当输入阶跃信号时，系统能够以可调的速率无冲击地到达给定值要求，从而获得平稳而迅速的启动、转换或停止，进而提高生产率（图 6-10）。

A slope signal generator is used to control the speed of rise change and decline change. When a step signal is input, it can reach given value at adjustable speed without impact, so as to obtain a smooth and rapid start, transmission or stop, and improve productivity（图 6-10）.

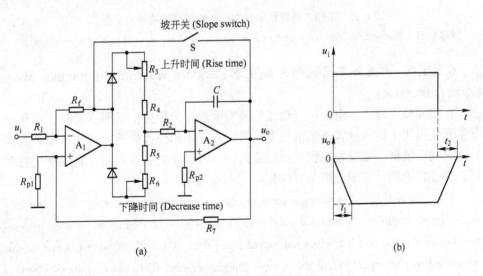

图 6-10 斜坡信号发生器电路

(Fig. 6-10 Slope signal generation circuit)

(a) 斜坡信号发生电路 (Slope signals generation circuit); (b) 斜坡信号发生电路的输入输出特性 (Input/output characteristics of slope signals occurred circuit)

6.3.1.2 正弦波信号发生器(Sine-wave signal generator)

正弦波信号发生器常用于产生颤振信号，以及用于差动变压器式位移传感器的测量放大电路。图6-11为采用二极管稳幅的文氏电桥正弦波发生器。

A sine wave signal generator is often used to generate a chatter signal, and it is also used as a measure amplifier circuit of a differential transformer displacement transducer. The Fig. 6-11 shows the Wien-bridge sine-wave generator in which the diode is used to stable amplitude.

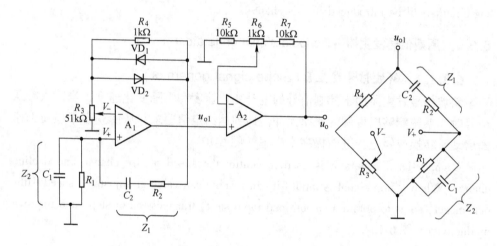

图 6-11 采用二极管稳幅的文氏电桥正弦波信号发生器
(Fig. 6-11 Wien-bridge sine-wave generator in which the diode is used to steady amplitude)

6.3.1.3 方波和三角波信号发生器(Square wave and triangular wave signal generator)

在比例放大器中，也常用三角波信号发生器作为颤振信号源，此外，三角波发生器还可用于组成脉宽调制信号发生器。在反相积分器的输入端加一串方波信号，求输出波形。如果是在积分器的输出端，可以获得一连串正、负斜波交替电压，即三角波形，它的周期与方波相同（图6-12）。

It is common to use a triangle wave signal generator as a chatter signal source in a proportional amplifier. In addition, the triangle wave generator can also be used in composition of a pulse width modulation signal generator. A bunch of square wave signals are added in the input end of an inverse integrator, and then the output waveform is found. A series of positive and negative alternating slope voltage can be got in the output end of an integrator, which is called triangle wave. Its cycle is the same as the cycle of square wave (Fig. 6-12).

6.3 控制信号发生电路 (Control signal generation circuit)

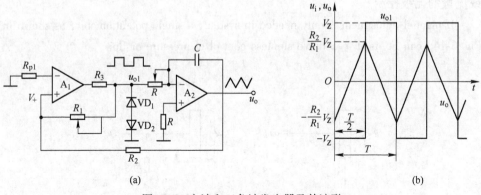

图 6-12 方波和三角波发生器及其波形

(Fig. 6-12 Square wave and triangular wave signal generator and its waveform)

(a) 电路 (Circuit); (b) 波形 (Waveform)

6.3.1.4 脉宽调制 (PWM) 信号发生电路 (Pulse-width modulation (PWM) signal occurring circuit)

采用 PWM 信号控制比例电磁铁有很多优点,阀芯的运动响应取决于 PWM 信号的平均值,使阀芯工作时处于微振动状态,可大大减小比例阀的滞环。另外,PWM 信号的电流放大电路为开关式功放电路,放大器只工作在导通和截止状态,节能效果好(图 6-13)。

There are many advantages of using a PWM signal control proportional solenoid. The movement of the valve core responds to the average of a PWM signal, so as to make the valve core work in micro vibration condition, and proportional valve hysteresis is decreased. In addition, the current amplifier of a PWM signal is switch type power amplifier circuit, which can only work in the condition of get-on and cut-off, so its energy saving effect is good (Fig. 6-13).

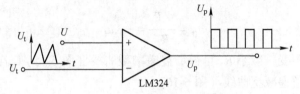

图 6-13 脉宽调制器原理图

(Fig. 6-13 Pulse-width modulation elementary diagram)

6.3.2 非周期信号发生器 (Nonperiodic signal generator)

6.3.2.1 手动电位器信号发生器 (Manual potentiometer signal generator)

如果系统仅需手动调节,就可采用如图 6-14 所示的单电位器来实现对压力或流量的无级控制。

If manual adjustment is only needed in system, a single potentiometer, as shown in Fig. 6-14, can be used to achieve stepless control of pressure or flow.

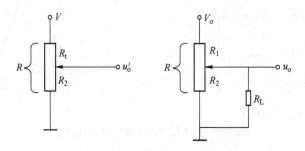

图 6-14 单电位器无级控制电路

(Fig. 6-14 Stepless control circuit with single potentiometer)

6.3.2.2 增益型函数发生器(Gain function generator)

当需要产生某种形状的曲线函数时,可采用分段线性化的办法来近似代替曲线。这样因为每一段是线性关系,从而可以用运放比例放大器来实现。各段之间的转换可采用稳压管等非线性元件转换。

When a function with a certain shape needs to be produced, a common method is piecewise linearization to replace the curve approximately. There is a linear relationship for each paragraph, thus we can use an op-amp proportional amplifier to achieve it. A stabilivolt and any other non linear elements can be used for conversion between each paragraph.

由反相比例放大的特性(According to inverse proportion magnified characteristics):

$$u_0 = \frac{-R_f}{R_1} = -Ku_i \tag{6-1}$$

若保持 R_f 不变,减小 R_1 可得增益递增型曲线(图6-15)。若保持 R_1 不变,减小 R_f 可得增益递减型曲线(图6-16)。

A curve of increasing gain can be obtained by fixing R_f and decreasing R_1 (Fig. 6-15), and a curve of decreasing gain can be obtained by fixing R_1 and decreasing R_f (Fig. 6-16).

支路愈多、分段愈细,折线与曲线就更接近,模拟就更准确。

If there are more branches, subsections are more careful, the broken line will get closer to the curve, and the simulation will be more accurate.

6.4 比例放大器电路 (Proportional amplifier circuit)

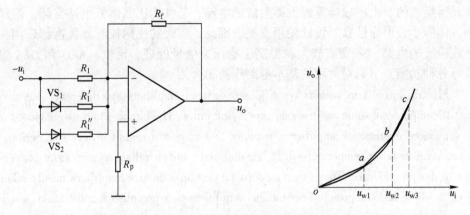

图 6-15 增益递增型函数发生器
(Fig. 6-15 Increase progressively gain function generator)

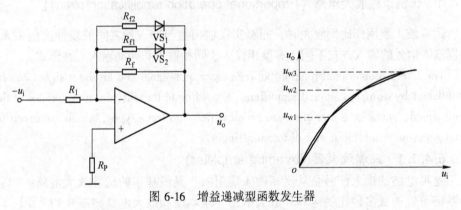

图 6-16 增益递减型函数发生器
(Fig. 6-16 Decrease progressively gain function generator)

6.4 比例放大器电路 (Proportional amplifier circuit)

比例放大器是一种专用的电子装置,用来对比例阀的控制电磁铁提供特定波形的控制电流,并对整个比例阀或系统进行开环或闭环控制。所以,比例放大器应具有相应的功能,以满足生产上的需要。

A proportional amplifier is a dedicated electronic device which is used to provide control current with specific waveform to the control solenoid of proportional valves and to control proportional valves or the whole system by way of open loop or closed-loop control. Therefore, the proportional amplifier should have corresponding function to meet the needs of production.

各种类型比例放大器的内部电路不尽相同,但它们都是由一些基本信号处理单元所组成的。由集成运算放大器组成的电路，其性能比晶体管电路优越，温漂小，体积小，可靠性高，设计使用方便。因此，现代的比例控制放大器都采用由运放组成的电路，来完成放大、振荡、稳压、信号处理、比较、A/D 和 D/A 变换等各种功能。下面对常用的基本电路作简单介绍。

Electric circuits in various types of proportional amplifiers are different, but they are all composed of some basic signal processing units. The circuit which is composed of an integrated operational amplifier is superior to a transistor circuit in terms of performance, such as lower temperature-drift, smaller size, higher reliability and more convenient to design and use. Therefore, modern proportional control amplifiers mostly adopt circuits which are composed of operational amplifiers to accomplish amplification, oscillation, voltage stabilization, signal process, comparison, A/D and D/A conversion and so on. A brief introduction of some basic circuits used in common is as follows.

6.4.1　比例运算放大电路 (Proportional operation amplification circuit)

运算放大器用作比例放大时，能够实现输出量与输入量之间的线性比例关系。根据输入信号的输入方式不同，有反相输入、同相输入和差动输入三种形式。

The linear proportional relation between an output signal and an input signal can be established by using operational amplifiers. According to the different input modes of the input signal, operational amplifier can be classified into three types, which are inverting input, non-inverting input and differential input.

6.4.1.1　反相放大器 (Inverting amplifier)

这种电路的输入信号是从反相输入端引入，是最基本的运算放大电路，以它为基础可以构成多种比例运算和放大电路。基本反相放大电路如图 6-17 所示。

An inverting amplifying circuit whose input signals are imported from the inverting input end is the most basic operational amplifier circuit. A variety of scaling operation and amplification circuits can be composed on the basis of it. A basic reverse amplification circuit is shown as Fig. 6-17.

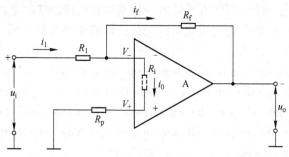

图 6-17　基本反相放大电路

(Fig. 6-17　Basic reverse amplification circuit)

6.4 比例放大器电路 (Proportional amplifier circuit)

电路输出电压为 (The output voltage of circuit is):

$$u_o = \frac{-R_f}{R_1} u_i \tag{6-2}$$

式中的负号表示输出电压与输入电压反向。由式可见，反相放大时，输出电压与输入电压之间的关系决定于比值 R_f/R_1，与放大器的放大倍数无关。

Minus means the output voltage and the input voltage are reverse. It can be seen that the relationship between output voltage and input voltage is decided by R_f/R_1, when using inverting amplification. And it has nothing to do with the amplification factor.

在图 6-17 中，如果 R_f/R_1，则电路输出电压为 (In Fig. 6-17, if $R_f = R_1$, then the output voltage of circuit is):

$$u_o = -u_i \tag{6-3}$$

在这种情况下，运算放大器作变号运算，称为反相器。反相器在三位四通比例方向阀的双通道电控器中得到应用。

Under the circumstance, the value will be reversed during the operation of operational amplifier, so we usually call an operational amplifier as an inverter. An inverter is used in the dual-channel electric controller of a three-position and four-way directional control valve.

在反相放大器的输入端安排多条支路并接，在输出端就可实现多路信号线性迭加，即反相加法器，如图 6-18 所示。

If branches are parallel connected at the input end of an inverting amplifier, a multiple signal linear superposition can be realized at the output end. At this time, the inverting amplifier is used as an inverting summator, which can be seen in Fig. 6-18.

如果 $R_1 = R_2 = R_3$，各路信号按相同比例迭加，如某一路输入信号为反相，则对该信号作减法运算。反相加法器的优点是，每增加一路信号输入，只需增加一个电阻，而且电阻的数值很容易确定。由于存在"虚地"，各路输入信号之间互不影响。

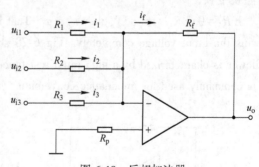

图 6-18 反相加法器
(Fig. 6-18 Inverting summator)

If $R_1 = R_2 = R_3$, multiple signals are added with same proportion, and are subtracted when one input signal is reversed. The advantage of an inverting summator is that only one resistance needs to be added when an input signal added, and the added numerical value is easy to determine. Because of the dummy ground, the in-

Chapter 6　Proportional Amplifier

put signals are not affected each other.

6.4.1.2　同相放大器(Non-inverting amplifier)

同相放大器的电路如图 6-19 所示。输入信号 u_i 从同相输入端引入，R_p 为平衡电阻，反相输入端经 R_1 接地。

A circuit of a non-inverting amplifier is shown as Fig. 6-19. An input signal u_i is imported from the non-inverting input end. R_p is a balance resistance, the inverting input is connected with the ground through R_1.

输出电压为 (The output voltage is):

$$u_0 = \left(1 + \frac{R_f}{R_1}\right) u_i \qquad (6-4)$$

故输出电压与输入电压成比例，比例系数为大于等于 1 的正数。同相放大器的优点是输入电阻很大，一般可达 10MΩ 以上，所以常用作电压跟随器。

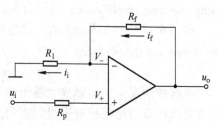

图 6-19　同相放大器
(Fig. 6-19　Non-inverting amplifier)

So the output voltage is proportional to the input voltage. A scale coefficient is a positive number equal to or bigger than 1. The advantage of a non-inverting amplifier is that its input resistance is quite large, which is generally more than 10MΩ, so it is often used as a voltage follower.

如果 $R_f = 0$ 或 $R_1 = \infty$(断开)，则 $u_o = u_i$。即输出电压完全跟随输入电压变化。图 6-20 为电压跟随器。其特点是输入电阻值 R_i 高，输出电阻值 R_o 低，常用于阻抗变换。

If $R_f = 0$ or $R_1 = \infty$(off), then $u_o = u_i$. That is to say the output voltage changes following the input voltage completely. Fig. 6-20 shows the voltage follower. The voltage follower is characterized by a high input resistance R_i and a low output resistance R_o. So it is commonly used for impedance conversion.

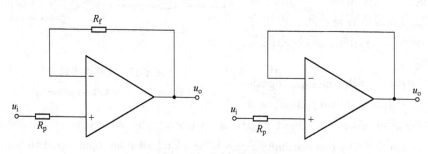

图 6-20　电压跟随器
(Fig. 6-20　The voltage follower)

6.4.1.3 差动放大器(Differential amplifier)

差动放大器两个输入端都有输入信号,如图 6-21 所示。

An input differential amplifier has two input signals, as is shown in Fig. 6-21.

设同相输入端的信号 $u_{i2} = 0$,反向输入端的输入为 u_{i1},对应的输出信号为 u_{o1},可得 (If the non-inverting input signal is $u_{i2} = 0$, the inverting input signal is u_{i1}, and the corresponding output signal is u_{o1}, so):

$$u_{o1} = -\frac{R_f}{R_1}u_{i1} \quad (6\text{-}5)$$

设 $u_{i1} = 0$,同向输入端的信号为 u_{i2},可得 (If $u_{i1} = 0$, and the non-inverting input signal is u_{i2}, so):

$$u_{o2} = \left(1 + \frac{R_f}{R_1}\right)\left(\frac{R_3}{R_2 + R_3}\right)u_{i2} \quad (6\text{-}6)$$

式中,$\left(\dfrac{R_3}{R_2 + R_3}\right)u_{i2}$ 是 R_3 和 R_2 把 u_{i2} 分压后在同向输入端得到的信号。

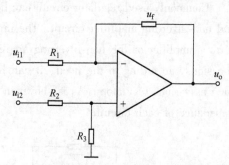

图 6-21 差动运算放大电路
(Fig. 6-21 Differential operational amplifier)

The formula $\left(\dfrac{R_3}{R_2 + R_3}\right)u_{i2}$ represents the input signal which is achieved after u_{i2} is divided by R_3 and R_2.

根据叠加原理,由以上两式可得 (According to the principle of superposition and the above two formulas, we can obtain that):

$$u_o = u_{o1} + u_{o2} = -\frac{R_f}{R_1}u_{i1} + \left(1 + \frac{R_f}{R_1}\right)\left(\frac{R_3}{R_2 + R_3}\right)u_{i2} \quad (6\text{-}7)$$

When $R_1 // R_f = R_1 // R_3$, or $R_2 = R_1$, $R_3 = R_f$

$$u_o = \frac{R_f}{R_1}(u_{i2} - u_{i1}) = K_f \Delta u_1 \quad (6\text{-}8)$$

称 $\Delta u_i = u_{i2} - u_{i1}$ 为差模,称 $K_f = \dfrac{R_f}{R_1}$ 为差模放大系数。当 $K_f = 1$ 时,可构成减法器。

$\Delta u_i = u_{i2} - u_{i1}$ is called differential mode. $K_f = \dfrac{R_f}{R_1}$ is called the magnification coefficient of differential mode. When $K_f = 1$, a differential amplifier can be used as a subtractor.

6.4.2 调节器电路 (Regulator circuit)

常用的调节器电路可称为一种广义反相型和同相型放大电路。反相放大器和同相放大器的输入回路和反馈回路均为纯电阻。在有些场合，根据需要，它们可以是电抗元件或阻抗网络。例如各种调节器就是这种类型的电路。如图 6-22 所示，用 Z_1、Z_f 和 Z_p 代表各回路的总阻抗。

Commonly used regulator circuit can be described as a kind of generalized inverting and non-inverting amplifier circuit. The input circuit and the feedback circuit of the inverting amplifier or the non-inverting amplifier are all pure resistance circuit. In some occasions, according to the need, it can be reactance element or impedance network, such as various regulators. As is shown in Fig. 6-22, Z_1, Z_f and Z_p represent the total impedance of each circuit.

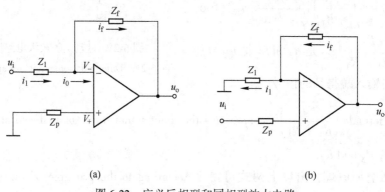

图 6-22 广义反相型和同相型放大电路

(Fig. 6-22 The generalized inverting and non-inverting amplifier)

(a) 广义反相型电路 (Generalized inverting circuit); (b) 广义同相型电路 (Generalized non-inverting circuit)

比例放大器中常采用各种调节器，用以优化各类阀的性能。主要任务是当给定值发生变化时，调节控制对象的实际值与新的设定值相对应；另一主要任务是消除或减小干扰量对调节量的影响。常用的调节器有比例调节器（P 调节器）、积分调节器（I 调节器）、微分调节器（D 调节器）、PI 调节器、PD 调节器和 PID 调节器。

A variety of regulators are often used in proportional amplifiers to optimize all kinds of valve performance. The main task is adjusting the object to correspond with a new changed given value, and another task is eliminating or reducing the effects of disturbance variable working on regulating variable. Commonly used regulators are proportional regulators (P regulator), integral regulators (I regulator), differential regulators (D regulator), PI regulators, PD regulators and PID regulators.

6.4.2.1 比例调节器（P）(Proportional regulator (P))

比例调节器电路与反相放大器的电路完全相同，输入量与输出量的关系是：
A proportional regulator circuit is completely the same with an inverting amplifier circuit. The relation between input and output quantity is：

$$u_o = \frac{-R_f}{R_1}u_i \tag{6-2}$$

令 $K_p = \dfrac{R_f}{R_1}$ 即得比例调节器的比例系数。

Order $K_p = \dfrac{R_f}{R_1}$, namely the proportionality coefficient of a proportional regulator.

P 调节器的优点是结构简单、容易调整和响应速度快，一旦出现调节偏差，输出量立即按比例产生变化。但 P 调节器工作时需要一个调节偏差，比例系数 K_p 越大，调节偏差越小。P 调节器可作为闭环控制系统中开环增益的调节环节。

P regulator has such advantages as simple structure, adjusting easily and fast response. Once a deviation of control appears, the output will change immediately in a certain proportion, but the P regulator needs a control deviation during the working time. The greater the proportion coefficient K_p is, the smaller the control deviation becomes. The P regulator can be used as an adjustment part of open-loop gain in the closed-loop control system.

6.4.2.2 积分调节器（Ⅰ）(Integral regulator (Ⅰ))

在 P 调节器的反馈电路中，用电容代替反馈电阻，就可获得积分调节器。积分调节器的开环传递函数是（An integral regulator can be acquired by way of using a capacitance instead of a feedback resistance in the feedback of a P regulator. The open-loop transfer function of an integral regulator is）：

$$W(s) = \frac{U_o(s)}{U_i(s)} = -\frac{Z_f}{Z_i} = -\frac{1}{TS} \tag{6-9}$$

式中　$T = R_1C$ ——积分时间常数（Integration time constant）。

将上式拉氏反变换，得（Through Laplace inverse transform）

$$u_o(t) = -\frac{1}{T}\int u_i(t)\,dt \tag{6-10}$$

设 $u_i(t)$ 为阶跃函数，$u_i(t) = A(t)$，若初始值为零，则输出电压为（Set up $u_i(t)$ as a step function, $u_i(t) = A(t)$, if the initial value is zero, then the output voltage is）：

$$u_o = -\frac{A}{T}t \tag{6-11}$$

由式（6-11）可知，$u_o(t)$ 将以斜率 $-\dfrac{A}{T}$ 线性递增，其过渡过程如图 6-23b 所示。

According to formula 6-11, $u_o(t)$ will linearlly increase at a slope $-\dfrac{A}{T}$, whose transition process is shown in Fig. 6-23b.

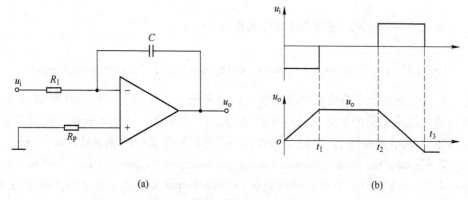

图 6-23 积分调节器及阶跃响应特性

(Fig. 6-23 The integral regulator and its step response characteristics)

(a) 积分调节器 (Integral controller); (b) 阶跃响应特性 (Step response characteristic)

积分调节器具有延缓、积累和记忆功能。所谓延缓功能，是指输入变化很快，但输出信号只能渐变。这一特性可用作斜坡上升、下降的调节电路。所谓积累功能，是指只要输入存在，积累就不停止。这一特性可用来消除系统误差。所谓记忆功能，是指输入突然消失后，其输出保持原值不变。

An integral regulator has functions of delay, accumulation and memory. The so-called delay function is that an output signal can only gradually changing in spite of the fast change of the input signal. This feature can be used as a ramp up-down regulating circuit. The so-called accumulation function is that the accumulation will not stop as long as the input exists. This feature can be used to eliminate systematic errors. The so-called memory function means that the input suddenly disappears, but the output keeps original value and remains the same.

6.4.2.3 微分调节器 (D) (Differential regulator (D))

微分是积分的逆运算，将积分电路中 R_1 和 C 对调，就构成纯微分调节器。但纯微分调节器在固有频率附近会出现谐振，使系统不稳定。为此，在输入端串联一个电阻 R_1（形成小阻尼），以提高稳定性。采取补偿措施后的微分调节器电路如图 6-24 所示。

6.4 比例放大器电路 (Proportional amplifier circuit)

It is known to all that differential is inverse of integral. A pure differential regulator can be achieved through putting R_1 and C reversed in an integral circuit. The pure differential regulator may appear resonance near the natural frequency, so that make the system unstable. For this reason, a resistance R_1 needs to be connected in series in the input end to promote stability. The differential regulator circuit after taking indemnifying measures is shown in Fig. 6-24.

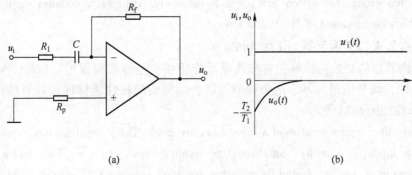

(a) (b)

图 6-24 微分调节器及其单位阶跃响应

(Fig. 6-24 The differential regulator circuit and its unit step response)
(a) 微分调节器 (The differential regulator); (b) 单位阶跃响应 (The unit step response)

微分调节器的传递函数为 (The differential regulator transfer function is):

$$W(s) = \frac{U_o(s)}{U_i(s)} = -\frac{z_f(s)}{Z_i(s)} = -\frac{T_2 s}{1 + T_1 S} \tag{6-12}$$

式中 $T_1 = R_1 C$ ——时间常数 (Time constant);

$T_2 = R_f C$ ——时间常数 (Time constant)。

设输入为单位阶跃信号, 即 $u_i(t) = l(t)$, 初值为零, 输出电压的时间响应函数为 (Supposing a unit step signal as the input signal, that means $u_i(t) = l(t)$, whose initial value is zero. The time response function of the output voltage is):

$$u_o(t) = \frac{-T_2}{T_1} e^{\frac{t}{T_1}} u_i(t) \tag{6-13}$$

其过渡特性如图 6-24b 所示。显然, D 调节器只会对变化着的信号产生响应, 对固定不变的输入不会有微分作用。所以微分作用是高通电路, 对高频干扰和噪声反应很灵敏, 容易产生自激振荡, 使用时应增加 R_1 以增大阻尼, 限制高频增益。在信号变化快时, 微分作用给出较大的控制量, 这有利于改善动态偏差和减少过渡时间;但微分作用不能克服静误差, 且微分作用过强时, 系统不易稳定。通常由 D 与 P、I 调节器构成复合调节器使用。

The transition characteristics are shown in Fig. 6-24b. Apparently, the D regulator will only respond to changing signals, but to a fixed input will not have differential effects. Therefore, the differential effect circuit is a high-pass circuit. It is sensitive to high-frequency interference and noise response, and it can produce self-oscillation easily. So R_1 should be increased to magnify damping to limit high-frequency gain. The differential action will produce a larger control variable which is helpful to improve dynamic deviation and reduce the time of transition when the signal is rapidly changing. However, differential effects cannot overcome static error, and when the differential effect is too strong, the system can not stabilize easily. Usually, a complex regulator is used which is composed of D, P and I regulator.

6.4.2.4　PI 调节器(PI regulator)

P 调节器的初始特性好，对输入量反应很快，但有残留误差；I 调节器的完成特性好，调节偏差为零。两个调节器联合所得到的特性能够保留两者的优点。PI 调节器如图 6-25 所示。

The initial characteristics of a P regulator are good. The P regulator has a quick response to an input quantity, but it has the residual error. An I regulator has a good characteristic of completion and its adjusting deviation is zero. A PI regulator which is a combination of the two regulators has both characteristics. The regulator is shown in Fig. 6-25.

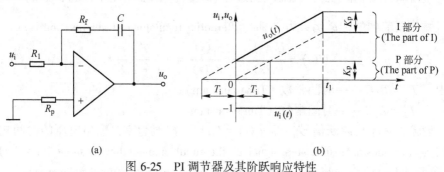

(a)　　　　　　　　　　　　　　(b)

图 6-25　PI 调节器及其阶跃响应特性

(Fig. 6-25　PI regulator and its step response characteristics)

(a) PI 调节器 (PI regulator); (b) 阶跃响应特性 (Step response characteristics)

PI 调节器的传递函数为 (The PI regulator's transfer function is):

$$W(s) = -\frac{U_0(s)}{U_i(s)} = -\frac{z_2(s)}{Z_1(s)} = -\left(K_p + \frac{1}{T_i s}\right) \quad (6\text{-}14)$$

式中　$K_p = \dfrac{R_f}{R_1}$——比例系数 (Proportional coefficient);

　　　$T_i = R_f C$——积分时间常数，也是比例环节的超前时间 (Integral time constant, which is also leading time of the proportional element)。

6.4 比例放大器电路 (Proportional amplifier circuit)

PI 调节器的特征常数为比例系数 K_p 及比例超前时间 T_i ,其时域响应特性曲线如图 6-25b 所示,时域响应方程为:

The characteristic constants of the PI regulator are a proportional coefficient K_p and a proportional leading time T_i. Its characteristic curve of time response is shown in Fig. 6-25b. The time response equation is:

$$-u_o(t) = K_p u_i(t) + \frac{1}{T_i}\int u_i(t)\,dt + u_{i0} \tag{6-15}$$

式中 u_{i0} —— $u_i(t)$ 的初值 (The initial value of $u_i(t)$)。

PI 调节器的性能与积分器相似,只是它开始工作的时间相当于提前了超前时间 T_n。

The PI regulator is similar to integrator in terms of performance, only the beginning is advanced for a leading time T_n.

6.4.2.5 PD 调节器 (PD regulator)

PD 调节器如图 6-26 所示 (A PD regulator is shown in Fig. 6-26)。

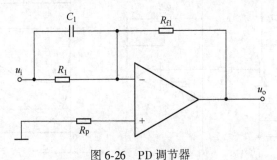

图 6-26 PD 调节器

(Fig. 6-26 The PD regulator)

令 (Set) $Z_1 = \dfrac{R_1}{(R_1 C_1 s + 1)}$, $Z_f = R_2$,则 PD 调节器的传递函数为 (The PD regulator transform function is):

$$W(s) = -\frac{U_o(s)}{U_i(s)} = -\frac{z_f(s)}{Z_i(s)} = K_p(T_d s + 1) \tag{6-16}$$

式中 $K_p = \dfrac{R_f}{R_1}$ ——PD 调节器的放大系数 (The proportional coefficient of the PD regulator);

$T_d = R_1 C_1$ ——D 调节器的时间常数 (The time constant of the D regulator)。

PD 调节器对快速变化的输入信号作出超前响应,因而可以加速调节过程。但由于没有积分 I 作用,仍存在一个静态偏差。

The PD regulator can make advanced response to a quickly changed input signal, so it can speed up the regulating process. Because the PD regulator lacks of the function of integral I, it still has a static deviation.

6.4.2.6 PID 调节器(PID regulator)

PID 调节器性能良好，使用广泛。将其放大系数设计成可调的，就可以适用于各种调节对象。构成 PID 调节器的方案很多，图 6-27 是比例放大器中广泛采用的一种方案。

A PID regulator has good performance, so it is widely used. A PID regulator can be used in all kinds of controlled plant, if its magnification coefficient is designed adjustable. There are many schemes of a PID regulator constitution. One of them which is widely used in a proportional amplifier is shown in Fig. 6-27.

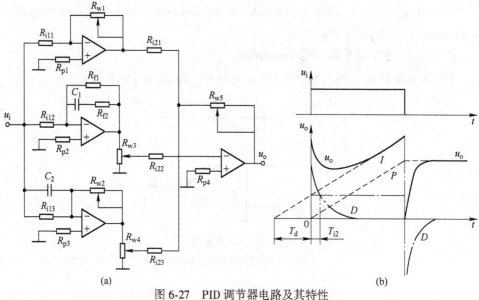

图 6-27 PID 调节器电路及其特性

(Fig. 6-27 The PID regulator circuit and its characteristics)

(a) 电路 (Circuit); (b) 特性 (Characteristics)

该方案的特点是可根据需要方便地取舍 P、I、D 中的任一个环节，每一个环节的时间常数都独立可调，互不影响。该方案通过加法电路将独立的 P、I、D 电路叠加在一起。

The characteristic of this scheme is that we can accept or reject any element of P, I and D conveniently as needed. The time constant of each element is adjustable independently and without interference. The P, I and D circuit are added together by using added circuit shown in this scheme.

令 (Set) $R_{w5}/R_{i21} = 1$, $R_{i21} = R_{i22} = R_{i23}$，PID 控制器的传递函数为 (The PID regulator's transform function is):

6.4 比例放大器电路 (Proportional amplifier circuit)

$$W(s) = \frac{U_o(s)}{U_i(s)} = K_p + K_i \frac{1+T_1 s}{1+T_2 s} + K_d(T_d s + 1) \qquad (6\text{-}17)$$

式中 K_p ——比例环节的比例放大系数（The proportion amplification coefficient of the proportional link）, $K_p = R_{w1}/R_{i11}$;

K_i ——积分环节的比例放大系数（The proportion amplification coefficient of the integral link）, $K_i = R_{f1}/R_{i12}$;

T_{i1} ——积分环节中惯性环节的时间常数（The time constant of the inertia element in integral link）, $T_{i1} = R_{12}C$;

T_{i2} ——积分环节的时间常数（The time constant of the integral link）, $T_{i2} = (R_{f1} + R_{f2})C_1$, 注意到 $T_{i2} > T_{i1}$, 故 $K_i \dfrac{1+T_{i1}s}{1+T_{i2}s}$ 相当于积分环节（considering $T_{i2} > T_{i1}$, so $K_i \dfrac{1+T_{i1}s}{1+T_{i2}s}$ is equivalent to the integral link）;

K_d ——微分环节的比例放大系数（The proportion amplification coefficient of the differential link）, $K_d = R_{w2}/R_{i13}$;

T_d ——微分环节的时间常数（The time constant of the proportional link）, $T_d = R_{i13}C_2$ 。

式中的各项系数是根据系统校正的需要配置的。

Every coefficient in the formula is configured according to the requirement of the system correction.

6.4.3 功率放大级 (Power amplifier stage)

功率放大级是比例控制放大器的核心单元。它的作用是对各种控制信号进行综合和放大，向电-机械转换装置提供足够大的驱动电流。对于比例电磁铁，通常所需的驱动电流为 0.8~1.2A。此外，比例控制放大器还需要有良好的稳态和动态特性。为此，功率级的输入信号除给定控制信号（经斜坡函数发生器处理）外，通常还包括多种其他信号：颤振信号、初始电流设定信号、负载电流反馈信号和检测值反馈等信号。

A power amplifier stage is the core unit of a proportional control amplifier. Its function is synthesizing and amplifying various kinds of control signals, and providing enough current to drive an electro-mechanical conversion device. To a proportional solenoid, usually the required drive current is about 0.8~1.2A. In addition, the proportional control amplifier must have good stability and dynamic performance. To this end, the input signal of power level except for a given control signal (through slope function generator processing) usually includes a variety of any other signals which are flutter signal, initial current setting signal, load current feedback signal, feedback signal of detection value and so on.

根据控制信号综合后的处理方式不同，功率放大级有模拟式和脉宽调制（PWM）式两种。

According to the ways which are used for the comprehensive treatment of the control signal, the amplifier stage can be classified into two types-analog type and pulse width modulation (PWM) type.

6.4.3.1 模拟式功率放大电路(Analog power amplifier circuit)

模拟式功率放大器将叠加后的各种信号直接传送给功放管，如图 6-28 所示。

An analog power amplifier will transmit all kinds of stacked signals to the power amplifier tube directly, as is shown in Fig. 6-28.

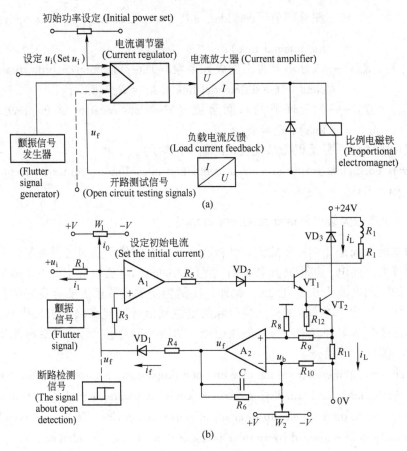

图 6-28 模拟式功率放大电路

(Fig. 6-28 Analog power amplifier circuit)

(a) 结构框图 (Structure diagram); (b) 电路原理图 (Schematic circuit diagram)

6.4 比例放大器电路 (Proportional amplifier circuit)

目前，模拟式比例放大器使用较普遍，具有控制精度高，响应速度快，结构简单的优点，但功放管工作在线性区，功耗大，温升高，效率较低。

At present, an analog proportional amplifier is widely used with advantages of high control accuracy, fast response and simple structure. But the power amplifier tubes working in linear area is characterized by high power consumption, high temperature rise and low efficiency.

6.4.3.2 脉宽调制 (PWM) 式功率放大电路 (PMW power amplifier circuit)

在各种控制信号和功放管之间加入 PWM 环节，就得到脉宽调制式功率放大电路，如图 6-29 所示。

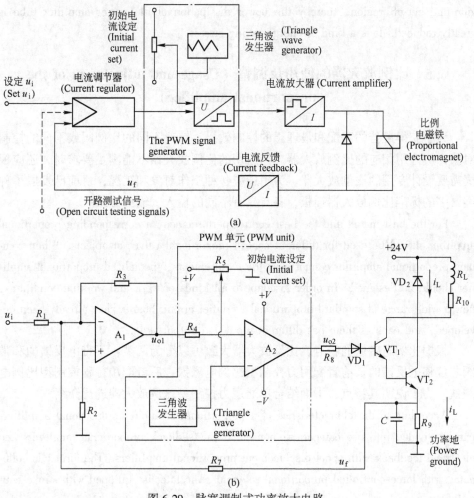

图 6-29 脉宽调制式功率放大电路

(Fig. 6-29 PWM power amplifier)

(a) 原理框图 (Structure diagram); (b) 电路原理图 (Schematic circuit diagram)

A PMW power amplifier can be achieved by adding a PWM link between all kinds of control signals and power amplifier tubes. As shown in Fig. 6-29.

在 PWM 信号的控制下，功放管工作在开关状态。开状态时，功放管导通，压降约等于零；关状态时，功放管截止管的电流约为零，功放管只工作在饱和区和截止区，因而使功放管的功耗大为降低。这是一种节能的电路。

Under the control of a PWM signal, the power amplifier tube works in the on-off state. In power-on state, the power amplifier tube is conducted, and pressure drop approximately equal to zero. In power-off state, the power amplifier tube is closed, and current in the tube is about zero. The power amplifier tube only works in saturation region and cut-off region, thereby the power dissipation of the power amplifier tube is greatly reduced. It is a kind of energy-saving circuit.

6.5　比例放大器的使用及调整（Usage and adjustment of the proportional amplifier）

为了获得最佳的匹配和最理想的控制效果，针对不同的比例阀或不同的控制对象，应选用相应的比例放大器。当通用的比例放大器不能满足要求时，还应根据需要设计专用的比例放大器。为了适应各种工作对象和工况，目前已发展了种类繁多的标准比例放大器插板，各自具有不同的输入、输出方式。

For the best match and the best control performance, a corresponding proportional amplifier should be used for different control proportional valves or objects. When common proportional amplifier cannot meet the requirements, specialized proportional amplifier should be designed. In order to adapt to all kinds of work and working conditions, now a wide range of standard proportional amplifier circuit boards have already been developed, and each of them has different input and output style.

根据比例电磁铁的特点，比例放大器大致可以分为两类：不带电反馈的和带阀芯位移电反馈的。前者配用力控制型比例电磁铁，后者配用行程控制型比例电磁铁。为了说明其特点，下面结合单通道力控制型比例放大器进行介绍。

According to the characteristics of a proportional solenoid, proportional amplifiers can be divided into two categories: non-electrical feedback proportional amplifiers and electrical feedback with spool displacement proportional amplifiers. The former is collocated with force-controlled proportional solenoid. The later is equipped with stroke-controlled proportional solenoid. We will take the single-channel force-controlled proportional amplifier for example to illustrate its features.

单通道力控制型比例放大器是最简单的一种比例放大器，主要用于控制先导

6.5 比例放大器的使用及调整
(Usage and adjustment of the proportional amplifier)

式溢流阀和减压阀、比例节流阀、比例泵和比例马达等由单个力控制型比例电磁铁驱动的比例元件。它的内部电路比较简单,所需的外部接线也较少。图 6-30 为单通道比例放大器的框图。

The single-channel force-controlled proportional amplifier is the simplest type of proportional amplifiers. It is mainly used to control pilot operated relief valve, pressure reducing valve, proportional throttling valve, proportional pump, proportional motor and any other elements which are controlled by a single force-controlled proportional solenoid-driven component. Its internal circuits are relatively simple, and less external wiring is required. Fig. 6-30 shows the block diagram of the single-channel proportional amplifier.

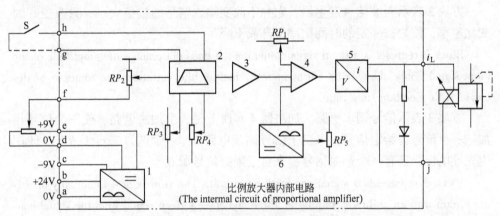

图 6-30 单通道开环比例放大器框图

(Fig. 6-30 Proportion of single-channel open-loop amplifier block diagram)

1—电源装置(Power device);2—斜坡信号发生器(Slope signal generator);3—信号调理电路(Signal conditioning circuit);4—加法器(Summator);5—功率放大器(PA);6—颤振信号源(Flutter source);RP_1—零点电流调节电位器(Zero current adjustment potentiometer);RP_2—限制最大电流幅值电位器(Limit the maximum current amplitude potentiometer);RP_3—斜坡上升时间调节电位器(Ramp up time adjustment potentiometer);RP_4—斜坡下降时间调节电位器(Slope fall time adjustment potentiometer);RP_5—颤振信号频率调节电位器(Flutter frequency adjustment potentiometer)

图中点划线内的为比例放大器的内部回路(The internal circuit of the proportional amplifier is shown in dash dot line of the figure)。

方块 1 表示滤波和稳压器,它的功能是对电源所提供的 24V 直流电压作进一步的平滑滤波,滤波后一部分供给功放级作功率放大,另一部分经稳压后分离出正负电源电压供内部运放电路使用。另外还通过接线柱 c(-9V)和 e(+9V)向外提供稳定的正、负参考电压,供产生给定电压用。电位器的一只脚接于端子 e(+9V),另一只脚接于端子 d(0V),电位器的动臂连接端子 f,这样通过转

第6章 比例放大器
Chapter 6　Proportional Amplifier

动电位器便可得到 0~9V 的指令信号。

Block 1 represents a filter and a voltage regulator. Its function is to provide a 24V DC voltage for power source to further filtering. The part of the filtered power is supplied for power amplification by power amplifier stage, and the other part of the filtered power is detached to positive and negative power supply for internal op-amp circuit. Moreover, through the terminal c (-9V) and e (+9 V), it provides a stable positive and negative reference voltage to generate a given voltage. One foot of a potentiometer is connected to terminal e (+9 V), the other foot is connected to terminal d (0V), the movable arm of the potentiometer is connected to terminal f, so that a command signal of 0 to 9V can be obtained by turning the potentiometer.

方块 2 表示斜坡信号发生器，它并不改变输入信号的幅值大小，只改变其的变化速率，使之按一定的时间比率上升或下降。

Block 2 represents a ramp signal generator, it does not change the amplitude of the input signal, only changes its variation rate, making the variation rate increase or decrease with a certain time ratio.

方块 3 表示信号调理电路。加法器 4 实质上是一个加法电路，使三个信号相加：一个信号为给定信号，来自于信号调理电路 3；一个信号来自于 RP_1，使产生先导电流；还有一个颤振信号，来自于颤振信号源 6。

Block 3 represents a signal conditioning circuit. The summator 4 is actually an adder circuit with an ability of adding the three signals—a given signal from the signal conditioning circuit 3, a signal from the RP_1, to produce pilot current and a flutter signal from the flutter source 6.

功率放大器 5 有两个主要功能：它把控制信号放大到驱动电磁铁所需的功率水平；提供电流反馈，使通过线圈的电流稳定，与线圈温度、接线损失等无关。

The power amplifier 5 has two main functions: it amplifies the control signal to a required level for driving solenoid, and it provides current feedback, so that it make the current through the coil stable without interference of coil temperature and wiring losses.

为了获得最理想的效果，每一比例放大器都有若干调整，用于匹配不同的比例元件或满足控制要求。这些调整通常称为预置或设定，是通过比例放大器内的一些多圈电位器来进行的。对力控制比例放大器，通常有 4~5 个电位器，即 $RP_1 \sim RP_5$ 需要进行设定。

In order to obtain optimal results, each proportional amplifier has a number of adjustments to match different proportional components or control requirements. These adjustments which are carried out with some multi-turn potentiometers in the proportional

6.5 比例放大器的使用及调整
(Usage and adjustment of the proportional amplifier)

amplifier are often referred as preset or setting. For a force control proportional amplifier, there are usually 4 ~ 5 potentiometers, namely RP_1 ~ RP_5, which need to be set.

A 初始电流设定 (Initial current setting)

由于比例阀的特性中都存在一定的死区特性,只有越过这个死区范围,其特性才是线性的。为了克服这个死区,特向比例电磁铁提供一个初始电流(或称为先导电流)。调整或预置初始电流是利用电位器 RP_1 来进行的,其调整范围为 0~300mA。由于各个阀的死区范围不同,使用前应根据具体情况进行调整。

As the characteristics of proportional valves have certain dead zone, only over the dead zone, its characteristic is linear. To overcome this dead zone, an initial current (or pilot current) is needed to provide for the proportional solenoid. Adjusting or presetting the initial current is carried out by using a potentiometer RP_1, the adjustment range of the potentiometer RP_1 is 0 ~ 300mA. Because dead zone of each valve is different, adjustment should be made depending on the specific situation.

B 最大工作电流设定 (Maximum operating current setting)

为了防止系统有过大的输出量,确保安全,最大允许工作电流的设定是必需的。这时要使用电位器 RP_2。为便于调整,RP_2 通常放在面板上。RP_2 的调节用于限制给定输入的水平。

In order to prevent the system from having too much output, and ensure safety, a maximum allowable operating current setting must be required. Then we need a potentiometer RP_2. In order to be adjusted conveniently, RP_2 is usually placed on the panel. The RP_2 adjustment is used to limit the level of given input.

在调整 RP_1 和 RP_2 时应注意的问题是:RP_2 的调整值会加到 RP_1 的设定值上。换句话说,RP_1 和 RP_2 的调整次序将对最终结果产生影响。

The question in adjustment of RP_1 and RP_2 should be paid attention: The adjustment value of RP_2 will be added to the settings of RP_1. In other words, the order of the adjustment of RP_1 and RP_2 will impact final outcome.

综上所述,RP_1 和 RP_2 的设定方法如下:置输入信号为零,调整 RP_1 至系统将要开始有输出为止,以消除死区;置输入信号为 100%,逐步增加 RP_2,使其达到希望的输出值为止。

In summary, RP_1 and RP_2 is set as following: set the input signal as zero, and adjust the RP_1 until the system begins to output, so as to eliminate the dead zone; set the input signal as 100%, and gradually increase the RP_2 until it achieves the desired output value.

C 斜坡上升、下降时间的调整（Time adjustment of slope rise and fall）

带有单通道力控制型电磁铁的比例放大器有两个斜坡时间调节器，一个（RP_3）用于调节上升时间，另一个（RP_4）用于调节下降时间。它们是互相独立调节的。这样利用斜坡信号发生器可获得加速、减速、升压和卸压等调整功能。放大器的斜坡调整时间范围为 0.1~5s。

A proportional amplifier with single channel force-controlled solenoid has two slope time regulators. The one (RP_3) is used to adjust rise time, and the other (RP_4) is used to adjust fall time. They are independent of each other in regulation. We can acquire acceleration, deceleration, pressure boosting, pressure relief and any other adjustment features by using a slope signal generator. The amplifier's slope adjustment time is 0.1 ~ 5s.

还应指出，如果对斜坡上升、下降时间不作要求，可以采用一根导线或开关使斜坡信号发生器短路，这时意味着对 RP_1 和 RP_2 的设定值或对输入信号作出直接的响应。

It should also be pointed out that if ramp up and down time isn't required, a wire or a switch can be used to make the slope signal generator shorten, which means that the generator will make a direct response to the set value of RP_1 and RP_2 or the input signal.

D 颤振信号参数的调整（Flutter signal parameter adjustments）

颤振信号的参数一般在工厂制造时已经调整到最佳范围，因而无需调整。通常比例阀中采用的颤振信号幅值为额定控制信号的 10% ~ 20%，频率为100~200Hz。

The general chatter signal parameters have been adjusted to the best range during the manufacturing in factory, and therefore need no adjustments. The flutter signal amplitude commonly used in proportional valves is 10% ~ 20% of the control signal's, the frequency is 100 ~ 200Hz.

单通道行程控制型比例放大器与力控型放大器相比较，主要的区别是，前者加入了断路检测器、位置反馈通道。反馈通道上有四个基本环节：比例积分微分（PID）调节器、比例放大器、振荡器和解调器。来自斜坡发生器的信号与反馈信号在PID调节器上进行比较，得出差值信号经过 P、I、D 调节，以获得最佳的控制效果，使阀芯保持在准确位置。

The main difference between a single-channel stroke-controlled proportional amplifier and a force-controlled proportional amplifier is that the former has an open circuit detector and a position feedback channel. There are four basic elements in the feedback channel: a PID regulator, a matched amplifier, an oscillator and a demodulator. In the PID regulator, signals from the ramp generator are compared with feedback signals. The obtained differential signals through P, I and D regulating, and then to get

6.5 比例放大器的使用及调整
(Usage and adjustment of the proportional amplifier)

the best control effect, so as to let the valve core at an accurate position.

图 6-31 所示为浙江大学流体传动及控制研究所为适应比例控制技术向高动态、高精度方向发展的需要而开发的,图 6-32 是该比例控制放大器的结构框图。其主要特点是:

As is shown in Fig. 6-31 BKT-12 fast switch type proportional amplifier is developed by the institute of fluid transmission and control of Zhejiang University to meet the need of development to high dynamic and high precision. Fig. 6-32 is the structure diagram of this proportional control amplifier. Its main characteristics are:

(1) 采用开环放大技术,降低了功耗;

Using an open loop amplification technique to reduce power dissipation;

(2) 采用高频脉宽调制,使颤振分量的频率和幅度单独可调,改善了稳态控制性能;

Adopting high-frequency pulse width modulation to make flutter component's frequency and amplitude adjustable independently;

(3) 采用反接卸荷功率驱动电路,提高了比例电磁铁电流动态响应的快速性。

Adopting a reverse connect unloading power drive circuit to improve dynamic response of the proportional solenoid current.

这种比例控制能在较高的环境温度下工作,且能满足较高的动态性能要求。

This kind of proportional control can adapt to high temperature environment, and can meet high dynamic performance requirements.

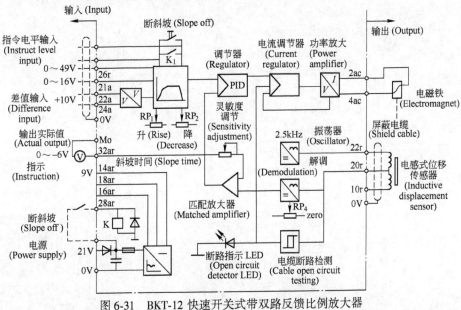

图 6-31 BKT-12 快速开关式带双路反馈比例放大器

(Fig. 6-31 BKT-12 fast switch type proportional amplifier with two feedback channels)

第 6 章 比例放大器
Chapter 6　Proportional Amplifier

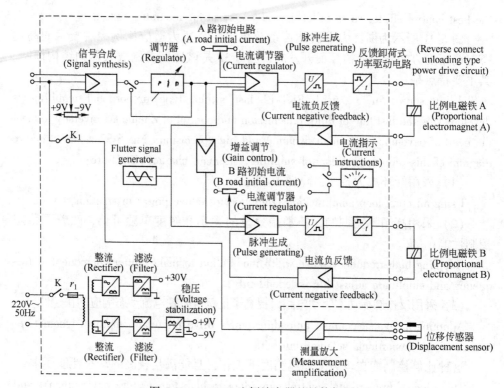

图 6-32　BKT-12 比例放大器的结构框图
(Fig. 6-32　The structure diagram of BKT-12)

第 7 章 电液比例控制系统的分析与设计
Chapter 7　Analysis and Design of Electro-hydraulic Proportional Control System

电液比例控制技术已获得广泛应用。根据所采用的控制元件,电液比例控制系统可分为:

Electro-hydraulic proportional control technology has been widely applied. According to the control components, electro-hydraulic proportional control system can be divided into:

(1) 采用比例压力阀(溢流阀和减压阀)压力(力)控制系统;

Pressure (force) control system using proportional pressure valve (pressure relief valve and pressure reducing valve);

(2) 采用比例节流阀和比例调速阀的速度控制系统;

Speed control system using proportional throttle and proportional speed regulating valve;

(3) 采用比例方向阀的位置、速度或压力控制系统;

Position, speed or pressure control system using proportional directional control valve;

(4) 采用比例控制泵的容积调速、压力或功率控制系统;

Volume Speed control, pressure or power control system using proportional control pump;

(5) 采用压力、流量、方向比例控制元件组成的复合控制系统。

Composite control system using pressure, flow, direction proportional control component.

在以上每一种控制系统中,又有开环控制和闭环控制之分。

Each of the above-mentioned control systems can also be divided into open-loop and closed-loop control system.

7.1　电液比例控制系统的设计内容与步骤 (Contents and steps of designing an electro-hydraulic proportional control system)

(1) 明确产品生产过程和工艺对电液比例控制系统的要求;

Be clear on the requirements of the process and the technology of production on the electro-hydraulic proportional control systems;

(2) 明确主机（设备）的结构特点、布置方式和控制要求；

Be clear on structural features, layout and control requirements of host (equipment);

(3) 确定电液比例控制系统的工况及其特点；

Be clear on the working conditions and characteristics of electro-hydraulic proportional control system;

(4) 估算各执行元件运动参数和动力参数；

Estimate the motion parameters and dynamic parameters of the actuators;

(5) 拟定系统方案，元件选型，绘制比例控制系统原理图；

Prepare system solutions, select components and draw schematic diagrams of proportional control system;

(6) 建立系统数学模型（包括静态和动态的），计算系统数学模型中的重要参数；

Establish mathematical model (including static and dynamic), calculate the important parameters in the mathematical model;

(7) 校核系统静态性能和动态性能；

Check the static performance and dynamic performance of the system;

(8) 绘制系统装配图和非外购件的零件图。

Draw the assemble drawings of system and the part drawings of non-purchased items.

7.2　电液比例控制系统的方案拟订（Programming of electro-hydraulic proportional control system）

拟订系统方案就是决定系统中的重大问题，其主要内容包括：

Programming of system solution is to determine the system's major problems, whose contents mainly include:

(1) 确定电液比例控制系统的控制方式；

Determine the control mode of the proportional control system;

(2) 确定电液比例控制系统的控制系统类型；

Determine the type of proportional control system;

(3) 确定电液比例控制系统的控制信号类型；

Determine the type of control signal in proportional control system;

(4) 确定电液比例控制系统的控制元件类型；

7.2 电液比例控制系统的方案拟订
(Programming of electro-hydraulic proportional control system)

Determine the type of control elements in the proportional control system;

(5) 确定电液比例控制系统的检测元件类型。

Determine the type of testing components in the proportional control system.

7.2.1 确定比例控制系统的控制方式 (Determine control mode of the proportional control system)

确定系统采用开环控制还是闭环控制。

Be sure on whether the system is open loop control or closed loop control.

开环控制系统不具备抗干扰能力，其控制精度取决于组成系统各元件或环节的精度，不存在稳定性问题。采用闭环控制主要基于以下四个方面的考虑：

Open-loop control system does not have anti-interference capability, whose control accuracy depends on the accuracy of components and parts of the system, and there is no problem with stability. Closed-loop control is adopted mainly based on the following four considerations：

(1) 提高系统自动化水平，避免外部干扰的影响；

Improve the automation level of the system, to keep it from external interference;

(2) 提高电液比例系统的控制精度；

Improve the control accuracy of the electro-hydraulic proportional system;

(3) 提高系统的动态性能（快速响应性），开环控制系统的快速响应性能完全是由系统的极限加速度 α_{max} 来决定的，当比例控制系统的快速响应要求超出极限加速度 α_{max} 时，开环控制失效；

Improve the dynamic performance (fast response). The fast response performance of open loop control system is entirely determined by the limit acceleration α_{max} of the system, when the requests of the rapid response of the proportional control system is more than α_{max}, the open-loop control system cannot do anything;

(4) 提高阻尼，保证系统稳定性。

Improve the damping to ensure the system stability.

7.2.2 确定比例控制系统的控制系统类型 (Determine the type of proportional control system)

电液比例技术在各种类型的工业设备中获得了广泛应用。闭环控制系统按照它所控制的物理参数，可分为电液比例位置控制系统、电液比例速度控制系统、电液比例加速度控制系统和电液比例力（压力）控制系统。在拟订系统方案时，首先要确定所设计的比例控制系统属于哪一种类型。通常情况下，电液比例系统控制的物理量（即系统类型）是不能随意确定和改变的，它是由设备用途、工艺要求和控制过程所决定的。

Electro-hydraulic proportional technology has been applied widely in all types of industrial equipments. Closed loop control system in accordance with physical parameters controlled by itself, can be divided into electro-hydraulic proportional position control system, electro-hydraulic proportional speed control system, electro-hydraulic proportional acceleration control system and electro-hydraulic proportional force (pressure) control system. When preparing the system scheme, the type which the proportion control system belongs to should be determined firstly. Generally, the physical quantity controlled by the electro-hydraulic proportional control system, namely the type of system, cannot be determined and changed optionally. It is determined by the equipment applications, process requirements and control process.

电液比例控制系统的类型一经确定，所采用的控制元件和检测元件的类型也就基本上确定了。

Once the electro-hydraulic proportional control system's type is determined, the type of control components and testing components is also basically determined.

7.2.3 确定比例控制系统的控制信号类型（Determine the type of control signals in proportional control system）

可供电液比例控制系统选用的控制信号有模拟式和数字式两种类型。

There are two kinds of control signals in the hydraulic proportional control systems which are analog and digital.

数字式检测元件的分辨率很高，用它构成数字式电液比例控制系统，可直接与计算机连接。另外，以数字方式给定的信号具有相当高的精度。

Digital detection devices have high resolution and are used to constitute a digital electro-hydraulic proportional control system. It can be directly connected to the computer. In addition, the signal given digitally has very high precision.

在控制精度要求高，设备有多个执行元件需要控制，且控制关系较复杂（如多缸同步控制系统），以及设备已经采用计算机控制方案时，推荐采用全数字控制系统，以简化控制系统设计。

Full digital control system is commanded to simplify the design of control system in the case of high control precision requirements, needing to control multiple executive elements on the device, complex relationship of control (such as multi cylinder synchronization control system) and computer control plan used by device.

采用全数字电液比例控制系统还有一个突出的优点，即控制信号可以在总线上传输，便于计算机读取和存储。

7.2 电液比例控制系统的方案拟订
(Programming of electro-hydraulic proportional control system)

Fully digital electro-hydraulic proportional control system also has a prominent advantage that the control signal can be transmitted on the bus, easy to be read and to be stored by the computer.

尽管如此,目前工业上采用的绝大多数比例阀仍然是模拟信号控制的,相配套的比例放大器和检测元件也是模拟式的。模拟式控制系统理解起来较简单,调试和处理也较直观,是目前普遍采用的方案,但其分辨率和控制精度较低,抗干扰能力稍差。

Nonetheless, at present, the overwhelming majority of proportional valves which are used in industry as well as the matched amplifiers and detecting elements are still controlled by analog signal. Analog control system is simple to understand and intuitional to debug and handle, which is a commonly used scheme. But its resolution ratio and control precision is quite low, and its anti-interference capability is relatively poor.

7.2.4 确定比例控制系统的控制元素类型 (Determine the type of control elements in proportional control system)

电液比例控制系统有阀控和泵控两种控制方案,究竟是采用阀控系统还是泵控系统,要根据控制精度和响应速度的要求、功率的大小、所要求效率及成本的高低等综合确定。阀控系统的控制精度和响应速度高,但效率低。泵控系统的效率高,可用于大功率控制场合,但控制精度和响应速度低。

There are two control schemes of electro-hydraulic proportional control systems, whether to take pump control system or valve control system depends on the requirements of control precision and response speed, the level of the power, the level of efficiency and the cost to determine and so on. Valve control system has higher accuracy and response speed, but lower efficiency. Pump control system has high efficiency, can be used in big power control situations, but the control accuracy and the speed of response are low.

7.2.5 确定比例控制系统的检测元件类型 (Determine the type of detection components in proportional control system)

检测环节包括传动机构、传感器及其二次仪表。当被控对象做直线运动,所采用的传感器为测速机、光码盘、圆形感应同步器或圆光栅一类旋转式传感器时,需要采用传动机构将直线运动转换成旋转运动。常用的传动机构有传送钢带、钢性绳、齿轮-齿条和摩擦轮系等。

Detection unit includes transmission mechanisms, sensors and its secondary instruments. When the controlled object moves rectilinearly and sensors like tachometer, optical encoder, circular inductosyn or such rotating sensors as circular grating etc are applied in the system, the linear motion needs to be transformed into rotary motion by the transmission mechanism. There are several common transmission mechanisms such as driving belt or steel rope, rack and pinion and friction gear train etc.

传感器的类型应根据所检测的物理量类型、要求的检测精度及动态响应速度、工作环境及成本来确定。位置控制系统采用位置传感器,速度控制系统采用速度传感器,力或压力控制系统采用压力传感器。

The type of sensor should be based on the type of physical quantity required testing accuracy and dynamic response speed and working environment and the cost. Position control system uses position sensor, speed control system uses speed sensors, force or pressure control system uses pressure sensors.

检测元件输入信号的量程要与被控制信号的最大变化范围相一致,不能小于被控制信号的最大范围,否则会损坏传感器件,但也不能比被控制信号的最大变化范围大得太多,否则会影响传感器的分辨率,引起过大的控制误差。

The range of input signal of detecting component should be consistent with the one of the biggest variation of the control signal, which cannot be less than the maximum range of controlled signals, otherwise it will damage the sensor parts. However, it cannot be wider than the maximum range too much, otherwise the resolution of the sensor will be affected and cause an excessive control error.

根据控制工程的经验,检测元件的精度等级必须大于系统控制精度的4倍以上。检测元件的响应速度应达系统所要求频宽的8~10倍以上。

According to the experience of control engineering, the grade of accuracy of the detecting components must be more than 4 times greater than the system's control accuracy. The response speed of detection component should be 8 to 10 times greater than the bandwidth required by the system.

检测元件的输出信号应采用标准的控制信号,如 0~10V、0~±10V、0~20mA、4~20mA、0~5V、0~±5V 等,以便与下一级元件的输入信号相匹配。

The output signal detection devices should use standard control signals, such as 0~10V, 0~±10V, 0~20mA, 4~20mA, 0~5V, 0~±5V, in order to make test components and input signal of the next level of components matched.

7.3 电液比例控制系统的静态分析 (Static analysis of electro-hydraulic proportional control system)

静态设计的主要内容是确定液压动力元件参数,选择系统的组成元件。液压

7.3 电液比例控制系统的静态分析
(Static analysis of electro-hydraulic proportional control system)

动力元件是比例伺服系统的关键部件。它在整个工作循环中拖动负载按要求的速度运动的同时,其主要性能参数应能满足整个系统所要求的动态特性。此外,动力元件参数的选择还必须考虑与负载参数的最佳匹配,以保证系统的功耗最小,效率最高。

The main contents of static design are to determine the hydraulic power units and to choose system components. Hydraulic power units are the critical components of proportional servo system. They drag the load on requested speed to motion in the whole working cycle meanwhile their specification can meet the dynamic characteristics that the whole system required. In addition, the power component parameters must also be considered for the best match with the load parameters, in order to ensure that the system's power dissipation is minimal, while the efficiency is maximal.

液压动力元件参数选择包括系统的供油压力 p_s,液压执行器的主要规格尺寸(如液压缸的有效面积 A_p),比例阀的规格(最大空载流量 q_{0m} 及额定流量 q_n)。

The choices of hydraulic power components include the oil pressure of system p_s, the main size of the hydraulic actuator, the effective area of hydraulic cylinder A_p, the size of the proportional valve (the biggest flow of the carrying idler q_{0m} and rating flow q_n).

7.3.1 比例控制系统供油压力的选择 (Selection of oil pressure of the proportional control system)

选用较高的供油压力,在相同输出功率条件下,可减小执行元件-液压缸的活塞面积,从而使泵和动力元件尺寸小、重量轻,设备结构紧凑;同时油腔的容积减小,容积弹性模数增大,有利于提高系统的响应速度。但是,随供油压力的增加,由于受材料强度的限制,液压元件的尺寸和重量也有增加的趋势,元件的加工精度也需提高,系统的造价也随之提高。同时,高压时,泄漏大,发热高,系统功率损失增加,噪声加大,元件寿命降低,维护也变得困难。所以,条件允许时,通常还是选用较低的供油压力。

To choose the higher oil pressure supplied, under the same condition of output power, can decrease the piston area of actuator—hydraulic cylinder. Thereby it can make the size of the pump and power components small and light, and make structure of equipment compact, meanwhile decreasing the volume of oil chamber and increasing the elastic modulus, which are conducive to improve response speed of the system. But with the oil pressure increasing, because of the limited strength of raw material, the size and weight of the hydraulic components tend to increase, process precision of components needs to be improved as well, the cost of the system is also increasing. At the same time, when the system is at a high pressure, the leakage and heat productivity is large, the power loss and the noise increases, and the component life will get shorter, which makes the system difficult to maintain. So the lower oil pressure is more likely to be chosen under allowable conditions.

常用的供油压力等级为 7~28MPa。

The normal oil supply pressure level is between 7~28MPa.

7.3.2 确定液压执行器规格尺寸及阀规格 (Determine the size of hydraulic actuators and the specification of valve)

由于执行器和比例阀是整个液压伺服系统的关键部件，直接影响到系统性能的优劣，故应十分注意其选型和设计问题。在按下述方法确定出液压缸的有效面积的最佳值 A_p，计算出相应的缸筒内径 D、活塞杆直径 d 等结构尺寸并确定比例阀的流量后，进行选型及结构设计。

Because the hydraulic actuators and the proportional valves are the key parts to the whole hydraulic servo system, which would directly influence the performance, so much attention should be paid on the components selection and design. According to the following method, after determining the optimum value A_p of the effective area of hydraulic cylinder, figuring out the corresponding cylinder inside diameter D, piston rod diameter d and other structuredimensions, as well as determining the flow of proportional valve, then the components can be selected and the structure can be designed.

7.3.2.1 按负载匹配确定(Determine according to load matching)

A 负载特性（负载力与负载速度之间的关系）(The characteristic of the load (the relation between the carrying force and the load speed))

负载特性曲线（负载轨迹图）是根据负载的情况，以横轴为负载力（可转化为负载压力）、纵轴为负载速度（可转化为负载流量）作出的曲线（如图 7-1 所示），其方程为负载轨迹方程。

Load characteristic curve (load path chart) is on the basis of the condition of load to take the horizontal axis as carrying (it can translate into load pressure) and to take the vertical axis as the load speed (it can translate into load flow), which can make a curve (as shown in Fig. 7-1). This equation is load trajectory equation.

负载工作的每一个工况都应在负载特性曲线内。较为简单的运动过程，其负载特性为特殊点，一般为最大功率点或最大速度（转速）点和最大负载力（转矩）点，如图 7-1 中 A、B、C 三点分别为最大速度点、最大负载力点和最大功率点。

7.3 电液比例控制系统的静态分析
(Static analysis of electro-hydraulic proportional control system)

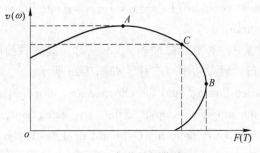

图 7-1 负载特性曲线
(Fig. 7-1 Curve of the character of the load)

Every working condition of the load working should be in the load characteristic curve. The less simple movement process whose load characteristics are the special dot, which generally is the dot of the most maximum power or the maximum speed (swiveling speed) or the maximum load bearing. As is shown in Fig. 7-1, A, B and C are respectively the maximum speed point, the maximum load bearing point and the maximum power point.

B 液压动力元件的输出特性 (The export character of hydraulic power units)

液压动力元件的输出特性可由比例阀的流量-压力曲线经坐标变换得到。在速度（转速)-力（转矩）平面上，绘出的液压动力元件输出特征曲线为抛物线。图 7-2 所示为动力元件（阀控缸）的输出特征曲线。由图 7-2 可知：

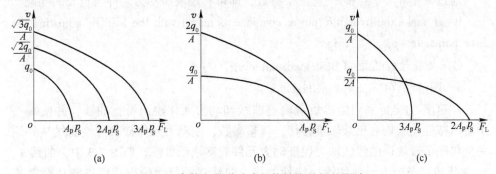

图 7-2 参数变化对动力机构输出特性的影响
(Fig. 7-2 Influence by parameter changing to output character of power mechanism)
(a) 供油压力变化 (Oil pressure changing); (b) 阀容量变化 (Valve capacity changing);
(c) 液压缸面积变化 (Cylinder area changing)

The export character of hydraulic power units can be obtained by the coordinate transformation of the flow-pressure curve of the proportional valve. On the planar of speed (rotate speed)-force (torque), we can draw the characteristic curve of the hydraulic power components as a parabola. The characteristic curve of a hydraulic component (hydraulic cylinder controlled by servo valve) is shown in the following figure. We can draw conclusion from Fig. 7-2:

(1) 当空载流量 q_{0m} 和液压缸面积 A_p 不变时，提高供油压力 p_s，曲线向外扩展，最大功率提高，最大功率点右移，如图 7-2a 所示；

When the unloaded flow q_{0m} and the effective area of cylinder piston A_p are fixed, with the increase of the supply pressure p_s, the curve extendsout, the maximum power raises and the maximum power point moves toward right, as is shown in Fig. 7-2a;

(2) 当供油压力 p_s 和液压缸面积 A_p 不变，加大空载流量 q_{0m}，曲线的顶点 $A_p p_s$ 不变，最大功率提高，最大功率点不变，如图 7-2b 所示；

When the supply pressure p_s and the effective area of cylinder piston A_p are fixed, with the increase of the unloaded flow q_{0m}, the curve and the peak $A_p p_s$ remain the same, the maximum power raises and the maximum power point remains the same, as is shown in Fig. 7-2b;

(3) 当供油压力 p_s 和空载流量 q_{0m} 不变，加大液压缸面积 A_p，曲线变低，顶点右移，最大功率不变，最大功率点右移，如图 7-2c 所示。

When the supply pressure p_s and the unloaded flow q_{0m} are fixed, with the increase of the effective area of cylinder piston A_p, the curve drops, with its peak moves towards right, the maximum power stays the same and the maximum power point moves toward right, as is shown in Fig. 7-2c.

通过调整 p_s、q_{0m} 和 A_p 这三个参数，即可实现液压动力元件与负载的匹配。

It can make the hydraulic power components match with the load by adjusting the three parameters p_s, q_{0m}, A_p.

C 负载最佳匹配 (Best load matching)

a 图解法 (Graphical method)

在速度-力坐标系内绘出负载轨迹曲线和动力元件输出特性曲线，并使每一条输出特性曲线均与负载轨迹相切，调整参数，使动力元件输出特性曲线从外侧完全包围负载轨迹曲线，即可保证动力元件能够拖动负载。如图 7-3 中，曲线 1、2、3 代表三条动力元件的输出特性曲线。曲线 3 的最大输出功率点与负载轨迹最大功率点 c 相重合，满足负载最佳匹配条件；曲线 1 和 2 的最大功率输出点（a 点和 b 点）大于负载的最大功率点（c 点），虽能拖动负载，但动力元件的功率未得到充分利用，故效率都较低。

Draw the load trajectory curve and the output characteristic curve of the power units in the speed-force coordinate system, and make each output characteristic curve tangent to the load trajectory curve, then adjust parameters so that the output characteristic curve of the power units can completely surround the load trajectory curve, in this way the load can be dragged by the power units. As is shown in Fig. 7-3, curve 1, 2, 3 represent three power units output characteristic curves. The maximum output power points of the curve 3 and the maximum power point c of load trajectory coincide, which meet the best matching conditions load. The maximum output power points (point a and point b) of the curve 1 and 2 are greater than the maximum power point (point c). Though in this case the load can be dragged, yet it will lead to the underutilization of the power, so the efficiency is relatively low.

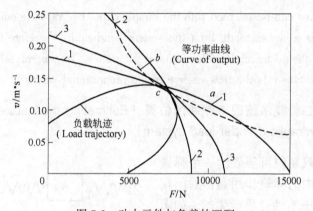

图 7-3 动力元件与负载的匹配

(Fig. 7-3 Match of power units and load)

b 近似计算法 (Approximate calculation method)

在工程设计中，设计动力元件时常采用近似计算法，即按最大负载力 F_{Lmax} 选择动力元件。在动力元件输出特性曲线上，限定 $F_{Lmax} \leq p_L A = \frac{2}{3} p_s A$，并认为负载力、最大速度和最大加速度是同时出现的，这样液压缸的有效面积可按下式计算：

In engineering design, the power units are often designed by using the approximate calculation method, namely the power units are chosen according to the maximum load bearing F_{Lmax}. In the output characteristic curve of the power units, it is limited that $F_{Lmax} \leq p_L A = \frac{2}{3} p_s A$, and it's also considered that the load, the maximum speed and the maximum acceleration appear at the same time. In this way, the effective area of hydraulic cylinder can be figured out according to the following formula:

$$A = \frac{F_{Lmax}}{p_L} = \frac{m\ddot{x} + B\dot{x} + kx + F_L}{\frac{2}{3} p_s} \tag{7-1}$$

7.3.2.2 确定比例阀规格(Determine the size of proportional valve)

根据所确定的供油压力 p_s 和由负载流量 q_L（即要求比例阀输出的流量）计算得到的比例阀空载流量 q_{0m}，即可由比例阀样本确定比例阀的规格。因为比例阀输出流量是限制系统频宽的一个重要因素，所以比例阀流量应留有余量。通常可取15%左右的负载流量作为比例阀的流量储备。

According to the selected supply pressure p_s and the load flow q_L (the output flow of proportional valve), the load flow q_{0m} can be figured out, and the specifications of proportional valves can be selected with the sample data. Because the output flow of proportional valve is a key factor to limit the system bandwidth, therefore the flow of proportional valves should be made with a suitable allowance. In general, about 15% of the load flow can be taken as the flow reserve of the proportional valves.

7.3.3 液压缸-负载系统固有频率的估算 (Estimating the nature frequency of the hydraulic cylinder-load system)

液压缸负载系统可等效为一个弹簧质量系统。其中的弹簧作用是由封闭在液压缸中的可压缩性介质产生的，如图7-4所示。

Hydraulic cylinder load system can be regarded as a spring mass system. One spring effect is produced by compressible medium which is closed in hydraulic cylinder. As shown in Fig. 7-4.

液压缸-负载系统的固有频率是电液比例控制系统极为重要的参数。

The natural frequency of the hydraulic cylinder-load system is an extremely important parameter of the electro hydraulic proportional control system.

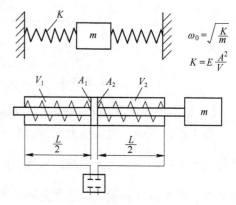

图7-4 液压缸的等效弹簧质量系统
(Fig. 7-4 Equivalent spring mass system of hydraulic cylinder)

在不考虑液压缸-负载系统与基座连接刚度的情况下，固有频率的计算公式为 (Without considering the connection stiffness of hydraulic cylinder-load system and the base, the formula for calculating the natural frequency is):

$$f_0 = \frac{1}{2\pi}\sqrt{K/m} \tag{7-2}$$

式中　f_0——液压缸-负载质量系统的固有频率 (The natural frequency of hydraulic cylinder-load system);

7.3 电液比例控制系统的静态分析
(Static analysis of electro-hydraulic proportional control system)

K——油液压缩性形成的弹簧（称为液压弹簧）刚度（The formation of oil compression spring (known as the hydraulic spring) stiffness）;

m——液压缸驱动的质量（Hydraulic cylinder-driven quality）。

其中，液压弹簧刚度主要由"受压缩"的油液体积决定。液压缸的单侧液压弹簧刚度为：

Among them, the hydraulic spring stiffness is mainly decided by the oil volume which is compressed. Side hydraulic cylinder hydraulic spring stiffness is:

$$K = E\frac{A^2}{V} \tag{7-3}$$

式中 E——液压油的弹性模量，取值范围较大，在 $(1\sim1.4)\times10^9$ Pa 之间 (Modulus of hydraulic oil, a large range between $(1\sim1.4)\times10^9$ Pa);

V——液压缸的单侧容积（The volume of side hydraulic cylinder）。

液压缸的结构形式不同，f_0 的计算略有不同。

Hydraulic cylinder structure style is different, f_0 has slightly different calculation。

7.3.3.1 双出杆液压缸(Double-rod piston hydraulic cylinder)

这种结构形式的液压缸具有对称的面积。当活塞处于全行程的中位时，其固有频率的计算简图如图 7-4 所示。

This structure has symmetrical area of this hydraulic cylinder. When the piston is in the middle, the natural frequency calculation diagram is shown in Fig. 7-4.

这时，总的液压弹簧刚度是液压缸两腔液压弹簧的刚度之和，即

At this time, the total hydraulic spring stiffness is the sum of the two cavities, that is

$$K = E\left(\frac{A_1^2}{A_1\frac{L}{2}} + \frac{A_2^2}{A_2\frac{L}{2}}\right) \tag{7-4}$$

由于 $A_1 = A_2 = A$，故（As $A_1 = A_2 = A$, so）

$$K = \frac{4EA}{L} \tag{7-5}$$

式 (7-5) 代入式 (7-2) 即可求得（Put equation (7-5) into equation (7-2) can get）

$$f_0 = \frac{1}{2\pi}\sqrt{\frac{4EA}{mL}} \tag{7-6}$$

式中 L——活塞行程（Piston stroke）。

7.3.3.2 单出杆液压缸(Single-rod piston cylinder)

这种结构形式的液压缸在工程实际中应用广泛，如图 7-5 所示。

This structure type of cylinder is widely used in the practical engineering, as shown in Fig. 7-5.

其最低固有频率 $f_{0\min}$ 的计算是先以双出杆液压缸 f_0 计算为基础，然后采用系数 $\beta = \left(1 + \sqrt{\alpha}\right)/2$ 进行修正获得。

The calculation of the lowest natural frequency $f_{0\min}$ is based on the calculation of double-rod piston hydraulic cylinder f_0, and then, is got by using the coefficient $\beta = \left(1 + \sqrt{\alpha}\right)/2$ to amendment to get.

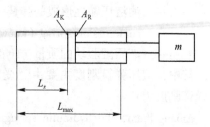

图 7-5 单出杆液压缸固有频率估算简图
(Fig. 7-5 Diagram to estimate single-rod piston cylinder nature frequency)

Where

$$\alpha = \frac{A_R(\text{环形面积, circular-area})}{A_K(\text{活塞面积, piston-area})} \tag{7-7}$$

Thus

$$f_{0\min} = \frac{1}{2\pi}\sqrt{\frac{4EA_K}{mL}\left(\frac{1+\sqrt{\alpha}}{2}\right)} \tag{7-8}$$

7.4 电液比例控制系统的动态分析（Dynamic analysis of electro-hydraulic proportional control system）

建立系统数学模型的目的在于对系统的性能进行定量分析，以了解系统的技术指标是否满足要求。电液比例控制系统的数学模型也是系统分析和校正的依据。

The purpose of establishing system mathematical model is to get quantitative analysis of the performance of the system, and thus to know whether the technical specifications of the system meet the requirements or not. The mathematical model of the electro-hydraulic proportional control system is the basis of systems analysis and correction.

由于闭环控制系统涉及的问题很多，并且有文献对液压伺服控制进行了详细深入的分析，而电液比例闭环控制系统的建模又与伺服控制系统的建模方法相同，故本书以位置闭环控制系统为例，讨论电液比例闭环控制系统在建模和分析过程中需要注意的问题。

Since closed-loop control system involves a lot of problems that have been analyzed in depth in monographs on hydraulic servo control, and the modeling of proportional electro-hydraulic closed-loop control system is same with that of servo control system.

7.4 电液比例控制系统的动态分析
(Dynamic analysis of electro-hydraulic proportional control system)

Therefore, this book takes the position loop control system for example to discuss issues to note in the modeling and analysis of proportional electro-hydraulic closed-loop control system.

7.4.1 闭环控制系统开环增益的组成 (Composition of open loop gain of closed loop control system)

闭环控制系统有输出量的检测和反馈，当把反馈输出端断开，则可得到闭环控制系统的开环回路，如图 7-6 所示。

There are the detection and feedback of the output in the closed-loop control system, when the feedback output is disconnected, the open-loop circuit of the closed-loop control system can be get as shown in Fig. 7-6.

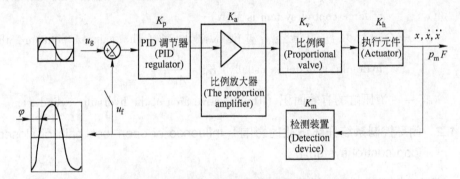

图 7-6 闭环控制系统回路增益的组成

(Fig. 7-6 Gain composition of closed-loop control system)

该开环回路对应的开环增益为 (Corresponding gain of the open loop is)

$$K_c = K_p K_a K_v K_h K_m \tag{7-9}$$

式中　K_c——闭环控制系统的开环增益 (The open loop gain of the closed loop control system);

K_p——PID 调节器的增益，其中的比例环节是必不可少的; PID 调节器的参数在系统校正后确定，在建模的初级阶段，暂不考虑 PID 调节器环节 (PID regulator gain, the proportion links is essential; PID regulator parameters are determined only after the system calibration, in the initial stages of modeling, the PID regulator link will not be considered);

K_a——比例放大器的增益 (The proportion amplifier gain);

K_v——比例控制阀的增益，其量纲因采用不同类型的比例控制阀而异：比例压力控制阀增益的量纲为 MPa/A，比例流量阀和比例方向阀增益的量纲为 $m^3/(s \cdot A)$ (Proportional control valve gain, its dimension is different according to different types of control valves. Proportional pressure control valve dimension is MPa/A, proportional flow control valve and the proportional directional valve dimensions are $m^3/(s \cdot A)$);

K_m——检测装置的增益,压力控制系统采用压力传感器,增益量纲为 V/MPa;位置控制系统采用位置传感器,增益量纲为 V/m;速度控制系统采用速度传感器增益量纲为 Vs/m;力控制系统增益的量纲为 V/N (The gain of the detection device, the pressure control system uses pressure sensor whose dimension is the V/MPa; position control system uses a position sensor whose dimension is V/m; speed control system uses the speed sensor whose dimension is Vs/m; the dimension of force control system is V/N);

K_h——执行元件的增益,对液压缸通常 (The gain of actuator, the hydraulic cylinder is usually) $K_h = \dfrac{1}{A_h}$;

A_h——液压缸的有效面积 (The effective area of the hydraulic cylinder)。

7.4.2 闭环控制系统开环增益的影响 (Influence of open-loop gain of closed-loop control system)

闭环控制系统的开环增益是建立自动控制系统数学模型时需要确定的一个极为重要的参数。该参数对闭环控制系统的性能(稳定性、控制精度和动态响应)会产生重要影响,如图 7-7 所示。

The open-loop gain of the closed-loop control system is a very important parameter to be determined in building mathematical models in the automatic control system, this parameter will have an important impact on the closed-loop control system performance (stability, control accuracy, dynamic response), as shown in Fig. 7-7.

图 7-7 显示,开环增益的变化对系统的稳定性有着直接的影响,值的增加或减小将导致开环波德图上幅频特性曲线向上或向下平行移动,但相频特性曲线保持不变。

Fig. 7-7 shows changes in open-loop gain have a direct impact on the stability of the system, the amplitude-frequency characteristic curve upward or downward parallelly in the open-loop bode plot due to the increase or decrease of the value of the open-loop gain, but the phase frequency characteristics curve remains unchanged.

为确保系统参数的变化不致造成系统不稳定,对于位置控制系统,K_c 值应保证系统获得足够的增益裕量(相位移为-180°时,幅值比为 3~6dB)和相位裕量

7.4 电液比例控制系统的动态分析
(Dynamic analysis of electro-hydraulic proportional control system)

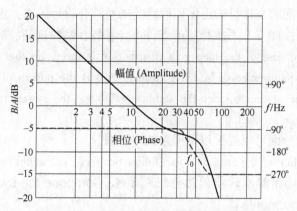

图 7-7 开环增益对闭环控制系统性能的影响

(Fig. 7-7 Influence of closed-loop control system characteristic by open-loop gain)

(幅值比为 0dB 时,相位移小于 135°)。

To ensure that the changes of system parameters will not cause system instability, as for the position control system, the value of K_c should ensure that the system obtain sufficient gain margin (when the phase shift is $-180°$, amplitude ratio is 3~6dB) and phase margin (amplitude ratio is 0dB, the phase shift is less than 135°).

一般来说,系统的结构一经确定,元件也已选定,K_a、K_v、K_h 和 K_m 的值就不容易再改变,而回路的总增益又影响到系统的稳定性和控制精度等性能指标,故在设计闭环系统时,总是把 K_p 设计成可调节的。在闭环控制系统的前向通道里预留一个增益可调的环节,就可达到这一目的。

In general, the structure of the system has been determined, the components have been selected, the value of K_a、K_v、K_h、K_m is not easy to change, and the total loop gain will affect the performance of system stability and control accuracy, so for the closed-loop system the K_p is always designed to be adjustable. It can be achieved in closed loop control system links to add an adjustable gain to the channel reserved.

7.4.3 闭环控制系统开环增益的分配 (Distribution of the open loop gain of closed-loop control system)

设计过程中,是先按式求得闭环控制系统的开环增益,而后才去根据系统的结构和组成系统的环节确定每个环节的增益。因此,存在一个回路增益的分配问题。

In the design process, an open-loop gain of the closed-loop control system is obtained according to the formula, and then gains of each link can be determined according to the system structure and the links. So there is an issue of distribution of the loop gain.

回路增益分配的直接目的是保证回路的增益符合设计要求,最终目的是追求系统的最佳控制性能(在系统稳定的条件下,系统控制精度高、响应速度快)。

The direct purpose of the loop gain distribution is to ensure that the gain of loop meets the design requirements, and the ultimate goal is the pursuit of the best control performance of the system (under the stable conditions, the system have high control precision and fast response).

回路增益的分配可以遵循以下两条技术路线。

The distribution of the loop gain can follow two technical routes as follows.

7.4.3.1 选用标准的控制元件组成系统(Compose the system by standard control components)

采用这种方法进行回路增益的分配,具体做法是:

This approach to the distribution of loop gain, which would be:

(1) 先确定 K_h: 一般情况下, K_h 是在进行系统的参数估算时,根据设备负载和速度要求就已经确定了的参数。由于输入执行元件的参数是流量,故在确定了 K_h 的同时,必须明确执行元件所需要的最大流量。该流量是选用比例控制元件的依据。

As certain K_h: Under normal circumstances, K_h has been determined during estimation the parameter of the system, according to the equipment load and speed requirements. The input parameter of the actuator is the flow, so when K_h has been determined, the components required maximum flow must be determined, which is the basis for the selection of the proportional control components.

(2) 选择比例控制元件:确保获得理想的 K_c 值,是闭环控制系统与开关型液压系统设计内容的重要区别之一。切忌采用增大比例控制元件规格的做法来提高安全系数,即盲目提高 K_v 值。

Select the proportional control components: Make sure to get the optimal value K_c, which is one of the important differences between closed-loop control system and switching-type hydraulic system in terms of design. Never improve the safety factor by increasing the specification of proportional control components, which means improving the value of K_v without destination.

(3) 确定 K_a: 一般情况下,生产厂家提供配套的比例放大器,故当比例控制元件选定之后,比例放大器也就随之确定,即 K_a 也就确定了。

Determine K_a: Under normal circumstances, the manufacturer will provide the matching proportional amplifier. Since the proportional control element is selected, the proportional amplifier is also determined, so does K_a.

(4) 确定 K_m: 检测环节的输入是闭环系统的输出,其量程要与系统输出的

7.4 电液比例控制系统的动态分析
(Dynamic analysis of electro-hydraulic proportional control system)

最大范围一致或略大即可。检测环节的输出是与系统给定信号同量纲、同幅值的物理量（实现减法的需要），即

Determine K_m: The input of the detection element is the output of the closed-loop system, its range should be consistent with the maximum extent of the system output or slightly larger than it. The output of the detection link is a physical quantity which is same with the dimension of a given signal and the amplitude (the subtraction required), namely

$$K_m = x_{max}(t)/y_{max}(t) \tag{7-10}$$

式中　$y_{max}(t)$——检测环节的输入信号（The input signal of detection part）；

　　　$x_{max}(t)$——检测环节的输出信号（The output signal of detection aspects）。

（5）设计 PID 调节装置中的比例环节 K_p：K_p 是回路增益中的最后一个未知环节，按 $K_p = K_c/K_a K_v K_m K_h$ 的关系来确定。初步设计时，是先给 K_p 一个范围。该范围的下限不受限制，上限应能调整。

Design the proportion link K_p of PID adjustment device: K_p is the last unknown link in the loop gain, and is determined by the relationship $K_p = K_c/K_a K_v K_m K_h$. The preliminary design is to give K_p a range. The lower limit of the range is not restricted and the ceiling should be able to adjust.

在系统中的实际值由调试结果确定。在调试现场，K_p 的合理值可取为

The actual value in the system is determined by the debugging results. At the scene of debugging, a reasonable value of K_p is

$$K_p = \frac{2}{3} K_{pmax} \tag{7-11}$$

式中　K_{pmax}——调试过程中缓慢增加 K_p，到系统刚刚出现振荡时的值。系统刚出现振荡的状态可用示波器观察确定（During the process of debugging, slowly increase K_p to the value when system just appear the oscillation, which can be observed by the oscilloscope）。

7.4.3.2 为控制系统设计特定环节的增益 (Design the specific links gain for control system)

当采用标准元件难以获得理想的系统性能时，就要考虑为比例控制系统设计特定的控制元件，以获得所要求的增益。

When using standard components is difficult to obtain the desired system performance, the specific control elements for the proportional control system should be considered to obtain the desired gain.

这种方案一般在批量大，或性能有特殊要求的比例控制系统上采用。也有生产厂为了加强技术垄断，在所生产的比例控制系统中设计专用的控制元件。

Such programs are generally used in large quantities, or in proportional control system with special performance requirements. There are also some manufacturers designing specialized control components in the proportional control system, in order to strengthen the technological monopoly.

建立闭环控制系统的动态数学模型,就是求出系统的传递函数,是闭环控制系统设计的关键内容。传递函数是分析与校核系统静态、动态、稳定性性能和进行系统校正的依据。

The establishment of a dynamic mathematical model of the closed-loop control system is to calculate the transfer function of the system, which is an important part in the design process of closed-loop control system. The transfer function is the basis of analysis, checking static characteristic, dynamic characteristic and stability of the system, as well as system compensation.

7.4.4 闭环控制系统的动态数学模型 (Dynamic mathematical model of closed-loop control system)

建立比例控制系统的动态数学模型可按如下步骤进行:

To establish the dynamic mathematical model of the proportional control system can be in accordance with the following steps:

(1) 根据拟定的系统方案和参数估算的结果确定组成系统的各个环节。

According to the results of the proposed system solution and parameter estimation, the composition of the various segments of the system can be determined.

(2) 按信号在系统中的传递过程将各个环节依次连接起来,构成系统框图(图7-9)。

According to the signal transfer process in the system to link up all aspects, a system block diagram is formed (Fig. 7-9).

(3) 根据系统和选定元件对应的技术参数,求出每个环节的传递函数,并填入系统框图,得到初步的系统传递函数方框图。

According to the system and selected components corresponding to the technical parameters, find the transfer function of each link, fill the system block diagram, and gain the initial system transfer function block diagram.

(4) 对初步获得的传递函数方框图进行分析与校核。

Analysis and check the initial transfer function block diagram.

本章节以闭环位置控制系统为例,介绍采用比例方向阀的位置控制系统的数学模型。

This section takes closed loop position control system for example, introduces the

7.4 电液比例控制系统的动态分析
(Dynamic analysis of electro-hydraulic proportional control system)

proportional directional valve position control system mathematical model.

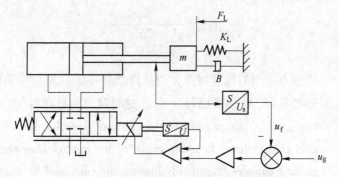

图 7-8 采用比例方向阀构成的位置控制系统

(Fig. 7-8 Position control system using of proportional directional valve)

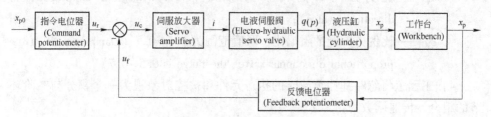

图 7-9 位置控制系统框图

(Fig. 7-9 Block diagram of position control system)

动态数学模型可按两种情形考虑：

Dynamic mathematical model can be considered according to two situations:

(1) 只有惯性负载的位置控制系统 (Position control system with inertia load only)

1) 比例放大器：由于其转折频率比系统的频宽高得多，故可近似为比例环节。

Proportional amplifier: For its corner frequency is much higher than the bandwidth of the system, it can be approximated as a proportional element.

2) 位置检测传感器：其频宽也比系统频宽高得多，亦可近似为比例环节。

The position detection sensor: For its bandwidth is much higher than the system bandwidth, it also can be approximated as a proportional element.

3) 比例方向阀：根据测试结果，工程中将比例方向阀视为一个二阶环节。

The proportional direction valve: Based on test results, the proportional direction valve can be seen as a second order element in the project.

传递函数为：

第7章 电液比例控制系统的分析与设计
Chapter 7 Analysis and Design of Electro-hydraulic Proportional Control System

The transfer function is:

$$W_{pv}(s) = \frac{K_q}{\dfrac{s^2}{\omega_v^2} + \dfrac{2\delta_v s}{\omega_v} + 1} \tag{7-12}$$

式中　K_q——比例方向阀的流量增益,当从产品样本的控制特性曲线中计算流量增益时,K_q 通常已包含 K_a ,这时对应的量纲为 $m^3/(s \cdot V)$ (The flow gain of the proportional directional control valve, when the flow gain is calculated from the curve of the control characteristics of the product samples, K_q usually has already contained K_a, then the corresponding dimension is $m^3/(s \cdot V)$);

ω_v——比例方向阀的相频宽(The phase bandwidth of proportional directional valve);

δ_v——比例方向阀的阻尼比,取值范围为 0.5~0.7(Damping ratio of the proportional directional valve, the range is 0.5~0.7)。

4) 工程上将忽略弹性负载时的执行元件和被控对象视为一个积分与二阶环节的组合。传递函数为:

When the elastic load is ignored, actuator and the controlled object are considered as a combination of an integral and second order element in the project. Transfer function is:

$$W_h(s) = \frac{1/A_h}{s\left(\dfrac{s^2}{\omega_h^2} + \dfrac{2\delta_h s}{\omega_h} + 1\right)} \tag{7-13}$$

式中　A_h——液压缸的有效作用面积,根据运动方向,分别取 A_K 或 A_R (The effective area of the hydraulic cylinder, according to the direction of movement, A_K or A_R is taken respectively);

δ_h——液压缸-负载质量系统的阻尼比,取值范围为 0.1~0.2(Damping ratio of hydraulic cylinder-load quality system, the range is 0.1 to 0.2);

ω_h——液压缸-负载质量系统的固有频率(Natural frequency of hydraulic cylinder- load quality system),

$$\omega_h = \sqrt{\frac{K_h}{m_t}} = \sqrt{\frac{4EA_h^2}{m_t V_t}} \tag{7-14}$$

其中　E——液压油的体积弹性模量(Bulk elastic modulus of hydraulic oil);

7.4 电液比例控制系统的动态分析
(Dynamic analysis of electro-hydraulic proportional control system)

m_t ——包含负载和液压执行元件运动部分的总质量(The total mass which contains the load and the moving parts of hydraulic actuators);

V_t ——比例控制阀至液压缸两腔的总容积(Volume from Proportional control valve to Cylinder chamber)。

图 7-10 中,K_a 是比例放大器的增益,由于其转折频率比系统的频宽高得多,故可视为比例环节;K_m 是位置传感器的增益,其频宽比系统频宽高得多,可视为比例环节。

In the Fig. 7-10, K_a is the gain of proportional amplifier, which can be regarded as the proportional links, due to its corner frequency is much higher than the bandwidth of the system; K_m is the gain of the position sensor, whose bandwidth is much higher than the system bandwidth, so it can be regarded as a proportional link.

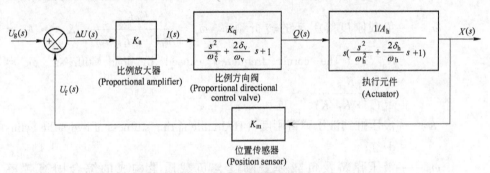

图 7-10 只有惯性负载的位置控制系统传递函数框图
(Fig. 7-10 Position control system transfer function block diagram with inertia load only)

由图 7-10 求得系统的开环传递函数为

The open-loop transfer function can be obtained according to Fig. 7-10 that

$$G(s) = \frac{K_a K_q K_m / A_h}{s\left(\dfrac{s^2}{\omega_v^2} + \dfrac{2\delta_v s}{\omega_v} + 1\right)\left(\dfrac{s^2}{\omega_h^2} + \dfrac{2\delta_h s}{\omega_h} + 1\right)} = \frac{K_c}{s\left(\dfrac{s^2}{\omega_v^2} + \dfrac{2\delta_v s}{\omega_v} + 1\right)\left(\dfrac{s^2}{\omega_h^2} + \dfrac{2\delta_h s}{\omega_h} + 1\right)}$$

(7-15)

(2) 带有弹性负载的位置控制系统 (Position control system with elastic load)。

带有弹性负载的位置控制系统中,除惯性负载外,还有弹性负载。以弹性负载为主的位置控制系统,称为弹性负载位置控制系统,其框图与图 7-9 相同,从控制放大器输出电流到液压缸位置输出的开环传递函数为:

Position control systems with elastic load: position control system has the inertia load, as well as flexible load. Position control system based on elastic load is called elastic load position control system. Its block diagram is same with Fig. 7-9, the open loop transfer function from the control amplifier output current to the hydraulic cylinder output position is:

$$\frac{X(s)}{I(s)} = \frac{\dfrac{K_p A_h}{K_L}}{\left(\dfrac{s}{\omega_r}+1\right)\left(\dfrac{s^2}{\omega_v^2}+\dfrac{2\delta_v s}{\omega_v}+1\right)\left(\dfrac{s^2}{\omega_0^2}+\dfrac{2\delta_0 s}{\omega_0}+1\right)} \tag{7-16}$$

式中 K_p——比例方向阀的压力增益,$K_p = K_q/K_c$ (The pressure gains of the proportional directional control valve, $K_p = K_q/K_c$);

K_L——负载弹簧刚度 (The spring stiffness of the load);

K_c——比例方向阀的流量-压力系数 (The flow-pressure coefficient of the proportional directional valve);

A_h——液压缸的有效面积 (The effective area of the hydraulic cylinder);

ω_r——负载刚度等引起的转折频率,$\omega_r = \dfrac{K_L K_{ce}}{A_h^2(1+K_L/K)}$, $K_{ce} \approx K_c = K_q/K_p$ (The corner frequency caused by load stiffness, $\omega_r = \dfrac{K_L K_{ce}}{A_h^2(1+K_L/K)}$, $K_{ce} \approx K_c = K_q/K_p$);

K——液压缸的液压弹簧刚度 (Hydraulic spring stiffness of hydraulic cylinder);

ω_0——液压弹簧及负载弹簧刚度与负载质量构成的综合固有频率 (rad/s), $\omega_0 = \sqrt{\omega_h + \omega_m} = \omega_h\sqrt{1+K_L/K}$ (The natural frequency formed by stiffness of load spring and hydraulic spring and quality of load (rad/s), $\omega_0 = \sqrt{\omega_h + \omega_m} = \omega_h\sqrt{1+K_L/K}$);

δ_0——液压缸的阻尼比,一般取 0.1~0.2 (Damping ratio of hydraulic cylinder is typically 0.1 to 0.2)。

因此,包含有弹性负载位置控制系统的传递函数框图如图 7-11 所示。
Therefore, position control system transfer function block diagram with the elastic load is shown in Fig. 7-11.

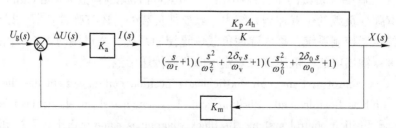

图 7-11 包含弹性负载的位置控制系统传递函数框图

(Fig. 7-11 Position control system transfer function block diagram with elastic load)

7.5 电液比例控制系统的静态性能分析
(Static performance analysis of electro-hydraulic proportional control system)

根据系统对输入信号的响应过程，系统中的误差有动态误差和稳态误差两种形式。系统在响应过程中存在的误差动态分量称为动态误差。如果系统是稳定的，则当系统动态响应过程结束进入稳态后，系统输出的实际值与期望值的差值（也称为误差的稳态分量）构成了系统的稳态误差。稳态误差的大小反映了系统控制精度的高低。

According to the response process of the input signal, the system errors can be divided into dynamic errors and steady-state errors. The dynamic component of the system error in the response process is called dynamic error. If the system is stable, when the system dynamic response process end into the steady state, the actual value of the system output with the expected difference (also called the steady-state component of the error) constitute the steady state error. The size of the steady-state error reflects the level of control precision.

闭环系统的误差来源包括（The error sources of closed-loop system include）：

(1) 输入误差 (Input error);

(2) 干扰误差 (Interference error);

(3) 元件误差 (Component error)。

7.5 电液比例控制系统的静态性能分析 (Static performance analysis of electro-hydraulic proportional control system)

7.5.1 输入误差 (Input error)

闭环控制系统典型框图及误差的定义方式如图 7-12 所示（Typical closed loop control system block diagram and the definition of error are shown in Fig. 7-12）。

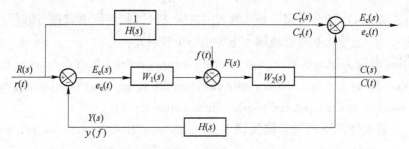

图 7-12 闭环控制系统典型框图及误差的定义方式

(Fig. 7-12 Typical closed loop control system block diagram and the definition of error)

根据定义，系统误差是指在系统的输出端，期望输出与实际输出的差值，即
According to the definition, a system error is the difference between expected

output and actual output in the system output end, that is

$$e_c(t) = c_r(t) - c(t) \tag{7-17}$$

闭环系统通常用系统偏差来反映系统误差的大小,并定义

Closed-loop system always use systematic bias to reflect the size of the system error, and it defines

$$e_e(f) = r(f) - y(f) \tag{7-18}$$

$e_e(t)$ 的物理含义虽然不同于 $e_c(t)$,但因其便于测量,且间接反映了 $e_c(t)$ 的大小,故闭环系统通常就用 $e_e(t)$ 表示 $e_c(t)$,两者之间的关系为

Although the physical meaning of $e_e(t)$ differs from $e_c(t)$, for it is easy to be measured and indirectly reflects the size of $e_c(t)$, closed-loop systems often use $e_e(t)$ to show $e_c(t)$, and the relationship between them is

$$E_c(s) = \frac{1}{H(s)} E_e(s) \tag{7-19}$$

由式可知 (It is known that):

(1) 当 $H(s) = 1$,即系统为单位反馈时,误差与偏差的数值相等,两种定义下的误差是一致的。

When $H(s) = 1$, namely when the feedback of system is a unit, the value of error is equal to that of deviation, and the errors of two definitions are same.

(2) 当系统反馈环节 $H(s)$ 的频宽足够高,$H(s)$ 可视为比例环节 (其增益为 K_m 时)。

When the bandwidth of system feedback link $H(s)$ is high enough, $H(s)$ can be regarded as proportional part (when the gain is K_m).

$$e_c(t) = e_e(t)/K_m \tag{7-20}$$

当扰动 $F(s) = 0$,且误差是从输入端来定义时,利用终值定理可方便地计算阶跃输入、等速输入和等加速输入下的稳态位置误差:

When the disturbance $F(s) = 0$, and the error is defined from the input, then the final value theory can be used to easily calculate the steady-state position error of step input, constant speed input and constant acceleration input:

(1) 阶跃输入下的稳态位置误差 (Steady-state position error under step input is) e_{ep}。

(2) 等速输入下的稳态速度误差 (Steady-state rate error under constant velocity input) e_{ev}。

(3) 等加速输入下的稳态加速度误差 (Steady-state input error under constant acceleration input) e_{ea}。

(4) 任意输入下的稳态误差(The steady-state error under any input)。

可见,任意输入信号可看成是阶跃、等速和等加速信号的合成,因而任意输入下总的稳态误差为

It can be seen that any input signal can be seen as to mix step, constant velocity and acceleration signals together, thus the total steady-state error under any input is

$$e_e = e_{ep} + e_{ev} + e_{ea} \tag{7-21}$$

7.5.2 指令输入引起的稳态误差(Steady-state error caused by command input)

由上可知(It is can be seen from the above analysis):

(1) 系统的型次越高,其稳态误差越小,即控制精度可提高。但系统的型次越高,越容易引起系统的不稳定。从确保闭环系统的稳定性要求出发,实际系统的型次一般不超过Ⅱ型,即开环传递函数中积分环节的数量不会超过2个。

The higher the type times of system are, the smaller the steady-state error is, which can improve the control accuracy. But the higher the type times of system are, the easier to lead to system instability. In terms of ensuring the stability of the closed loop system, the actual type of system views are generally not more than Ⅱ, the number of integral element in open loop transfer function will be not more than 2.

(2) 稳态位置误差、稳态速度误差和稳态加速度误差都是指输出跟踪指令与输入相比较存在的偏差。

Steady-state position error, steady-state speed error and acceleration error are all the bias, which are caused by comparing the output signal with input one.

(3) 稳态误差与系统的结构类型及输入形式有关。

The steady-state error has relationship with system structure type and form of the input.

(4) 开环增益大时稳态误差变小。

When the open loop gain is larger, steady state error becomes smaller.

在图7-12中,令$R(s) = 0$,可求得干扰$F(s)$作用下的系统输出如图7-13所示。

In Fig. 7-12, set $R(s) = 0$, the system output under interference $F(s)$ can be obtained as shown in Fig. 7-13.

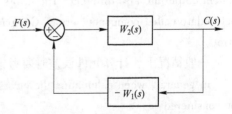

图7-13 干扰作用下的系统框图
(Fig. 7-13 System block diagram under interference)

7.5.3 扰动输入引起的稳态误差（Steady state error caused by interference input）

$$E_c(s) = - C(s) \tag{7-22}$$

该输出全部形成了干扰误差，即

All the output compose the interference error, that is,

由图 7-13 可得

From Fig. 7-13 we can conclude that

$$\frac{C(s)}{F(s)} = \frac{W_2(s)}{1 + W_1(s)W_2(s)} \tag{7-23}$$

由式（7-22）和式（7-23）可得

From formula (7-22) and (7-23) we can conclude that

$$E_c(s) = \frac{- W_2(s)}{1 + W_1(s)W_2(s)} F(s) \tag{7-24}$$

由式（7-24）可见，干扰误差的大小也与闭环系统的结构（由误差传递函数的特征方程体现）和干扰信号的类型有关。

From formula (7-24) we can see that the size of the error also relates to the closed-loop system structure (expressed by the characteristic equation of the error transfer function) and the interference signal types.

7.5.4 元件误差（Components error）

构成闭环控制系统的各个元件本身存在一定的误差，由于元件在系统中的具体位置不同，对系统输出端造成的误差大小也是不同的。将组成系统的元件在系统输出端造成的误差称为元件误差。元件误差是造成系统误差的重要原因。

The components of closed-loop control system exit some errors, and since the components have different specific positions in the system, the size of errors caused by the system output are also different. The errors of the components caused at the system output endare called component error. Component error is an important reason for system errors.

一般情况下，计算元件误差时要考虑以下因素：

In general, when calculating the errors of components the following factors need to be considered:

（1）检测环节的误差（由分辨率引起），用 e_m 表示。

Detecting elements' error (caused by the resolution), expressed with e_m.

（2）控制元件的滞环、灵敏度和死区引起的误差，用 e_H 表示。

7.5 电液比例控制系统的静态性能分析
(Static performance analysis of electro-hydraulic proportional control system)

Control elements' hysteresis, sensitivity and errors caused by dead zone, expressed with e_H.

(3) 温漂引起的误差，用 e_T 表示。

Errors caused by temperature drift, expressed with e_T.

(4) 比例阀的压力增益引起的误差，用 e_P 表示。

Errors caused by pressure gain of proportional valve, expressed with e_P.

负载变化引起位置控制系统稳态误差的原理如图 7-14 所示。

The principle of the error of position control system that caused by load changes is shown as Fig. 7-14.

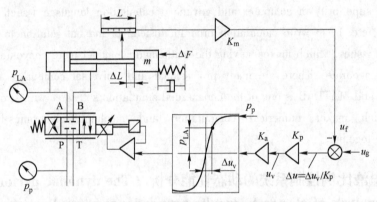

图 7-14 负载变化引起位置控制系统稳态误差示意图

(Fig. 7-14 Principle diagram of the error of position control system that caused by load changes)

负载的变化一方面由执行元件在一个工作周期中的不同阶段所驱动的负载不同造成，另一方面就是负载为弹性负载（负载大小与位移有关）或机架存在弹性造成的。元件误差是上述因素引起的误差之和，即

On one hand, changes in load are caused by the difference of load actuator in different periods of a working cycle, and on the other hand, the load is elastic load (the load size is relative to the displacement) or there exists elasticity in rack that leads to the changes in load. Component error is the sum error of the above factors

$$e_e = e_m + e_H + e_T + e_p \tag{7-25}$$

控制系统的仿真分析集中体现两个步骤：建模和仿真。其基本思想就是建立物理的或数学的模型来模拟现实的过程，以研究过程和规律。实物仿真比较形象、直观，如飞机和导弹外形在风洞中的模拟实验；利用沙盘模型作战；以及汽车的道路实验等。利用数学的语言、方法来描述实际问题，并用数值计算方法对这一问题进行分析，这一过程称为数字仿真。人们利用计算机在数值计算上的优势，采用高级计算语言（如 FORTRAN 语言、C 语言等），编制计算程序替代人工求解，这使得数学模型的求解变得更加方便、快捷和精确。有许多专业性和通

用性的计算仿真软件，MATLAB 是通用性较强的数值计算、机电液综合仿真商业软件之一。

Control system simulation analysis is intensively reflected in two steps: modeling and simulation. The basic idea is to build a physical or mathematical model to simulate the real process to research the process and the law. Physical simulation is intuitive and vivid, such as simulation experiments of aircraft and missiles in the wind tunnel; operations by using the sand table model; and the car's road tests. The process which uses the mathematical language and methods to describe an actual problem, and then uses numerical methods to analyze the issue, is known as digital simulation. People use the numerical superiority of computer and advanced calculation language (such as FORTRAN.C, etc.), to write calculation program instead of manual solution to solve the numerical issues, which makes solving the mathematical model more convenient, faster and more accurate. There are many professional and universal computing simulation software, and MATLAB is one of the commercial simulation softwares with strong generality which makes numerical computation and mechanics-electronics-hydraulics integrated together.

7.6 电液比例控制系统的动态性能分析 (The dynamic performance analysis of electro-hydraulic proportional control system)

MATLAB 1.0 版于 1984 年由 MATHWORKS 公司推出，其名称为由 Matrix Laborator 缩写而来，主要的优势在于它强大的矩阵处理和绘图功能。这一点非常适合于现代控制系统的计算机辅助设计。它刚推出就立刻引起国际控制学术界的重视。MATLAB 把计算、可视化、编程等基本功能都集中在一个易于使用的环境中，并且公式的表达和求解与日常数学运算相似，这一特点，使工程技术人员很容易地熟悉其使用环境，缩短学习和编程时间，为此 MATLAB 语言也被亲切地称为"演算纸式的语言"。

The edition of MATLAB 1.0 was launched by the MATHWORKS in 1984, whose name came from the abbreviation of Matrix Laborator, and its main advantage lies in its powerful matrix processing and graphics capabilities. This is very suitable for computer-aided design of modern control systems. Once launched, it drew great attention from the international control academia immediately. In MATLAB, computing, visualization, programming and other basic functions are integrated in a simple environment, and the expression and the solution of formulas is similar to the normal mathematical operations. This characteristic makes it easy for engineers to familiar with its using environment,

7.6 电液比例控制系统的动态性能分析
(The dynamic performance analysis of electro-hydraulic proportional control system)

shortening the time of learning and programming. For this, MATLAB language is also affectionately called "the language based on calculation paper".

7.6.1 MATLAB 仿真工具软件介绍 (Introduction of MATLAB simulation tool)

随着 MATLAB 的不断完善和功能的开发, 1993 年在 MATLAB 中集成了具有动态系统建模、仿真的工具 SIMULINK (图 7-15)。

With the continuous improvement and development of the function of MATLAB, in 1993, it integrated a dynamic system modeling, simulation tool, called SIMULINK (Fig. 7-15).

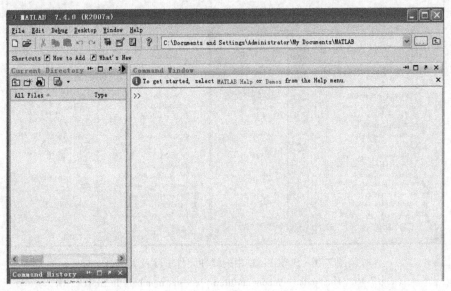

图 7-15 MATLAB 工作界面
(Fig. 7-15 MATLAB workspace)

SIMULINK 是图形仿真工具包,能对动态系统进行建模、仿真和综合分析,可处理线性和非线性方程,离散的、连续的和混合系统,进行单任务和多任务仿真分析。工程技术人员不需要编制任何程序,甚至不必编写一行代码,即可完成相当复杂控制系统的模型构建、仿真和分析校正,能直观、快捷地得到希望的参数 (图 7-16)。

SIMULINK is a graphical simulation tool kit, it can model, simulate and analyze dynamic systems, handle linear and nonlinear equations, discrete, continuous and hybrid systems, and simulate for single-task and multi-task. Engineers do not need the preparation of any program, even without writing a single line code to complete the very complex control system modeling, simulation and analysis of calibration, can directly and quickly achieve the desired parameters (Fig. 7-16).

在 SIMULINK 下进行控制系统仿真，分两步进行：首先是系统建模，其次是系统仿真和分析。

The control system simulation is divided into two steps in SIMULINK: modeling and then simulating and analyzing on the system.

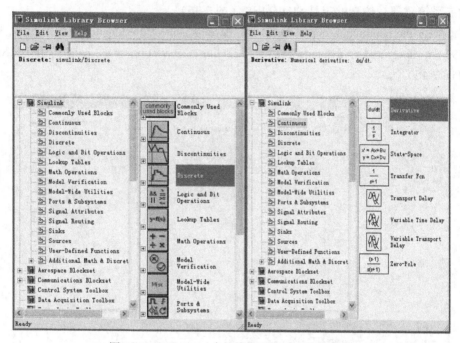

图 7-16　SIMULINK 中的线性元件子库及其成员

(Fig. 7-16　The linear components library and its members in SIMULINK)

7.6.2　闭环位置控制系统仿真实例（Example for closed-loop position control system simulation）

7.6.2.1　Simulink 建模（Simulink Modeling）

A　新建模型窗口（New model window）

点出"Simulink Library Browser"下的图标" ▯ "按钮新建 Untited * 模型窗口（图7-17），或者选择 MATLAB 命令窗口中的 File \ New \ Model 菜单选项。

Click out the icon " ▯ " button under the "Simulink Library Browser" to build a new model window Untited * (Fig. 7-17), or select the File \ New \ Model menu option under MATLAB command window.

7.6 电液比例控制系统的动态性能分析
(The dynamic performance analysis of electro-hydraulic proportional control system)

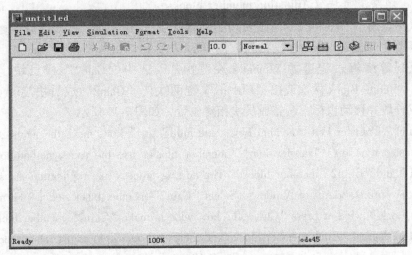

图 7-17 新建 Untited*模型窗口
(Fig. 7-17 New model window)

B 拖入成员块 (Drag member blocks)

从 Simulink 元件库中浏览窗口的"Continuous"子元件库中点击"Transfer Fcn"、"Intergrator"成员块,并拖到模型窗口。同样方法把"Commonly Used Blocks"子元件库中的"Sum"、"Gain"成员块拖到模型窗口(图 7-18)。

Click "Transfer Fcn", "Intergrator" members blocks and drag to the model window from browsing window "Continuous" sub-component library of the Simulink component library. Using the same method, drag "Sum", "Gain" member blocks of "Commonly Used Blocks" sub-component library into the model window (Fig. 7-18).

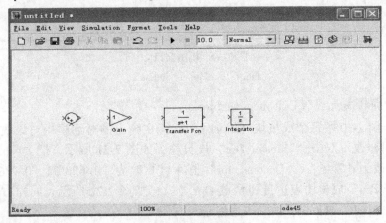

图 7-18 模型窗口 1
(Fig. 7-18 Model window 1)

C 调整成员块 (Adjusting member blocks)

鼠标点出"Transfer Fcn"成员块的同时按住"Ctrl"键，移动鼠标，复制一传递函数"Transfer Fcn1"成员块，同样方法复制"Gain1"和"Gain2"成员块。复制等编辑过程遵循 Windows 规范操作。选中"Gain2"成员块，点出 Format \ Rotate Block 两次或按"Ctrl+R"键两次使"Gain2"成员块旋转180°；单击成员块并移动鼠标，调整成员块相对位置，如图7-19所示。

Click "Transfer Fcn" member block and hold "Ctrl" key, move the mouse, copy one transfer function "Transfer Fcn1" member block, use the same method to copy "Gain1" and "Gain2" member blocks. The editing process like replication etc should follow the work standards of Windows. Select "Gain2" member block, click "Format \ Rotate Block" twice or press "Ctrl + R" key twice to make "Gain2" member block rotating 180 degrees. Click member blocks and then move mouse to adjust relative position of the member blocks. It is shown as Fig. 7-19.

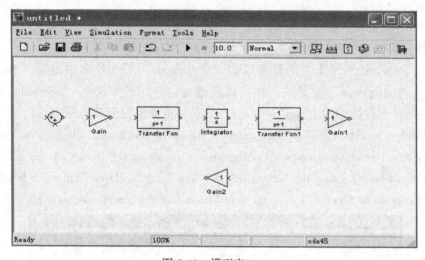

图 7-19 模型窗口 2

(Fig. 7-19 Model window 2)

D 编辑成员块 (Edit member blocks)

双击图 7-20 中每个成员块 (Block)，均能在弹出的对话框中对该块的参数进行编辑修改。双击"Transfer Fcn"成员块，如图 7-21 所示，把"Numerator"值 [1] 改为 [7e-5]；"Denominator"值 [11] 改为 [1/628^22 * 0.7/628 1]，"Transfer Fcn"成员块表达传递函数 $G_{pv}(s)$，如图 7-22 所示。同样方法修改"Transfer Fcn1"成员块，使它表达传递函数 $G_h(s)$。

Double click each member block in Fig. 7-20, and then its parameters can be edited in a pop-up dialog. Double click "Transfer Fcn" member block, as Fig. 7-21,

7.6 电液比例控制系统的动态性能分析
(The dynamic performance analysis of electro-hydraulic proportional control system)

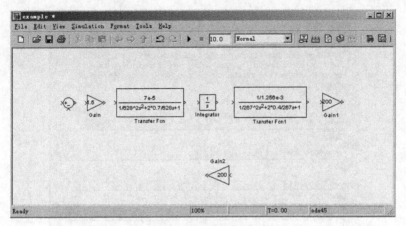

图 7-20 模型窗口 3
(Fig. 7-20 Model window 3)

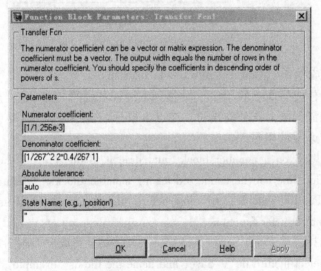

图 7-21 打开的成员块
(Fig. 7-21 Opened block)

change "Numerator" value [1] to [7e-5], and change "Denominator" value [11] to [1／628^22*0.7/628 1], thus making the "Transfer Fcn" member block express transfer function Gpv (s), as Fig. 7-22. Use the same method to modify "Transfer Fcn1" member block to make it express transfer function Gh (s).

对其他成员块进行相应的修改单击成员块后,用鼠标拖动成员块的任一角点,可改变成员块尺寸大小,使函数表达式显示完整。

After modifying other member blocks, drag any corner of the member blocks to change their size, so as to fully display the function expression.

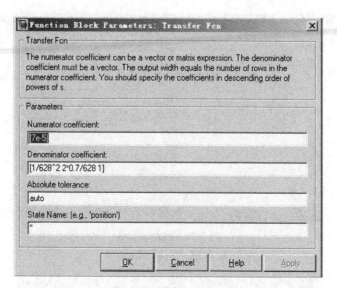

图 7-22 修改后成员块

(Fig. 7-22 Modified block)

E 连接成框图 (Connect into block diagram)

用鼠标点住成员块上的 ">",并拖到下一成员块的 ">" 处,成员块间就连上流程线。从流程线上作分支线时,在点击鼠标前需按住 "Ctrl" 键,其结果和通常书写的传递函数相同。

Click the ">" on a member block, and drag it to the ">" of the next member block, so the flow line can be connected among member blocks. When making branch line from the flow line, hold "Ctrl" key before clicking mouse, and its results is the same with normal transfer function.

最后,选择菜单 File \ Save,取文件名 example,完成后如图 7-23 所示。

Finally, select menu File \ Save, and name the file as "example". The finished file is shown as Fig. 7-23.

7.6.2.2 仿真 (Simulation)

A 检测:加入信号类型和输出显示类型 (Test: Add Signal type and output display type)

从 Simulink 元件库中浏览窗口的 "Sources" 子元件库中点击 "Sine Wave" 成员块,并拖到模型窗口。同样方法把 "Sinks" 子元件库中的 "Scope" 成员块拖到模型窗口,并复制 "Scope1" 成员块。连接构成一个正弦波输入、示波器显示输出的仿真图。

Click the "Sine Wave" member block from sub-component library of the "Sources" library under the Simulink library browser window, and drag them into the model win-

7.6 电液比例控制系统的动态性能分析
(The dynamic performance analysis of electro-hydraulic proportional control system)

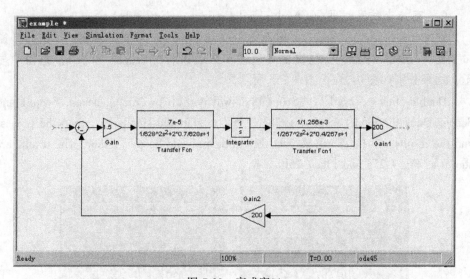

图 7-23 完成窗口
(Fig. 7-23 Block chart)

dow. In the same way, drag the "Scope" member block of the "Sinks" sub-component library into the model window, and copy the "Scope1" member block. Connect them to form a simulation diagram with a sine wave input and an oscillographic displayed output.

B 设置观测点 (Set the observation point)

用编辑成员块的方法为 K_p 赋值，或者在 MATLAB 命令窗口输入 $K_p = 1.5$，仿真框图如图 7-24 所示。

Assigned the K_p with the method of editing members block, or enter $K_p = 1.5$ in the MATLAB command window. The simulation block is shown as Fig. 7-24.

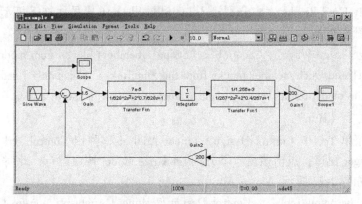

图 7-24 仿真框图
(Fig. 7-24 Simulation block)

C 直接查仿真过程（Check simulation process directly）

双击"Scope1"、"Scope"将弹出 Scope1 和 Scope 两个对话框，单击模型窗口工具栏"▶"开始进行仿真，其过程和结果分别在 Scope1 和 Scope 窗口中显示，结果分别如图 7-25a，图 7-25b 所示。

Double click "Scope1", "Scope", it will pop up two dialog boxes, Scope1 and Scope. Click the model window toolbar "▶" to start the simulation, then the process and the results will be shown respectively in Scope1 and Scope window. The results are shown as Fig. 7-25a and Fig. 7-25b.

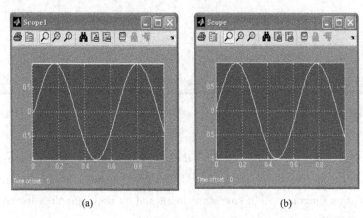

图 7-25 仿真结果

(Fig. 7-25 Simulation results)

(a) Scope1 仿真结果（Simulation result of Scope1）；(b) Scope 仿真结果（Simulation result of Scope）

D 加入输入输出点（Add input and output points）

去除模型窗口中的"Sine Wave"、"Scope1"、"Scope"成员块，从 Simulink \ Commonly Used Blocks 库中分别输入 In1 和输出 Out1 成员块拖入模型窗口，并连接，如图 7-26 所示。

Remove "Sine Wave", "Scope1", "Scope" members blocks from model window, and drag In1 and Out1 member blocks from the Simulink \ Commonly Used Blocks library into the model output window, and connect them. It is shown as Fig. 7-26.

E 运行（Run）

选择菜单 Tools \ Control Design \ Linear Analysis，弹出 Control and Estimation Tools Manager 窗口，点击该窗口下方的 Linearize Model 按钮运行，此时 $K_p = 1.5$，运行结果如图 7-27 所示。

Select the menu Tools \ Control Design \ Linear Analysis, pop-up Control and Estimation Tools Manager window, click on Linearize Model button below window to run. Now $K_p = 1.5$. The result is shown as Fig. 7-27.

7.6 电液比例控制系统的动态性能分析
(The dynamic performance analysis of electro-hydraulic proportional control system)

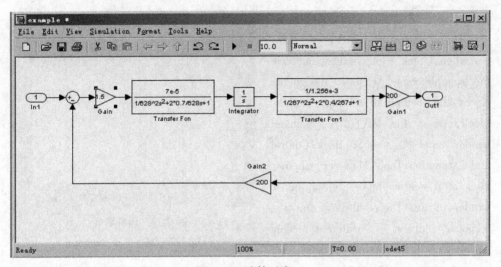

图 7-26 连接后窗口
(Fig. 7-26 Connected window)

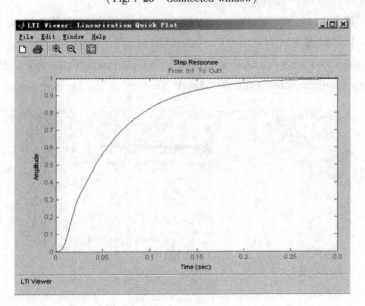

图 7-27 运行结果
(Fig. 7-27 Result of running)

F 改变传递函数的参数(Change the parameters of the transfer function)

不关闭 LTI Viewer：Linearization Quick Plot 窗口，激活 example 模型窗口。用编辑成员块的方法为 K_p 赋值，或者在 MATLAB 命令窗口输入 K_p = 4.5。返回 Control and Estimation Tools Manager 窗口，点击该窗口下方的 Linearize Model 按钮运行。结果便以不同颜色绘出响应曲线（蓝色 K_p = 7.5，绿色 K_p = 6.5），运行

结果如图 7-28 所示。

Do not close the LTI Viewer: Linearization Quick Plot window, activate the example model window. Assign the K_p by the method of editing members block, or in the MATLAB command window enter K_p = 4.5. Back Control and Estimation Tools Manager window, click on Linearize Model button below window to run. The result is to draw the response curves in different colors (Blue K_p = 7.5; green K_p = 6.5), the result is shown as Fig. 7-28.

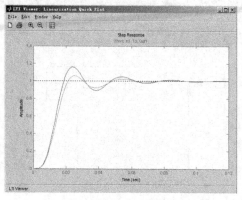

图 7-28 运行结果

(Fig. 7-28 Result of running)

G 改变坐标值 (Change coordinates)

选取 LTI Viewer: Linearization Quick Plot 绘图区鼠标右键 \ Preferences，在 Limits 选项卡里设置 X、Y 轴的坐标，设置窗口如图 7-29 所示。

Select the LTI Viewer: Linearization Quick Plot plot area, right click \ Preferences, set X, Y axis coordinates in the Limits tabbed. The setting window is shown as Fig. 7-29.

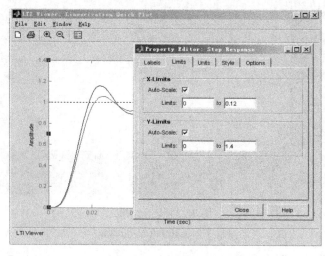

图 7-29 设置窗口

(Fig. 7-29 Setting window)

H 获取性能指标 (To obtain the performance index)

在 LTI Viewer: Linearization Quick Plot 绘图区鼠标右键 \ Characteristics \ 选

7.6 电液比例控制系统的动态性能分析
(The dynamic performance analysis of electro-hydraulic proportional control system)

项,将对已绘出的曲线标记特征值:如过渡过程时间(Rise Time)、进入稳态时间(Settling Time)、峰值点(Peak)等,点击并按住标记点,将显示该点特征值,如图 7-30 所示。

In the LTI Viewer: Linearization Quick Plot plotting area right clicks \ Characteristics \ option, and it will mark the curve which has been drawn with characteristic values, such as Rise Time, Settling Time, Peak and so on. Click a marker and hold down, then its characteristic value will be displayed, which are shown as Fig. 7-30.

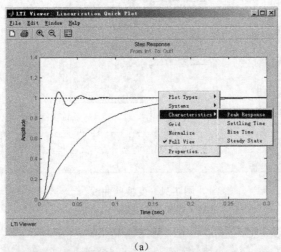

(a)

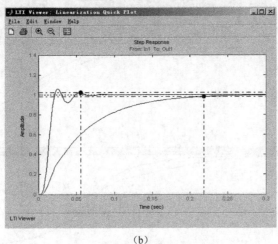

(b)

图 7-30 特征值标记

(Fig. 7-30 Mark the characteristic value)

(a) 标记前(Before marked); (b) 标记后(After marked)

I 变换响应曲线类型（Transform the type of response curve）

在 LTI Viewer：Linearization Quick Plot 绘图区鼠标右键 \ Plot Type \ Bode，进行曲线类型变换。变换前后曲线分别如图 7-31、图 7-32 所示。

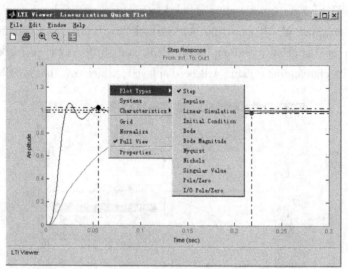

图 7-31 变换前曲线

(Fig. 7-31 Curve of before transformation)

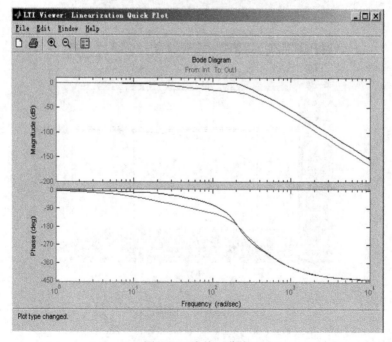

图 7-32 变换后曲线

(Fig. 7-32 Curve of after transformation)

7.7 电液比例控制系统的液压能源选择
(Choice of hydraulic energy of electro-hydraulic proportional control system)

In the LTI Viewer: Linearization Quick Plot plotting area, right click \ Plot Type \ Bode to change the curve type. The curves of before and after transformation are shown as Fig. 7-31 and Fig. 7-32.

J 显示多种类型曲线 (Display many types of curve)

在 LTI Viewer: Linearization Quick Plot 窗口选择 Edit \ Plot Configurations, 设置同时显示响应曲线的种类和布局, 如图 7-33 所示。

In the LTI Viewer: Linearization Quick Plot window, choose Edit \ Plot Configurations, set type and layout of response curves corresponding in one time. It is shown as Fig. 7-33.

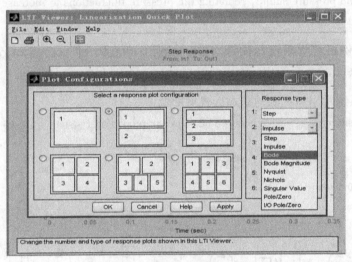

图 7-33 种类和布局的设置

(Fig. 7-33 Setting of type and layout)

7.7 电液比例控制系统的液压能源选择 (Choice of hydraulic energy of electro-hydraulic proportional control system)

合理选择和配置液压能源及相关辅件是保证液压系统长期可靠工作的重要因素。液压系统对液压能源的要求较液压传动系统严格,通常要求选用独立的液压能源。其要求有:

Reasonable choice and configuration of hydraulic energy and related accessories are important factors to ensure long-term reliable operation of hydraulic system. Hydraulic system has stricter requirements than hydraulic transmission about hydraulic energy, and it often requires independent hydraulic energy. The requirements are:

(1) 液压能源应能提供系统足够的压力和流量;

Hydraulic energy system should be able to provide enough pressure and flow;

(2) 保证油液的清洁度,以提高伺服可靠性;

Ensure the cleanliness of oil in order to improve the reliability of servo systems;

(3) 防止空气混入,以免影响系统的稳定性和快速性;

Prevent air mixing so as not to affect the system stability and fast;

(4) 保持油温恒定,以提高系统性能并延长元件寿命;

Maintain the oil temperature constant to improve system performance and extend component life;

(5) 保持压力稳定,减小压力波动,以提高响应能力。

Keep the pressure stable, reducing the pressure fluctuations to increase the responsiveness.

7.7.1 比例系统常用液压油源 (Common hydraulic oil source of proportional system)

比例系统常用恒压源的基本形式见图 7-34。

The Basic forms of steady pressure oil source of proportional system is shown as Fig. 7-34.

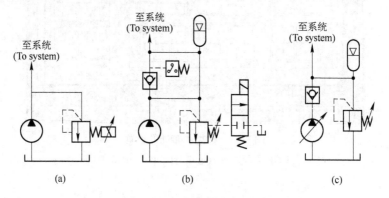

图 7-34 恒压源的基本形式

(Fig. 7-34 Basic forms of steady pressure oil source)

7.7.2 液压油源与负载的匹配 (Matching of hydraulic energy and load)

图 7-35 表示了液压能源与负载的匹配情况,负载特性曲线 a 和 b 分别表示液压能源的压力和流量不足。为了充分发挥能源的作用以提高系统效率,能源装置的最大功率点应尽量接近负载特性曲线的最大功率点。

Fig. 7-35 indicates the matching situation of the hydraulic energy and the load, the load characteristic curve a and b respectively show the shortage of pressure and flow of the hydraulic energy. In order to give full play to the role of energy to improve system efficiency, maximum power point of energy devices should be as close as to maximum power point of the load characteristic curve.

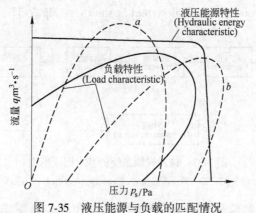

图 7-35 液压能源与负载的匹配情况

(Fig. 7-35 The matching of the hydraulic energy and the load)

7.8 液压伺服与比例控制系统设计举例(Instance of hydraulic servo and proportional control system design)

本节以闭环位置控制系统为例,介绍采用伺服阀的位置控制系统的数学模型,如图 7-36 所示。

To take a closed-loop position control system as an example, the mathematical model of position control system using servo valve is shown as Fig. 7-36.

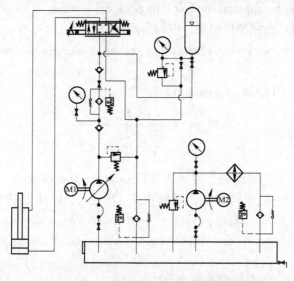

图 7-36 采用伺服阀构成的位置控制系统

(Fig. 7-36 Position control system composed by servo valve)

位置控制系统的组成框图如图 7-37 所示。

The diagram of position control system is shown as Fig. 7-37.

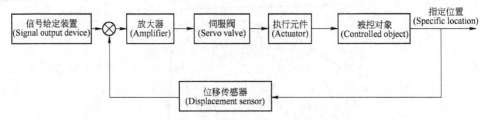

图 7-37　位置控制系统的基本组成框图

(Fig. 7-37　The diagram of position control system)

图 7-37 的动态数学模型按只有惯性负载的位置控制系统处理。这种位置控制系统应用相当广泛。图中各环节的传递函数按以下方法求出：

The dynamic mathematical model of Fig. 7-37 is dealt with as position control system with only inertial load. This kind of position control system is widely used. The transfer function of each element is calculated as follows：

(1) 放大器：由于其转折频率比系统的频宽高得多，故可近似为比例环节。

Amplifier：Because its corner frequency is much higher than the bandwidth of the system, it can be approximated as a proportional element.

(2) 位移传感器：其频宽也比系统频宽高得多，亦可近似为比例环节。

Displacement sensor：Because its bandwidth is much higher than that of the system, it can also be approximated as a proportional element.

(3) 伺服阀：根据测试结果，将伺服阀视为一个二阶环节。

Servo valve：it can be regarded as a second order element according to the test result.

传递函数为 (Transfer function is)：

$$W_{sv}(s) = \frac{K_{sv}}{\dfrac{s^2}{\omega_{sv}^2} + \dfrac{2\zeta_{sv}}{\omega_{sv}}s + 1} \qquad (7\text{-}26)$$

式中　$W_{sv}(s)$ ——伺服阀的传递函数 (Transfer function of servo valve)；

　　　K_{sv} ——伺服阀增益，以电流 I_N 为输入、以阀芯位移 X_v 为输出时 $K_{sv}=X_v/I_N$ (Gain of servo valve. When electric current I_N is the input, and displacement of valve core X_v is the output, $K_{sv}=X_v/I_N$)；

　　　ω_{sv} ——伺服阀固有频率 (Natural frequency of servo valve)；

　　　ζ_{sv} ——阻尼系数，无因次量；取决于伺服阀类型及规格，一般 0.5~

7.8 液压伺服与比例控制系统设计举例
(Instance of hydraulic servo and proportional control system design)

1 (Damping coefficient, which is a non-dimensional value. It depends on the type and the size of servo valve, and generally it is 0.5~1)。

输出液压缸位移 X_p 对输入阀芯位移 X_v 的传递函数为:

The transfer function of Output hydraulic cylinder displacement to Input core displacement is:

$$\frac{X_p}{X_v} = \frac{\dfrac{K_q}{A}}{s\left(\dfrac{s^2}{\omega_h^2} + \dfrac{2\zeta_h}{\omega_h}s + 1\right)} \quad (7\text{-}27)$$

$$\omega_h = \sqrt{\frac{4\beta_e A^2}{V_t m_t}} \quad (7\text{-}28)$$

$$\zeta_h = \frac{k_{ce}}{A}\sqrt{\frac{\beta_e m_t}{V_t}} + \frac{B_p}{4A}\sqrt{\frac{V_t}{\beta_e m_t}} \quad (7\text{-}29)$$

当 B_p 较小可以忽略不计时,ζ_h 可近似写成(When B_p is quiet small, ζ_h can be)

$$\zeta_h = \frac{K_{ce}}{A}\sqrt{\frac{\beta_e m_t}{V_t}} \quad (7\text{-}30)$$

$$K_q = \frac{\partial q_L}{\partial x_v} = C_d W \sqrt{\frac{2(p_s - p_L)}{\rho(1 + \eta^3)}} \quad (7\text{-}31)$$

式中　ω_h ——液压缸固有频率(Natural frequency of hydraulic cylinder);
　　　ζ_h ——液压阻尼比(Hydraulic damping coefficient);
　　　X_p ——液压缸位移(Displacement of hydraulic cylinder);
　　　K_q ——流量增益(Gain of flow);
　　　X_v ——阀芯位移(Displacement of core);
　　　β_e ——体积弹性模量(Bulk modulus of elasticity);
　　　V_t ——液压缸的容积(Volume of hydraulic cylinder);
　　　m_t ——活塞缸质量(Mass of piston);
　　　A ——活塞有效面积(Efficient area of piston);
　　　q_L ——负载流量(Load flow);
　　　C_d ——滑阀流量系数(Flow coefficient of slide valve);
　　　W ——滑阀开口面积梯度(Area gradient of the opening of slidevalve);
　　　p_s ——油源压力(Oil pressure);
　　　p_L ——负载压力(Load pressure);

η——液压缸两腔面积之比（Area ratio of the two chambers of hydraulic cylinder）；

ρ——液压油密度（Density of hydraulic oil）。

液压系统模型框图如图 7-38 所示（The diagram of hydraulic system is shown in Fig. 7-38）。

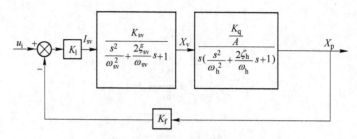

图 7-38　液压系统模型框图

(Fig. 7-38　Diagram of hydraulic system)

第8章 电液比例控制技术的工程应用

Chapter 8 Engineering Application of Electro-hydraulic Proportional Control Technology

8.1 电液比例控制技术在钢管水压试验机上的应用（Application of electro-hydraulic proportional control technology in pipe hydrostatic tester）

钢管水压试验机是制管生产线上的关键设备，用于试验钢管所能承受的压力，以使钢管满足输送石油、天然气等的要求。在进行钢管水压试验时，必须给钢管两端施加与管内水压值相适应的反作用力，以保证管内试验水压达到要求的值。该作用力太大，就有可能使钢管管体变形，甚至压弯钢管；力太小，管内的压力水就会从管端渗漏，甚至喷出伤人，同时还会破坏密封。因此要求施加在管端上的力在试压过程中随着管内水压值的变化而变化。为了满足此要求，在钢管水压机中应设有一套压力同步控制装置。以前国内大型钢管厂的试验机都是采用传统的机械杠杆式压力同步控制装置以实现钢管试压过程中保证水压力平衡的，如图 8-1 所示。

Pipe hydrostatic tester is a key device of pipe production line used to test pipe's bearing pressure, whether they can meet the requirements of transporting oil, natural gas etc. When conducting the pipe hydrostatic test, a counteracting force which is suitable for the water pressure in the pipes must be applied to the two ends of pipes to make sure that the water pressure in the pipe reaches the required level. If the force is too large, it may deform or even buckle the steel pipe. If too small, pressure fluid may leak from the pipe end, which will break the seal and even ejection outsides to cause severe damage. So the forces applied to pipe's ends are required to change with the variable water pressure in the pipe during the pressure testing process. In order to meet the requirement, a set of pressure synchro-control devices should be built into the pipe hydraulic tester. In the past, testers in large domestic steel pipe plants generally used traditional mechanical lever-type pressure synchro-control device to keep the balance of water pressure in the test process, as is shown in Fig. 8-1.

第 8 章 电液比例控制技术的工程应用
Chapter 8 Engineering Application of Electro-hydraulic Proportional Control Technology

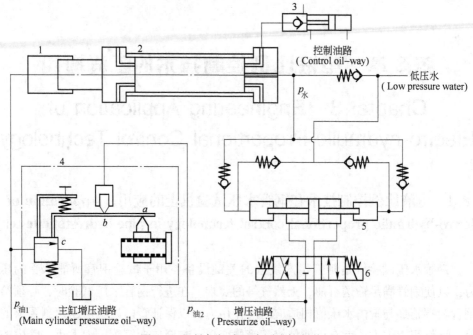

图 8-1 机械杠杆式压力同步控制装置

(Fig. 8-1 Mechanical lever-type pressure synchro-control device)

1—主加压油缸（Main pressure cylinder）；2—被试钢管（Test tube）；
3—放气装置（Deflating device）；4—压力同步装置（Pressure synchronous device）；
5—增压缸（Pressurized cylinders）；6—换向阀（Directional control valve）

工作原理（Working principle）：

(1) 增压部分：当电磁换向阀 6 右侧带电时，增压缸 5 的活塞右行，右侧输出的高压水经过单向阀进入钢管内，同时低压水经左侧的单向阀向增压缸补水；电磁换向阀 6 左侧带电时，增压缸左侧输出高压水。

Pressurization part: When right solenoid of electro-magnetic directional control valve is turned on, piston of pressurization cylinder 5 moves right. Output high pressure water flows into steel pipes through check valve, and at the same time, low pressure water supplies to pressurization cylinder through left check valve. When left solenoid of electro-magnetic directional control valve is turned on, left part of pressurization cylinder outputs high pressure water.

(2) 压力同步：为满足钢管不同口径规格的试压要求，通过调节杠杆支点 a 的位置来改变油压与水压值的比例关系。调节好后支点 a 不再动，b 点随压力油 $p_{油2}$ 的变化而变化，c 点随 b 点而动。c 点的上下移动改变了溢流阀的溢流压力，也就改变了 $p_{油1}$ 的压力。

Pressure synchronization: To meet pressure test requirements of different steel pipe

8.1 电液比例控制技术在钢管水压试验机上的应用
(Application of electro-hydraulic proportional control technology in pipe hydrostatic tester)

size, the ratio of oil and water pressure is changed by adjusting position of level pivot a. Pivot a remain stationary, pivot b varies with $p_{油2}$, pivot c moves following b. Vertical movement of c changes relief pressure of relief valve, so $p_{油1}$ is changed.

这种机械杠杆式压力同步控制装置存在诸多缺点：灵敏度差且难以精确调整；只适用于试管加压过程中的压力同步控制，不适用于压机卸压过程的压力同步；对于不同钢管口径规格的试压都要调整好该装置的杠杆支点，而且往往不能一次调整好平衡，给操作者带来很大不便，钢管试压效率很低。对此，我们考虑采用电液比例控制技术，如图 8-2 所示。

This kind of device has lots of disadvantages, e. g. poor sensitivity and difficulty of sensitive adjustment; only appropriate for pressure synchro-control, not appropriate for pressure synchronization of pressure relief etc. Level pivot has to be adjusted for pressure test of different steel pipe size always repeatedly. It is inconvenient for operators. Efficiency of pressure test is low. So the electro-hydraulic proportional control technology is considered, as is shown in Fig. 8-2.

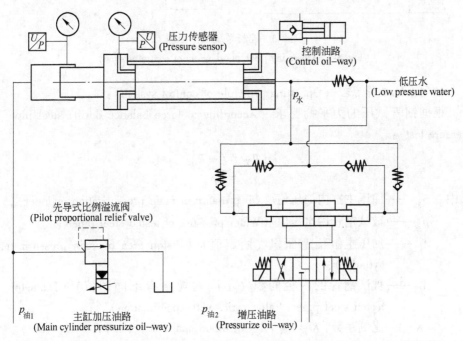

图 8-2 采用电液比例控制技术的液压原理图
(Fig. 8-2 Hydraulic schematic diagram of electro-hydraulic proportional control technology)

在图 8-2 中，采用电液比例溢流阀取代机械杠杆压力同步控制装置，并与比例放大器、压力传感器等组成压力反馈的闭环电液比例控制系统。其控制系统工作原理如图 8-3 所示。

As shown in Fig. 8-2, electro-hydraulic proportional relief valve instead of mechanical lever-type pressure synchronous control device, with proportional amplifier and pressure sensors etc, compose a closed-loop electro-hydraulic proportional control system with pressure feedback. The operating principle of the control system is shown in Fig. 8-3.

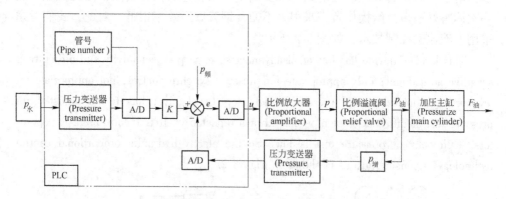

图 8-3 控制系统工作原理

(Fig. 8-3 Operating principle diagram of control system)

控制系统工作原理 (Operating principle of control system):

根据钢管试压的力平衡要求 (According to force balance during steel pipe's pressure test):

$$A_{缸} p_{油} = A_{管} p_{水}$$

$$p_{油} = K p_{水}$$

式中　$p_{油}$——加压主缸内油压值 (Oil pressure of main pressurization cylinder);

$p_{水}$——被试钢管内水压值 (Water pressure of tested steel pipe);

$A_{缸}$——加压主缸活塞面积 (固定值) (Piston area of main pressurization cylinder);

$A_{管}$——被试钢管的内径面积 (不同的规格有不同的值) (Diameter of tested steel pipe (Value varies with specification));

K——比例系数, $K = A_{管}/A_{缸}$ (Proportional coefficient, $K = A_{管}/A_{缸}$)。

不同的钢管管径规格对应不同的 K 值, 对于不同壁厚的钢管可适当调整 K 值。$p_{预}$ 是为了保证钢管在充低压水时, 不从管端渗漏而人为加向主缸的力, 一般凭经验来确定。

K changes with the size of specific diameter of steel pipe and can be adjusted slightly for different wall-thickness pipes. $p_{预}$ is added to main cylinder in order to prevent the pipe ends from leaking when pipes are filled with low pressure water. In general, it bases on empirical calculation.

可编程控制器（PLC）内存有不同钢管管径规格所对应的比例系数 K 值。在试压前，先将被试管的管号通过拨码开关输入给 PLC（管号对应于钢管的管径规格），PLC 可根据输入的管号从内存中找到相对应的 K 值。在整个试压过程中，PLC 不断对钢管内的水压（$p_水$）和加压主缸内的油压（$p_油$）进行采样，并通过内部运算输出相应值，控制比例溢流阀，调节加压缸内的油压值。在钢管试压的加压和卸压过程中，主缸内的油压值始终随钢管内水压值的变化而按一定的比例相应变化，从而保持油水压力的平衡关系。

PLC stores proportional coefficient K of steel pipe specifications of different diameters. Before pressure test, the number of tested pipe is input to PLC through dial switch (the number corresponds to diameter specification of the pipe), and PLC can find the corresponding K in memory according to the number. During the pressure test, PLC continuously takes samples from water pressure in pipes ($p_水$) and oil pressure in main pressurization cylinder ($p_油$). And calculates the values outputted to control proportional relief valve to adjust oil pressure in cylinder. During the processes of pressurization and relief, oil pressure in main cylinder varies proportionally with water pressure in the pipe to maintain the balance of oil and water pressure.

8.2　飞机拦阻器电液比例控制系统（Electro-hydraulic proportional control system of aircraft arresting system）

飞机拦阻器（图 8-4）是飞机着陆控制的辅助措施之一，主要用于陆基飞机的应急拦阻以及舰基飞机的自由飞着陆和舰基自由飞失败后的应急拦阻。目前，国外较先进的某飞机拦阻器是纯机液系统，以拦阻网或拦阻钢缆连接在一起沿机场跑道对称布置，将被拦阻飞机本身巨大的动能转化成液压能而生成使飞机制动的摩擦力。

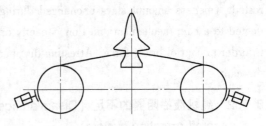

图 8-4　飞机拦阻器的简易模型
(Fig. 8-4　Simple model of aircraft arresting system)

Arresting system (Fig. 8-4) is one of auxiliary measures controlling aircraft landing, mainly used for the emergency arresting of land-based aircraft, free landing of ship-based jet and emergency arresting when free landing fail. At present, one more advanced foreign aircraft arresting system is the pure mechanical-hydraulic driven system, with arresting nets or connected wire rope laid symmetrically along the airport runway, thus the huge kinetic energy of aircraft is converted to hydraulic energy in order to produce the friction force which can brake aircraft.

8.2.1 改进前的液压系统 (Original hydraulic system)

图 8-5 为国外某飞机拦阻器的液压系统原理图，它包括静压和动压两部分。静压部分由 6、7、8、9、10 和 11 组成，由蓄能器 11 提供具有一定压力的油液，以保证飞机拦阻器在不工作时处于锁紧状态；手动泵可以补油。动压部分为能量转换装置，在飞机撞网带动尼龙带盘转动的同时，也带动液压泵 12 转动，输出压力油，产生使飞机制动的摩擦力。节流阀 18 为手动针阀，通过预先调节其开口量大小来适应不同重量、不同撞网速度飞机的拦阻，拦阻过程中节流口大小不变；节流阀 16 与一套凸轮机构连接在一起，靠凸轮型面的变化来改变节流阀的通流能力，以控制刹车压力。通过调节凸轮起始工作角来调节拦停距离。

What Fig. 8-5 shows is the schematic diagram of a foreign aircraft arresting hydraulic system that includes two parts of static and dynamic pressure. The static part (consists of 6, 7, 8, 9, 10, 11, and the accumulator provides the pressure to ensure that the system stays in a locking state off-time; The dynamic part is energy conversion device, which drives the hydraulic pump 12 to output pressurized oil that generates the friction force to brake the aircraft at the same time the aircraft crashes into the net and rolls the nylon reel. Relief valve 18 is a manual needle valve. By adjusting the amount of openess previously, aircrafts with different weight and net-crashing velocity can be arrested. Openess amount stays unchanged during the arresting. Relief valve 16 is connected to a cam mechanism and flow capacity can be changed based on the cam profile in order to control brake force. Arresting distance is regulated by initial working angle of the cam.

8.2.2 纯机液拦阻器的不足 (Disadvantages of pure mechanical-hydraulic aircraft arresting system)

纯机液拦阻器中流入摩擦片组件的油液受节流阀 16 及 18 的控制。节流阀 16 与凸轮机构 17 连接在一起，靠凸轮型面的变化来改变节流阀的通流能力，以控制刹车压力。通过调节凸轮起始工作角来调节拦停距离。对纯机液拦阻器，当飞机撞网状况发生变化时（例如飞机撞偏、飞机重量差异、刹车片摩擦系数变化等），拦阻不能达到预期效果。若凸轮型面受损，拦停效果也会受到极大影响，控制精度不高。

8.2 飞机拦阻器电液比例控制系统
（Electro-hydraulic proportional control system of aircraft arresting system）

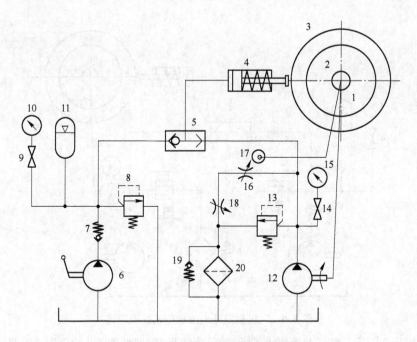

图 8-5 飞机拦阻器纯机液系统原理图

（Fig. 8-5 Schematic of pure mechanical-hydraulic aircraft arresting system）

1—转轴（Shaft）；2—尼龙带盘（Nylon reel）；3—摩擦制动盘（Friction brake disc）；
4—摩擦片组件（Friction components）；5—梭阀（Shuttle valve）；6—手动泵（Hand pump）；
7—单向阀（Check valve）；8，13—安全阀（Safety valve）；9，14—压力表开关（Pressure gauge switch）；
10，15—压力表（Pressure gauge）；11—蓄能器（Accumulator）；12—泵源（Pump source）；
16—节流阀（由凸轮机构调节）（Throttle（by the cam adjustment））；17—凸轮机构（Cam mechanism）；
18—节流阀（手动调节）（Throttle（manual adjustment））；19—单向阀（One-way valve）；
20—回油过滤器（Oil returning filter）

In the system, the oil flowing into the friction disk unit is controlled by flow control valves 16 and 18. A set of cam gear 17 is connected to the valve 16 so that the throttle valve's flow capacity can be changed by variation of the cam surface to control the brake pressure. The distance of arresting can be changed by regulating the initial working angle of the cam. As for the pure mechanical-hydraulic system, when the situation of striking nets varies (for instance, deviation of an aircraft, differences of weight of aircrafts, variation of brake block coefficient and so on), the blocking can't work as expected. If the cam surface is damaged, the blocking can be heavily affected with low control accuracy.

8.2.3 改进后的液压系统（Improved hydraulic system）

改进后的电液比例拦阻系统（图 8-6）仍包括静压和动压两部分。系统的静

Chapter 8　Engineering Application of Electro-hydraulic Proportional Control Technology

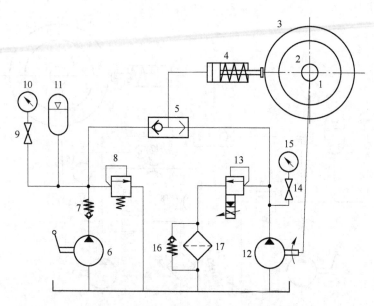

图 8-6　飞机拦阻器电液比例系统原理图

(Fig. 8-6　Schematic diagram of electro-hydraulic proportional aircraft arresting system)

压部分未做改动，起到拦阻前保持拦阻器稳定状态以及飞机刚撞网时避免对系统造成冲击的作用。动压部分将凸轮调节的节流阀改成电液比例溢流阀，这样通过调节比例溢流阀的输入电信号来改变系统输出压力，从而控制使飞机制动的摩擦力。引入电信号更便于用计算机构成自动控制系统。对于不同型号、不同拦停距离的飞机，只要事先计算出拦阻期间拦停位移与系统所需要达到的压力之间的关系，当拦阻飞机到达设定位移时，通过改变电信号的输入来改变系统输出压力（即刹车力）的大小，即可实现拦阻，达到拦停要求。

The improved electro-hydraulic proportional arresting system (Fig. 8-6) still consists of the static and dynamic pressure system. The static part remains the same, keeps the system stable before arresting and avoids the impact on system when engaging. As for the dynamic part, the cam throttle valve is replaced by electro-hydraulic proportional relief valve so that output pressure of the system changes with input electric signal of relief valve to control the brake force. Introduction of electric signal is more convenient to integrate an automatic control system with the help of computer. For aircrafts of different models and different arresting distances, if the relationship between pressure that system needs to achieve and arresting displacement during the arresting is calculated ahead of time, when the aircraft reach the set displacement, arresting requirement can be met by changing the input of electric signal to regulate system's output pressure (brake force).

8.3 电液比例控制技术在 CVT 中的应用 (Application of the electro-hydraulic proportional control technology in CVT)

随着电子技术和自动控制技术的快速发展，车用变速器的技术也越来越完善，形式也更加多样化，在越来越多的车辆上得到应用。

With the rapid development of electronic technology and automatic control technology, transmission technology of automobiles is becoming more and more improved and diversified, and it is widely used use in vehicles.

车用无级变速器 CVT（图 8-7，图 8-8）避免了齿轮传动比不连续和零件数量过多的缺点，能够实现真正的无级变速，具有一系列的优点，如：传动比连续、传递动力平稳、操纵方便，可使汽车行驶过程中经常处于良好的性能状态，节省燃油、改善汽车排放等。

Automobile Continuously Variable Transmission (CVT) (Fig. 8-7, Fig. 8-8) can realize real stepless speed variation in contrast to the traditional gear engagement, e.g. discontinuation and complicated design. CVT has a series of other advantages, such as continuous transmission ratio, steady transmission of power, convenient control, so automobiles can perform better with regard to energy efficiency and emission.

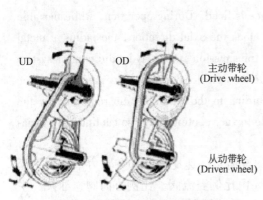

图 8-7 金属带 CVT 原理示意图
(Fig. 8-7 Schematic diagram of Metal V-Belt CVT)

图 8-8 无级变速器外观图
(Fig. 8-8 Outside view of CVT)

金属带式无级变速器属于摩擦传动式无级变速器，它主要利用两个锥形带轮来改变传动比，从而实现无级变速。从图 8-7 可见，发动机输出的动力经输入轴传到主动轮上，主动轮锥盘通过与金属带的 V 型摩擦片侧面接触所产生的摩擦力向前推动摩擦片，这样就使后一个摩擦片推压前一个摩擦片，在两者之间产生推压力。该压力形成于接触弧的始端，至终端逐渐加大。这种推力经金属带的摩擦片作用在从动轮上，由摩擦片通过与从动轮锥盘接触产生的摩擦力带动从动轮旋转，将动力传到了从动轴上。

Metal V-Belt CVT is one kind of friction driven continuously variable transmission and it uses two cone belt wheels to change transmission ratio. As the Fig. 8-7 shows, output power of engine is passed to drive wheel through input shaft, cone disc of which contact with side of V-shape friction disc of metal belt producing friction force to push the disc forward, so the back disc thrust the forward one to make forces. They are produced at the beginning of contacting arc, getting bigger gradually to the end, acting on the driven wheel through friction discs of metal belt. Friction force by friction disc contacting with cone disc of driven wheel makes driven wheel rotate and as a result, then power is passed to the driven shaft.

金属带的主动轮、从动轮皆由可动锥盘部分和不可动锥盘部分构成，它们的中心距是固定的。工作中，当主、从动轮的可动锥盘做轴向移动时，改变了金属传动带的工作半径，从而改变了传动比。可动锥盘的轴向移动量是根据发动机使用要求的变速比，通过液压控制系统分别调整主、从动轮上作用油缸的压力来调节的。由于工作节圆半径可连续调节，从而可实现无级变速。

Both drive wheel and driven wheel of metal belt consist of a movable cone disc and a non-movable one, and their centre distance is fixed. During operation, while movable cone discs of drive and driven wheel move along the axial direction, the radius of metal belting is changed resulting in variation of transmission ratio. Hydraulic control system respectively adjusts pressure of working cylinders on drive and driven wheels to change axial movement of movable cone discs according to the variable ratio requested by the engine. The radius of the pitch circle is able to vary continuously, so continuously variable transmission can be realized.

系统采用的电液比例控制系统如图 8-9 所示，速比控制和夹紧力采用同一压力源，由发动机带动泵运行。本系统主要由比例溢流阀、比例方向阀、油泵、主从动轮油缸和电子控制块组成。液压油从泵里排出来，由比例溢流阀根据控制系统的指令实时控制系统的压力。

The electronic-hydraulic proportional system is shown in Fig. 8-9. The control of the speed ratio and the clamping force are generated from the same pressure source, which is established by a pump driven by the engine. This system consists of proportional relief valve, proportional directional valve, pump, cylinders of drive and driven wheels, and the electronic control part. When oil is pressurized by the pump, the system pressure is under real-time control by the proportional relief valve according

8.3 电液比例控制技术在 CVT 中的应用
(Application of the electro-hydraulic proportional control technology in CVT)

to commands of the control system.

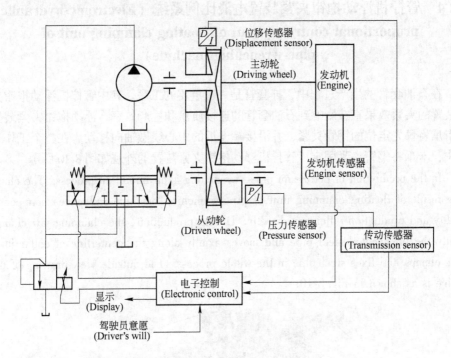

图 8-9 无级变速器控制原理图
(Fig. 8-9 Control schematic diagram of CVT)

液压油通过比例方向阀控制进入主动缸的流量，进一步控制主动轮可动锥盘部分前进的位移，通过金属带进一步改变从动轮可动锥盘部分的位置，从而间接地改变传动比。可以使用发动机转速传感器和转矩传感器，主动带轮的位移传感器，被动带轮的压力传感器，作为电子控制的输入信号。通过测量发动机的转速来调整速比，从而控制发动机转速满足工作要求。

The proportional directional valve controls flow rate of oil into the drive cylinder to further control the displacement of movable cone discs on the drive wheel. Then it will change the position of movable cone discs on the driven wheel through the metal belt, thereby change the transmission ratio indirectly. Input signals of electronic control can be determined from the speed and torque sensor of the engine, the displacement sensor of the drive belt wheel or the pressure sensor of the driven belt wheel. The speed ratio can be adjusted by measuring the engine's speed so as to meet the requirement of the engine speed.

8.4 管拧机浮动抱钳夹紧装置电液比例系统（Electronic-hydraulic proportional control system of floating clamping unit of pipe-wrenching machine）

在石油套管的生产过程中，拧套管是一道重要的工序。其中管拧机浮动抱钳夹紧装置的夹紧效果直接影响到石油套管的连接质量和生产效率。浮动抱钳夹紧装置的卡爪将预先定位的钢管夹紧，并沿接箍卡爪的中心线做轴向移动。在整个工作过程中，卡爪将钢管夹紧固定。管拧机浮动抱钳夹紧装置的外观如图 8-10 所示。

In the production of petroleum pipe, wrenching is an important process. The clamping result of floating clamping unit of pipe-wrenching machine influences connection quality and production efficiency directly. During production, the clamping jaw clamps tightly the pre-located steel pipe and moves axially along the centerline of collar jaw. Jaws clamps and fixes steel pipe in the whole process. The outside view drawing of the device is as shown in Fig. 8-10.

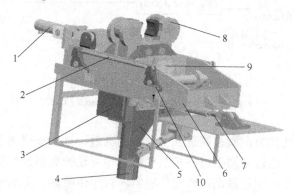

图 8-10 管拧机浮动抱钳夹紧装置的外观图

(Fig. 8-10 Outside view drawing of floating clamping unit of pipe-wrenching machine)
1—减振油缸（Damping cylinder）；2—水平传力机构（Horizontal force-transferring mechanism）；3—小箱体（Small box）；4—夹紧缸（Clamping cylinder）；5—竖直传力机构（Vertical force-transferring mechanism）；6—大滑台（Big slipway）；7—水平驱动缸（Horizontal drive cylinder）；8—夹爪体（Clamping draw body）；9—支撑轴（Suportting shaft）；10—凸轮随动轴承（Follow-up bearing）

8.4.1 工作原理（Operating principle）

管拧机是用于生产石油套管的关键设备，夹紧装置为其主要部件。首先，经步进梁输送过来的带有经过预拧管箍的钢管进入抱钳位置，夹紧油缸 4 活塞杆伸

8.4 管拧机浮动抱钳夹紧装置电液比例系统
(Electronic-hydraulic proportional control system of floating clamping unit of pipe-wrenching machine)

出,带动由夹爪体 8 等组成的四杆机构夹紧钢管。然后,水平油缸带动整个滑台向接箍卡盘移动。到达位置后,卡盘夹紧管箍并带动其绕卡盘中心旋转,由于钢管被夹紧装置夹紧固定,这样,管箍被拧紧。

Pipe-wrenching machine is a kind of key device of oil casing pipe production, and clamping device is the main part. Firstly, steel pipe with pre-tightening couplings are moved by the walking beam into holding-clamps position, then the piston rod of clamping cylinder 4 stretches out, driving the four-bar linkage made of clamping jaws 8 to clamp steel pipes. Secondly, the horizontal cylinder drives the slipway and couplings to move to the position where the chuck clamps the coupling and drives it to rotate round center of the chuck. Because pipes are clamped tightly, the coupling is screwed up.

在拧紧管箍过程中,拧箍力所产生的水平径向力通过竖直传力机构转化为竖直径向力,竖直径向力直接传递给大滑台。这样,钢管拧箍力均转化为竖直向下的力。凸轮随动轴承 10 为大滑台支撑导向轮。在水平传力机构的作用下,所有的竖直力均转化为水平力传递给液压减振油缸而被液压减振系统吸收。保证在夹紧钢管的同时,允许钢管存在一定的偏心。

During the process of screwing up, the horizontal radial force produced by the tightening force is transmitted vertically downward through vertical force-transferring mechanism and is transmitted to the slipway directly. In this way, tightening forces are all transmitted into vertical downward forces. The follow-up bearing of the cam 10 is supporting guide wheel of the giant slipway. Under the action of the horizontal force-transmitted mechanism, all vertical forces are transmitted into horizontal forces, passed to hydraulic damping cylinder and get absorbed. Some eccentricity is acceptable when they are clamped.

8.4.2 液压站 (Hydraulic station)

一用一备变量柱塞泵 4 和电磁溢流阀 5,用于泵的空载启动及工作时起安全保护作用。过滤冷却采用旁置式,模块化设计,系统不工作时仍可以单独进行过滤冷却(图 8-11)。

A variable piston pump 4 and an electromagnetism-flooding-valve 5 are used for non-load starting of the pump and pressure relief. The station has two variable piston pumps, a job, a standby. Filtration and cooling is carried out in an independent circuit with another pump (Fig. 8-11).

(1) 减振支路:调节减压阀 7 及换向阀 8,油液推动油缸 12 活塞杆伸出,推动水平传力机构调节滑台及钢管位置。到达理想位置后,换向阀 8 切换到中位。工作时,通过传力机构传递过来的力向后挤压活塞杆,无杆腔油液流向蓄能器,蓄能器内油压升高,油液的压力对活塞杆产生的推力与水平传力机构的力相平衡，此刻，活塞杆停止运动。在这个过程中，外力均被蓄能器吸收。

Damping branch: Through adjusting pressure reducing valve 7 and directional control valve 8, oil drives piston of cylinder 12 to stretch out, which drives the horizontal force-transmission mechanism to change the position of slipway and steel pipes. When reaching the desirable position, directional control valve 8 switches to the center-position. During the operating process, the force transmitted by force-transferring mechanism squeezes the piston rod backward, making oil in the cap end chamber flow into the accumulator, resulting in higher pressure in the accumulator. Finally, when the thrust force by oil acting on the piston rod balances the force of horizontal force mechanism, the piston rod stops. In the process, external forces are all absorbed by the accumulator.

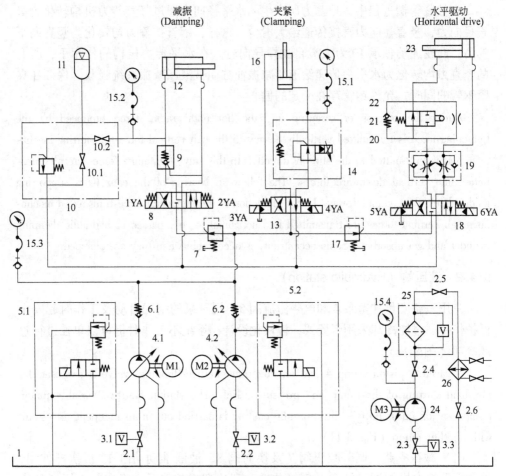

图 8-11 液压系统原理图

(Fig. 8-11 Schematic diagram of hydraulic system)

(2) 夹紧支路：用于夹紧钢管，保证拧箍时钢管不会发生打滑现象。通过调节两位四通电液换向阀来达到夹爪体松开和夹紧。活塞杆伸出，夹爪体夹紧钢管。夹持钢管的力通过调节比例减压阀来实现。这样，就可以根据不同规格的钢管设置不同的压力级别，避免因调定压力不变而在加工不同规格的钢管时，出现夹坏钢管或出现打滑现象。

Clamping branch: The branch is used to clamp steel pipes, to make sure of no slipping when tighten. Clamping jaws clamp and unfasten by adjusting the 2-position- 4 way electro hydraulic directional control valve. As the piston rod stretches out, clamping jaw clamps steel pipes once clamp force is generated by adjusting the proportional pressure reducing valve. So in that way, different pressure levels are set according to different specification of pipes, preventing damaging or slipping when processing different pipe size while pressure is unchanged.

(3) 水平支路：在接箍拧紧过程中，换向阀18处于中位，此时油缸为浮动状态。在接箍及钢管螺纹螺旋传动的作用下，钢管带动滑台向前移动，油缸被滑台驱动。

Horizontal branch: In the process of screwing up of joint couplings, directional control valve 18 is at the center-position, and the cylinder is free to float. Under the joint couplings and screw driving, the steel pipe drives slipway move forward, the cylinder is driven by the slipway.

8.5 带钢对中装置电液比例控制系统（Electro-hydraulic proportional control system of strip steel alignment device）

在冷轧带钢酸洗生产线中，带钢对中装置通过控制进入圆盘剪的带钢位置来控制圆盘剪剪切带钢废边的宽度。带钢对中装置起着非常重要的作用，国内有些冷轧薄板酸洗生产线原有的带钢对中装置采用人工目测、手工调节对中，宽度确定后带钢被夹紧的控制方式，很难保证产品的质量，以致经常出现废料，极大地影响了该机组乃至整个生产线的产能。该对中装置存在着一些固有的缺点：一是压力设定不连续，系统压力无法根据带钢厚度自动调节；二是带钢宽度变化引起的作用力通过液压缸反作用于液压系统，液压缸只起到一定的缓冲作用，而不能实现随带钢宽度自动调节。为了克服这些缺点，带钢对中装置的液压系统采用了电液比例控制技术，其工作原理如图8-12所示。

In the pickling line, strip steel alignment device controls rotary waste-shearing selvage's width by controlling the position of strip steels into rotary shears. The device plays a very important role. Some domestic original strip alignment device of cold-rolled sheet metal pickling production line depends on eye observation and manually adjusting, which is hard to ensure quality of products. It always results in unaccepted product and productivity of this unit and even the whole line is affected. Some inherent defects exist: firstly, pressure setting is discrete and system pressure is not able to be automatically adjusted. Secondly, forces caused by variation of strip width react on hydraulic system through cylinders. Cylinder only plays a role of damper, not able to automatically vary with strip width. To overcome these shortages, the hydraulic system uses eletro-hydraulic proportional control technology. Schematic diagram is shown in Fig. 8-12.

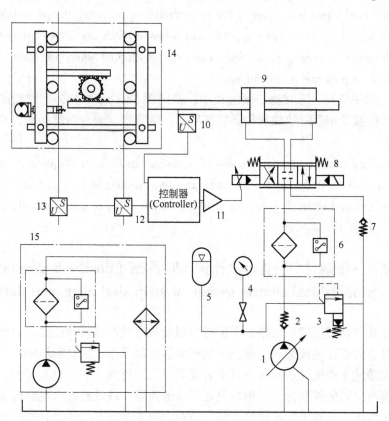

图 8-12 带钢对中装置电液比例控制系统原理图

(Fig. 8-12 Schematic diagram of electro-hydraulic proportional control system of strip steel alignment device)

1—恒压变量泵（Constant pressure variable pump）；2，7—单向阀（Check valve）；3—电磁溢流阀（Electro-magnetic relief valve）；4—压力表及其开关（Pressure gauge and switches）；5—蓄能器（Accumulator）；6—高压过滤器（High pressure precise filter）；8—电液伺服阀（Electro-hydraulic servo valve）；9—伺服缸（Servo cylinder）；10—位移传感器（Displacement sensor）；11—伺服放大器（Servo amplifier）；12—控制器（Controller）；13—测宽光栅（Width-measured grating）；14—带钢对中装置（Trip steel centering device）；15—过滤冷却装置（Filltering and cooling device）

8.5 带钢对中装置电液比例控制系统
(Electro-hydraulic proportional control system of strip steel alignment device)

主泵采用 Rexroth 公司生产的 A10VSO 系列恒压变量泵,为系统提供恒压油源,并能使输出流量随负载需要的流量而变化,使泵源无节流损失,最大限度地降低能耗。电磁溢流阀 3 在泵启动时卸荷,以保证泵空载启动;在泵正常工作时作为安全阀使用。蓄能器 5 采用反应灵敏的皮囊式蓄能器,具有减小柱塞泵输出压力脉动、稳定伺服阀入口压力的功能。泵的出口设置单向阀 2,用以防止蓄能器对泵造成反向冲击而损坏液压泵。高压精密过滤器用于保证进入伺服阀油液的清洁度。伺服阀 8 采用 MOOG 两级力反馈电液伺服阀,其输入额定信号为±10mA 的电流信号。控制器输出的电流信号用于控制伺服阀完成相应的阀口开口度,进而控制伺服缸 9 完成相应的动作。旁置过滤冷却系统 15 由抗污染能力强的螺杆泵、带有污染发讯装置的精密过滤器、冷却风扇和安全阀组成。冷却风扇的启、停由温控器控制,以使伺服系统工作在合适的温度范围内。

Main pump is A10VSO series pressure-compensated pump produced by Rexroth. It supplies constant pressure oil and its output flow rate can vary with load, preventing throttling loss of pump source and reducing energy consumption as possible. Electromagnetic relief valve 3 unloads the system in order to ensure a non-load starting of the pump. And it is used as a safety valve when the pump is normally operating. Bladder accumulator 5 is very sensitive which can reduce pressure fluctuation of piston pump and stabilize servo valve's inlet pressure. Check valve 2 is placed in the inlet of pump, used to prevent shock caused by reverse flow direction from accumulator to pump. Fine filter ensure oil cleanliness into servo valve. Servo valve 8 is a MOOG two-level force feedback electro-hydraulic servo valve, rated signal of which is ±10mA current. Current signal outputted by controller is used to control servo valve opens to a certain degree, making servo valve 9 complete corresponding movements. Side filtering and cooling system 15 consists of screw pump with strong anti-pollution ability, precise filter with pollution alarm device, cooling fan, and safety valve. Temperature controller controls the starting and stopping of cooling fan to make sure that the servo system works under a proper range of temperature.

在带钢对中装置的前部安装有带宽检测光栅 13,用以检测带钢的实际宽度 W_1。伺服缸 9 的活塞杆伸出端上安装 MTS 磁致伸缩式位移传感器,其检测精度可以达到±0.025mm,用以检测夹辊的实际位置 W_2。控制器采用 MOOG 公司的 MCS 伺服控制器,其具有柔性的硬件平台、高精度的模拟量输入输出通道及很高的闭环控制精度,很容易做到高精度的位置随动控制。

Detection grating 13 fixed at front of centering device is used to measure the actual width W_1 of strip steel real-timely. A MTS hysteresis telescope displacement sensor is set in extented end of piston rod. It provides a precision of ±0.025 mm, to detect real position W_2 of clamping roll. MCS servo controller of MOOG Co., Ltd. has a flexible hardware platform, high precise input and output channels of analog quantity and high precise closed-loop control, it is easy to meet requirements of high precision of position following control.

由位移传感器得到的位移信号经过一定的转换关系，转化为夹紧槽的宽度信号后反馈给控制器。在控制器内，W_1 和 W_2 经过运算后将偏差信号输入给数字调节器，经过调节器运算后的信号由控制器的模拟量输出端输出给伺服阀，以调节伺服阀的开度，进而控制伺服缸完成相应的动作，使夹辊随带钢的宽度变化产生随动动作，并保证夹紧带钢。

According to a certain transformation relation, displacement signal obtained by sensor is transformed into width signal of clamping groove as a feedback to controller. Inside the controller, after calculation of W_1 and W_2 error signal is inputted to digital regulator. Signal calculated by the regulator is transported to servo valve through the analog output end, to regulate opening degree of servo valve, controlling servo cylinder to complete corresponding movement, making clamping roll move with variation of strip steel width and clamp steels tightly.

工作原理（Operating principle）：

液压马达带动导板和导辊沿水平方向左右移动，来调整带钢中心线所在位置，并由制动器保证其位置不变。导板位置调整完毕、制动器锁死之后，带钢运动的中心线即被控制在该导板纵向中心线位置。伺服缸 9 可以带动齿条齿轮机构来完成夹辊夹紧及松开带钢动作。

Hydraulic motor drives guide plate and guide roller to move left and right horizontally, to regulate the position of strip steel's center line while the brake makes the position unchanged. After guide plate position completed and brake locked, the centerline of strip steel is controlled at the position of guide plate's vertical center line. The cylinder 9 can drive rack-and-pinion mechanism to clamp and unfasten strip steels.

对中装置前端安装的检测光栅 13 用来实时检测带钢的实际宽度 W_1，作为输入信号输送给 PLC。安装在油缸活塞杆上的位移变送器测量出来的信号，经过一定的转化关系转化为夹紧槽宽度信号反馈给 PLC。两个信号在 PLC 内部经计算后，输出信号给比例放大板，进而控制比例阀完成相应动作，使夹辊随带钢的宽度而改变，保证既夹紧带钢又不出现翻边。

Detection grating 13 fixed at front of centering device is used to measure the actual width W_1 of strip steel real-timely, which is an input signal transported to PLC. The signal obtained by displacement sensor fixed on the piston rod of the cylinder is transformed into clamping groove width signal with respect to some transforming relationship, which is transported to PLC. The two signals are calculated by PLC and are outputted to proportional amplifier, to control the proportional valve to move correspondingly. So clamping varies along with the width of strip steel by clamping tightly to prevent flanging.

8.6 电渣炉电极进给电液比例控制系统（Electro-hydraulic proportional control system of electrode feeding system for electroslag remelting furnace）

8.6.1 电渣重熔原理（Electroslag remelting principle）

电渣重熔是特种冶金的主要手段之一。经过电渣重熔的钢以其高纯度、组织致密和良好的力学性能成为各种高级钢和特殊钢的首选，被广泛应用在航空、航天、军工、能源工业等具有特殊要求的场合。

Electroslag remelting (ESR) is one of the main methods for special metallurgy. Steel through electroslag remelting is often preferred to various kinds of high-grade steel and special steel for its high purity, compact texture and good mechanical property. It has been widely used in occasions of special requirements, such as aviation, spaceflight, military and energy industry.

电渣重熔的基本原理如图 8-13 所示。首先在铜制的水冷结晶器中加入液态炉渣，将自耗电极的端部插入渣池中。当自耗电极、炉渣与底水箱通过短网和变压器形成供电回路时，电流从变压器输出并通过熔渣；由于熔渣具有一定的电阻并造成了绝大部分的电压降，从而在渣池中产生大量的热，进而将插入其中的金属电极熔化。熔化的金属液滴从电极端部滴落，穿过渣池进入金属熔池，并在水冷结晶器的强制冷却下逐渐凝固形成钢锭。钢锭由下而上逐渐冷却、凝固，金属熔池和渣池液面随之不断地上升。上升的渣池在结晶器内壁和钢锭之间形成一层能起保温作用的渣壳，不仅能促使钢锭表面平滑光洁，而且能减少钢锭的径向导热，有利于自下而上的

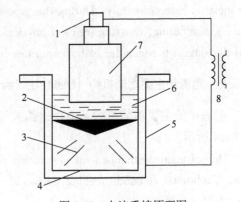

图 8-13 电渣重熔原理图
(Fig. 8-13 Schematic diagram of electrode remelting)
1—驱动电极（Drive electrode）；
2—金属熔池（Metal pool）；
3—钢锭（Ingot）；4—水冷却（Water cooling）；
5—结晶器（Crystallizer）；6—渣池（Slag basin）；
7—自耗电极（Consumbale electrode）；
8—变压器（Transformer）

顺序结晶，使钢锭形成没有缩孔的致密组织。自耗电极根据熔化的速度被缓慢地送进渣池，不断地补充正在凝固的金属熔池的体积。在电极熔化、金属液滴形成与滴落的过程中，金属熔池内的金属和炉渣发生着一系列的物理化学反应，从而除去了金属中的有害杂质元素和非金属夹杂物。

The basic principle of electroslag remelting is shown in Fig. 8-13. Firstly, add the molten slag into the coppery water-cooled crystallizer, and insert the end of the consumable electrode into the slag bath. The consumable electrode, slag, bottom water tank and transformer are connected by short network. When electric current passes the molten slag, the large resistance of slag causes huge voltage drop, thus heat is generated and the consumable electrode in the slag bath is melting. Then, the molten metal drops from the end of the electrode, goes through the metallic bath, freezes and becomes a steel ingot. The steel ingot cools down and freezes gradually from the bottom up, along with which the liquid level of the metallic bath and the slag bath rises. There comes out a slag shell playing a role of heat preservation between the mould wall and the steel ingot in the rising slag bath. It can not only make the surface of the ingot smooth and clean, but also reduce the radial heat conduction, thus being helpful to crystallize from the bottom up and form a compact texture without shrinkage cavity. The consumable electrode is gradually fed into the slag bath according to the melting rate, supplementing the freezing metallic bath constantly. During the processes of electrode melting and metal drops forming and falling, due to a series of physical and chemical reactions between the metal and the slag in the metallic bath, impurities in the metal are eliminated.

8.6.2　电渣炉主要功能机构 (Main mechanisms of electroslag remelting furnace)

电渣炉的主要功能机构是电极进给装置。传统电渣炉的电极进给装置机构如图 8-14 所示。

One of main mechanisms of electroslag remelting furnace is electrode feeding device. Traditional electrode feeding device of electroslag remelting furnace is shown in Fig. 8-14.

传统电渣炉的电极进给装置由电机+滑轮机构或伺服电机+滚珠丝杠组成，在机械结构上存在安装精度误差、机械磨损、机械效率降低等方面的问题，影响了横臂下降速度的控制精度，进而影响电渣重熔的熔速控制精度。

The electrode feeding device of traditional electroslag remelting furnace is composed of electromotor and pulley system, or servo motor and ball screw. Because of problems in mechanism, such as installation error, mechanical wear, and low mechanical efficiency, it impacts the control accuracy of cross-arm's falling speed, thus influencing

8.6 电渣炉电极进给电液比例控制系统
(Electro-hydraulic proportional control system of electrode feeding system for electroslag remelting furnace)

the control accuracy of melting rate in the electrode remelting process.

全液压传动电渣炉采用大尺寸液压缸代替滑轮机构和滚珠丝杠系统，采用电液比例调速阀对液压缸的下降速度进行控制。该种驱动与控制形式采用液压缸直接驱动导电横臂，避免了传动形式的转换，同时具有速比大、系统刚度大、频响高等优势，可以实现对导电横臂的快速、精确控制，进而实现电渣炉给料系统的精确控制。全液压电渣炉电极进给装置三维结构如图 8-15 所示，液压系统工作原理如图 8-16 所示。

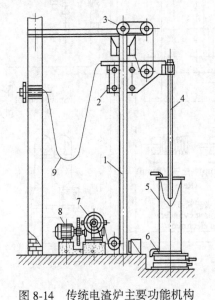
图 8-14 传统电渣炉主要功能机构
(Fig. 8-14 Main mechanisms of electroslag remelting furnace)
1—立柱 (Pillar)；2—横臂 (Cross arm)；
3—滑轮 (Pulley)；4—电极 (Electrode)；
5—结晶器 (Crystallizer)；
6—底部水箱 (Water tank in low position)；
7—钢丝绳 (Wire rope)；
8—电动机 (Electromotor)；9—铜排 (Copper bar)

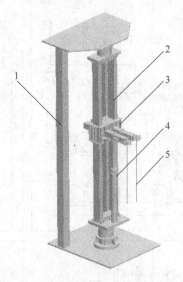
图 8-15 电极进给装置三维结构图
(Fig. 8-15 Triaxial structure of electrode feeding device)
1—支撑 (Support column)；
2—旋转立柱 (Split pillar)；
3—横臂 (Cross arm)；4—带有磁尺的液压缸
(Hydraulic cylinder with magnetic scale)；
5—电极 (Electrode)

Hydraulic driven electroslag remelting furnace uses hydraulic cylinder in large scale instead of pulley system or ball screw, and it uses hydraulic proportional speed regulating valve to control the rate of descend of the hydraulic cylinder. This form of driving and control uses hydraulic cylinder drive the current-conducting arm directly, thus avoiding the transformation of transmission type. Meanwhile, for its high speed ratio, large system stiffness and fast frequency response, it can achieve fast and accurate control of the current conducting arm, and then accurate control of feeding system in the electroslag remelting furnace. The electrode feeding device three-dimensional structure of hydraulic driven electroslag furnace are shown in Fig. 8-15, and its schematic diagram of the hydraulic system is shown in Fig. 8-16.

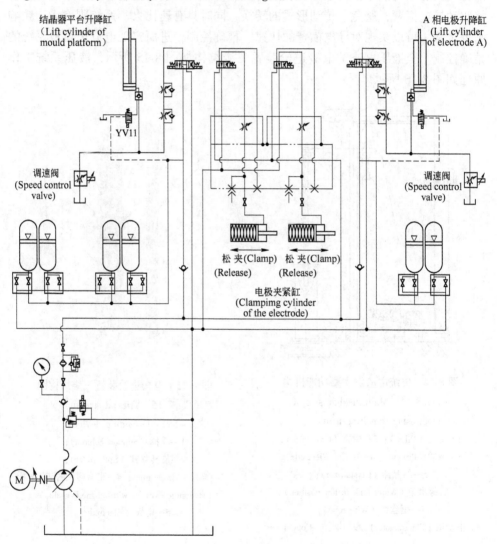

图 8-16 全液压传动电渣炉液压系统原理图
(Fig. 8-16 Hydraulic system schematic diagram of hydraulic driven electroslag remelting furnace)

8.6.3 电极进给装置电液比例控制系统 (Electro-hydraulic proportional control system for electrode feeding device)

全液压传动电渣炉的导电横臂采用大尺寸液压缸驱动,采用电液比例调速系

8.6 电渣炉电极进给电液比例控制系统
(Electro-hydraulic proportional control system of electrode feeding system for electroslag remelting furnace)

统进行液压缸的速度控制活塞杆的位置由高精度磁尺进行检测与反馈，进而形成闭环控制系统。其液压系统工作原理如图 8-17 所示。液压缸上升时，由电液换向阀 1 和蓄能器 8 提供动力源，电磁铁 YA1 得电。液压缸快速下降时，电磁铁 YA2、YA3 得电，液压缸在外负载和重力作用下快速下降，而不需要液压动力源。液压缸慢速下降时，YA3 失电，液压缸回油经由比例调速阀 5 流回油箱，液压缸的下降速度由比例调速阀调节。电极进给的过程即是液压缸慢速下降的过程，其电液比例控制原理如图 8-18 所示。

The current-conducting arm of hydraulic driven electroslag furnace is driven by large scale hydraulic cylinder, whose speed control is achieved by using electro-hydraulic proportional speed regulating system. The position of the piston rod is detected and fed back by high-precision magnetic scale, thus forming a closed-loop control system. The hydraulic system schematic diagram is shown in Fig. 8-17. When the hydraulic cylinder goes up, power is supplied by electro-hydraulic directional control valve 1 and accumulator 8, and electromagnet YA1 powers on. When the hydraulic cylinder descends quickly, electromagnets YA2 and YA3 power on, and the hydraulic cylinder goes down rapidly with no need for hydraulic power supply. When the hydraulic cylinder descends slowly, YA3 powers off, and return oil of the hydraulic cylinder runs back to the tank through proportional speed regulating valve 5, by which the descending speed of the cylinder is governed. The process of electrode feeding is namely the slowly descending process of the cylinder, whose electro-hydraulic proportional control principle is shown in Fig. 8-18.

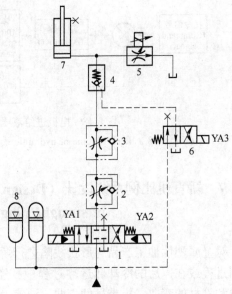

图 8-17 电极进给装置液压系统原理图
(Fig. 8-17 Schematic diagram of hydraulic system for electrode feeding device)
1—电液换向阀（Electro-hydraulic directional control valve）；
2, 3—单向节流阀（One-way throttle valve）；
4—液控单向阀（Hydraulic control one-way valve）；
5—比例调速阀（Proportional speed regulating valve）；
6—换向阀（Directional valve）；
7—液压缸（Hydraulic cylinder）；
8—蓄能器（Accumulator）

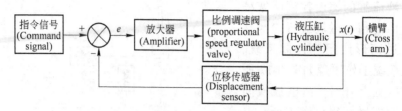

图 8-18 电极进给液压控制系统方框图

(Fig. 8-18 Block scheme of hydraulic control system for electrode feeding)

8.7 矫直机比例系统设计（Design of proportional control system of straightening machine）

对金属塑性加工产品的形状缺陷进行矫正，是重要的精整工序之一。轧材在轧制过程或在以后的冷却和运输过程中，经常会产生种种形状缺陷，诸如棒材、型材和管材的弯曲，板带材的弯曲、波浪、瓢曲等。通过各种矫直工序，可使弯曲等缺陷在外力作用下得以消除，使产品达到合格的状态。

Straightening of plastic treated metal productions' shape defects is one of the important finishing line processes. Rolling materials always generate shape defects in the process of rolling, cooling and transporting, such as bending of bars, section bars or pipe, bending, wave warps and buckling of plates and strips etc. Through finishing line processes, defects can be eliminated by external forces, to make productions qualified.

矫直机（图 8-19~图 8-21）是对金属棒材、管材、线材等进行矫直的设备。矫直机通过矫直辊对棒材等进行挤压使其改变直线度。矫直辊一般有两排，辊数量不等。也有两辊矫直机，依靠两辊（中间内凹，双曲线辊）的角度变化对不同直径的材料进行矫直。主要类型有压力矫直机、平衡辊矫直机、斜辊矫直机和旋转反弯矫直机等等。

Straightening machines (Fig. 8-19 ~ Fig. 8-21) are used to straighten metal bars, pipes, wire rods. By squeezing of rolls, straightness of workpiece is able to be changed. The straightening machines usually have two rows of rolls with various quantities. The two-roll-machine varies the angle of two rolls to straighten ma-

图 8-19 矫直机外观图

(Fig. 8-19 Outside view of straightening machine)

8.7 矫直机比例系统设计
(Design of proportional control system of straightening machine)

terials with different diameters. Main types are pressure straightening machines, parallel straightening machines, inclined roll straightening machines and rotary recurving straightening machines.

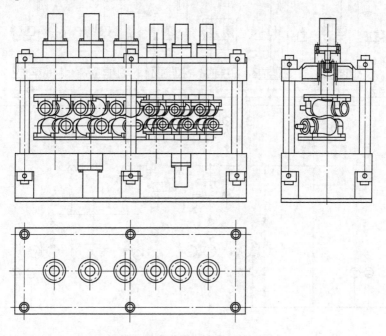

图 8-20 矫直机机械结构图
(Fig. 8-20 Structure of straightening machine)

这种矫直机的矫直过程为：辊子的位置与被矫直制品运动方向成某种角度，两或三个大辊的是主动压力辊，由电动机带动作同方向旋转；另一边若干个小辊为从动压力辊，它们是靠旋转着的圆棒或管材摩擦力使之旋转的。为了达到辊子对制品所要求的压缩量，这些小辊可以同时或分别向前或向后调整位置。一般辊子的数目越多，矫直后制品精度越高。当钢管头部到达第一对矫直辊后，该对辊的上辊快速压下，同理后面的矫直辊依次动作快速压下，钢管在矫直机里经过反复旋转弯曲变形后被矫直。当钢管尾部到达第一对矫直辊后，该对辊的上辊快速升起。同理后面的矫直辊依次动作快速升起。当钢管尾部离开最后一对矫直辊后，出口辊道中的辊道升起，将钢管输送至出料辊道的尾端。而后出料辊道降下，推料装置液压缸动作将钢管侧推出出料辊道，完成一根钢管的矫直。

The straightening process of this kind of straightening machine is as follows. Rolls' position is at an angle of original materials. Motors drive two or three driving rolls to rotate while small rolls in the other side are driven to rotate by the frictions of rotating pipes or bars. To meet the requirements of compression for products, these small rolls can simulta-

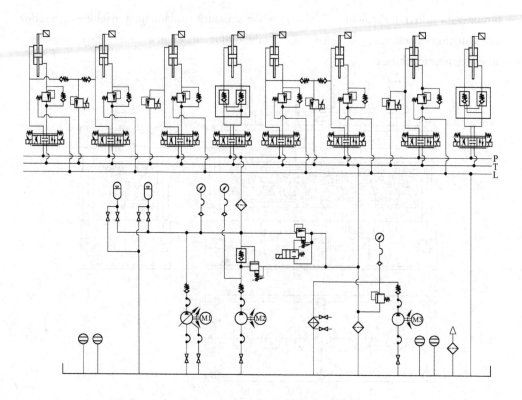

图 8-21 矫直机液压系统原理图
(Fig. 8-21 Hydraulic system diagram of straightening machine)

neously or separately move forward or backward to adjust positions. The more rolls, the more precision can be achieved after straightened. When the head of steel pipe reaches the first pair of straightening rolls, the top roll pushes down quickly. Similarly, the rear rolls pushes down in turn quickly. Steel pipe is straightened through repeat rotary and bending. When the end of steel pipe reaches the first pair straightening rolls, the top roll rises quickly. Similarly, the rear rolls rises quickly in turn. After the end of steel pipe leaves final pair of rolls, outlet roller table rises, and transports pipes to the end of discharging roller table, which falls off. Cylinder of material-pushing device pushes workpiece aside out of discharging roller table. So straightening process is completed.

8.8 比例技术在压铸机液压系统中的应用 (Application of proportional technology in die casting machine hydraulic system)

卧式冷室压铸机（图 8-22），是压铸有色金属（铝、镁、锌、铜等）中小型

8.8 比例技术在压铸机液压系统中的应用
(Application of proportional technology in die casting machine hydraulic system)

铸件的专用设备,广泛应用于航空、汽车、拖拉机、电讯电器、仪器仪表、照相机和家用电器等工业部门。

Horizontal cold chamber die casting machine (Fig. 8-22) is a special equipment for small and medium sized nonferrous metal (aluminum, magnesium, zinc, copper, etc.) casting. It's widely used in industrial fields such as aviation, automobile manufacturing, tractors, telecommunication and electrical appliances, instruments and apparatuses, cameras, and household appliances and so on.

图 8-22 卧式冷室压铸机外观图

(Fig. 8-22 Outside view of horizontal cold chamber die casting machine)

该机主要由合模、压射、液压和电气等部分组成。合模部分采用曲肘扩力机构,开合型平稳,速度快,效率高,性能可靠。压射部分采用压射、增压分控的四级压射系统,压射的各项参数可单独调节,简单可靠。特别是采用了无浮动活塞的内置单向阀增压装置,使建压时间更短,增压速度更快,完全可以满足镁、铝、锌、铜等有色金属压铸件的工艺要求。

This machine is mainly made up of mold clamping part, pressure casting part, hydraulic part and electrical part. The mold clamping part uses a crankshaft force expanding mechanism, which can open and close the mold smoothly, fast, efficiently, and reliably. The pressure casting part uses a four-stage system which separately controls pressure casting and pressurizing. Parameters of pressure casting can be independently regulated, which is simple and reliable. Especially, the built-in check valve pressurizing device with non-floating piston enables pressurization time shorter, speeding the pressurization process, which can completely meet the process requirements of nonferrous die casting, such as magnesium, aluminum, copper and so on.

机器液压部分工作原理 (Principle of hydraulic system):

(1) 压力调整:油泵输出压力油至系统各工作油缸 (图 8-23)。在机器运行过程中,系统压力低于 5MPa 时,高、低压泵同时向系统供油,机器则以较快的

第8章 电液比例控制技术的工程应用
Chapter 8 Engineering Application of Electro-hydraulic Proportional Control Technology

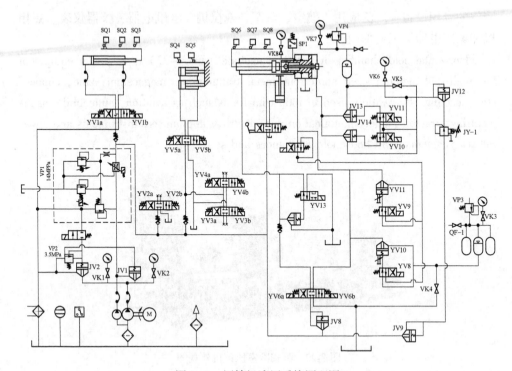

图 8-23 压铸机液压系统原理图

(Fig. 8-23 Hydraulic schematic of horizontal cold chamber die-casting machine)

速度运行；当系统压力达到 5MPa 时，JV2 打开，低压泵开始卸荷。系统最高压力为 14MPa，由厂方在出厂前完成设定，毋需用户自行调节。

Pressure setting: a pump outputs oil to working cylinders of this system (Fig. 8-23). During the working process, when pressure descends below 5 MPa, the high-pressure pump and the low-pressure pump output oil together so that the machine moves at a higher speed. When pressure reaches 5 MPa, JV2 opens and low-pressure pump begins to unload. The highest pressure of the system is 14 MPa, which is set before delivery and users don't need to regulate it themselves.

（2）机器的开合型：机器的开、合模动作由电液阀 YV1a、YV1b 控制。合模时，电磁阀 YV1b 得电，压力油进入合模油缸的无杆腔，进行合模；合模油缸的有杆腔油液回油箱，合模终止。开模时，电磁阀 YV1a 得电，压力油进入合模缸有杆腔；无杆腔油液回油箱，完成开模过程。

Die opening and clamping: the die opening and clamping movement is controlled by electro-hydraulic valve YV1a and YV1b. When clamping, valve YV1b's voltage rises and oil enters into the non-rod chamber of die clamping cylinder, starting die clamping. Oil in the rod chamber returns back to oil tank and die clamping stops. When

8.8 比例技术在压铸机液压系统中的应用
(Application of proportional technology in die casting machine hydraulic system)

opening, voltage of valve YV1a rises, pressure oil enters into the rod chamber, and oil in the non-rod chamber returns back to the tank. Thus the die opening is finished.

(3) 顶出、顶回：铸件的顶出、顶回由三位四通换向阀 YV5 控制。YV5a 得电，压力油推顶出缸活塞带动顶出板进行顶出，顶出时间由人机界面进行调整；反之，YV5b 得电，实现顶回。

Push-out and push-in: push-out and push-in processes are controlled by three-position four-way directional control valve YV5. When voltage of YV5a rises, oil pushes the cylinder piston with push-out plate. The push-out time is adjusted by human-computer interface. On the contrary, when voltage of YV5b rises, the push-in is done.

(4) 抽、插芯：本机器设有两个动芯和一个静芯，抽、插芯动作由 YV2、YV3 和 YV4 控制。抽、插芯程序按工艺要求选定后，按动触摸屏上的相应按钮，机器按选定程序运转。

Cores' pulling and inserting: this machine has two movable cores and a stationary core. Cores' pulling and inserting are controlled by YV2, YV3, and YV4. After the program is made by process requirements, once corresponding buttons are pressed, the machine will operate according to the given program.

(5) 压射与压射回程：慢压射和压射回程由三位四通电液换向阀 YV6 控制。按"压射"按钮，YV6a 得电，压力油自油泵进入压射缸右腔，进行慢压射；同时，YV13 得电，快排阀开启。慢压射时，由电液换向阀控制油缸供油，压射活塞杆带动接近开关触动杆，"快压启动"接近开关 SQ8 脱离；快压射阀 YV8 得电，快压射阀开启，压力油经 JV11 进入压射缸，进行快压射。金属液充型结束后，压射活塞停止，油腔压力上升，达到一定值后，压射腔压力继电器动作，电磁换向阀 YV11 得电，增压蓄能器的工作油液经 JV13 进入增压缸，即进行增压。增压起始时间由节流阀 JV13 调节。按"压回"按钮，YV6b、YV10 得电，压力油经单向阀进入压射有杆腔，增压缸回油阀 JV14 打开，增压活塞回程，打开内置单向阀，压射无杆腔接通回油，开始压射回程。

Injection and injection return stroke: slow injection and injection return stroke are controlled by three-position four-way electro-hydraulic directional valve YV6. If "Inject" button is pressed, voltage of YV6a rises, oil enters into the right chamber of injection cylinder from pump and slow injection is carried out. At the same time, voltage of YV13 rises and quick release valve opens. In the slow injection process, oil is supplied to control cylinder by electro-hydraulic directional valve. Injection piston rod drives the proximity switch bar, "Start quick injection" proximity switch SQ8 detaches, solenoid YV8 powers on, quick injection valve opens, oil through JV11 flows into injection cylinder, and quick injection is started. Liquid metal starts filling the mold, the injection piston stops. Pressure in the chamber rises until reaching a certain value to trigger the pressure relay in the injection chamber. Electro-hydraulic directional valve's solenoid YV11 powers on, then the working oil of pressurization accumulator enters pressurization cylinder through JV13 and pressurization is started. Pressurization starting time is controlled by relief valve JV13. If "Press Back" button is pressed, solenoid YV6b and YV10 power on. Subsequently, oil enters into injection rod chamber through check valve, pressurization cylinder's oil return valve JV14 opens, the piston returns, inner-set check valve opens, to connect injection non-rod chamber and injection return stroke starts.

8.9 风力发电机变桨距比例控制系统（Wind driven generator variable propeller pitch proportional control system）

变桨距控制系统的主要功能是通过调整桨叶节距，改变气流对叶片的攻角；在风力发电机（图 8-24）的启动过程提供启动转矩，在风速过高的时候改变风力发电机获得的空气动力转矩，从而保持功率输出恒定。

The main function of variable propeller pitch control system is changing the attack angle of the flow on the blade by adjusting the propeller pitch, supplying the starting torque in the starting process of wind driven generator (Fig. 8-24), and changing the aerodynamic torque attained by wind driven generator when the wind speed is too high, so as to keep the output power constant.

变桨距控制技术的发展经历了一段曲折的过程。最初研制的风力发电机组都被设计成可以全桨叶变距，但是由于设计人员对空气动力特性和运行工况考虑不足，可靠性远不能满足实际要求，灾难性事故不断发生，变桨距控制技术一度被摒弃。随着长期的实践和技术发展，在现代大型风力发电机组设计中，变桨距控制技术重新获得了重视和广泛应用。新型变桨距控制系统可以使桨叶和整机的负载受力状况大为改善。同时，变速恒频技术与变桨距控制系统相配合，可以提高机组的发电效率和电能质量。

The development of variable propeller pitch control technology has experienced a tortuous process. The initial wind driven generators were all designed as the whole blade pitch style, but because the designer had no sufficient consideration on the aerodynamic performance and operating conditions, its reliability was too far to satisfy the actual requirement, which usually leads to disastrous accidents. Therefore, variable propeller pitch control technology was abolished once. With the long development of practice and technology, this technology is reconsidered and used widely again in the modern big

8.9 风力发电机变桨距比例控制系统
（Wind driven generator variable propeller pitch proportional control system）

 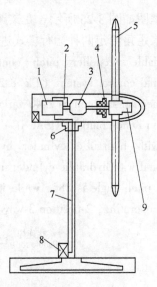

图 8-24　风力发电机实物及示意图

(Fig. 8-24　Wind driven generator physical and schematic diagram)

1—发电机（Generator）；2—控制柜（Control cabinet）；3—齿轮箱（Gear box）；
4—刹车（Brake）；5—叶片（Blade）；6—偏航系统（Yaw system）；
7—塔架（Tower frame）；8—电网引线（Power wire）；9—旋转头（Rotating head）

wind power generator design. The new variable propeller pitch control system can make the load stress condition of propeller and the whole body improved. At the same time, the match of variable speed constant frequency and variable propeller pitch control system can increase the efficiency of power generation and power quality.

风力发电中实现变桨距控制的液压系统工作原理如下：

The theory of variable propeller pitch control in wind generator is described as follows：

液压变桨距控制机构属于电液比例控制系统（图 8-25，图 8-26）。典型的变桨距液压执行机构如美国 Zond 公司的 Z-40 型变桨距风力发电机组，其工作原理如图 8-25 所示。桨叶通过机械连杆机构与液压缸相连接，节距角的变化同液压缸位移基本成正比。当液压缸活塞杆向左移动到最大位置时，节距角为 88°；而活塞向右移动到最大位置时，节距角为 -5°。在系统正常工作时，两位三通电磁换向阀 A、B、C 都通电，液控单向阀打开，液压缸的位移由电液比例换向阀进行精确控制。在风速低于额定风速时，不论风速如何变化，电液比例换向阀维持桨叶节距角为 3°，考虑到液压缸的泄漏，电液比例换向阀进行微调，保持节距角不变；当风速高于额定风速时，根据输出功率，利用电液比例换向阀精确改变输出流量，从而控制桨叶的节距角，使输出功率恒定。变桨距机构及其控制器的控制系统框图如图 8-26 所示。变桨距控制系统通常采用一个 FET 电子开关对功率给定和桨距角给定两种运行模式进行硬件电路的切换。

Variable propeller pitch control mechanism belongs to electric hydraulic proportional control system (Fig. 8-25, Fig. 8-26). Typical variable propeller pitch control mechanism such as Z-40 variable propeller pitch wind driven generator set which is made in Zond company USA. Its diagram is shown as Fig. 8-25. The propeller is connected with hydraulic cylinder by connecting rod mechanism, pitch change is proportional with hydraulic cylinder displacement. When the piston moves to the leftmost end, the pitch angle is 88°, while it is −5° at the rightmost end. When the system is normally operating, 2-position 3-way directional valves A, B, C all work, the hydraulic

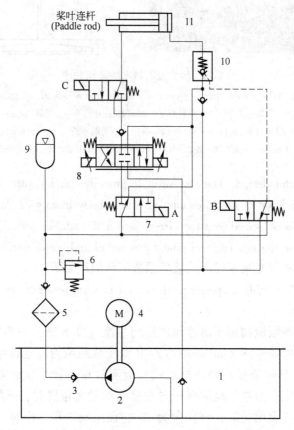

图 8-25 液压驱动变桨距机构工作原理简图

(Fig. 8-25 Principle schematic diagram of hydraulic driver variable propeller pitch mechanism)

1—油箱(Oil tank); 2—液压泵(Pump); 3—单向阀(Check valve); 4—电机(Motor);
5—过滤器(Filter); 6—溢流阀(Relief valve); 7—两位三通电磁阀(Two position three way directional valve);
8—电液比例阀(Proportional valve); 9—蓄能器(Accumulator);
10—液控单向阀(Hydraulic control unidirectional valve); 11—液压缸(Hydraulic cylinder)

operated check valve opens, and the displacement of hydraulic cylinder is controlled accurately by the proportional directional valve. When the wind speed is lower than the rated wind speed, no matter how the wind speed changes, the electro-hydraulic proportional direction valve can maintain the blade pitch angle at 3°. Considering the leakage of the hydraulic cylinder, the electro-hydraulic proportional direction valve can be fine-tuned to keep the pitch angle constant. When the wind speed is higher than the rated value, according to the output power, changing the output flow accurately by proportional directional valve can control the pitch angle to maintain the constant power. The variable propeller pitch mechanism and its controller system are shown in the Fig. 8-26. The usual variable propeller pitch control system use a FET electronic switch to switch between the two modes which are power setting and pitch angle setting.

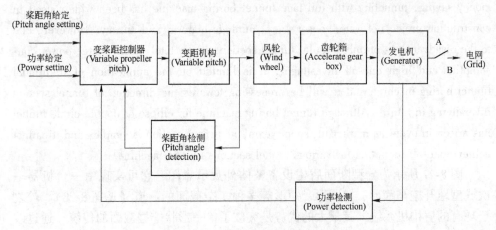

图 8-26 变桨距控制系统框图

(Fig. 8-26 Diagram of variable propeller pitch control system)

8.10 异型断面盾构设备电液比例控制系统（Electro-hydraulic proportional control system for special-section tunnel boring device）

8.10.1 异型断面盾构技术背景（Background of special-section tunnel boring technology）

随着交通的快速发展，在铁路、公路、水路等领域必然会出现不同的隧道，隧道的挖掘成为一项关键的工作。特别是在修建长隧道过程中，当隧道长度超过15km时，就要优先考虑采用隧道掘进机。

With the fast development of transport, there has come out various tunnels in construction of railway, highway and waterway, thus making tunnel boring quite pivotal. Especially in construction of long tunnels, which are more than 15km, tunnel boring machine will be given a prior consideration.

全断面隧道掘进机开挖施工具有高效、快速、衬砌量少、安全、经济等优点，已在世界各地广泛应用于水利、矿山、公路和铁路等建设工程。但是传统的全断面掘进机有一些使用局限性，开挖断面只能是圆形、整机重量大、消耗功率大、一次性投资高等，这大大限制了掘进机的使用范围。未来的掘进机将向断面多样化方向发展。虽然近年来国外市场上也出现了可以挖椭圆、双圆形的盾构机，但其可挖的形状都比较单一，无法由一台机器实现多种形状断面一次成型的功能。

Given the advantages, such as high efficiency, speediness, less lining, safety and money saving, tunneling with full-face tunnel boring machine has been widely used in construction projects of water conservancy, mining, highway and railway. However, due to the disadvantages of only circular tunnel section, heavy total weight, large power consumption and high one-off investment, it is limited on the application of traditional tunnel boring machine and it will be processing towards the direction of tunnel section diversifying in future. Although tunnel boring machine for ellipse or double circle tunnel has arisen in overseas market in recent years, its tunnel section is simplex and it cannot achieve one-off tunneling for various tunnel sections by one machine.

图 8-27 所示为异型断面盾构设备整体外观示意图，它可实现由一台机器一次成型地开挖马蹄形、椭圆形、矩形等多种复杂截面的隧道（见图 8-28）。较之于现有的盾构机系统，其最大的优势是突破了单一形状挖掘截面的传统，通过改变控制程序即可实现各种复杂形状断面的挖掘，并且一次挖掘成型。

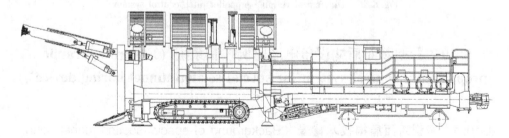

图 8-27 异型断面盾构设备外观图

(Fig. 8-27 General view of special-section tunnel boring machine)

Fig. 8-27 shows the general view of special-section tunnel boring machine, which can excavate special-section tunnels by one machine at a time, such as horseshoe, ellipse and rectangle etc. shown in Fig. 8-28. Compared with traditional tunnel boring ma-

8.10 异型断面盾构设备电液比例控制系统
(Electro-hydraulic proportional control system for special-section tunnel boring device)

chine, its biggest advantage is that it breaks though the restriction of simplex tunnel section and it can excavate various section tunnel by changing the control program.

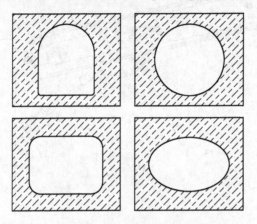

图 8-28 异型断面盾构设备可实现的断面形状

(Fig. 8-28 Tunnel sections available for special-section tunnel boring machine)

8.10.2 异型断面盾构设备主要功能机构 (Main mechanisms of special-section tunnel boring device)

如图 8-29 所示，异型断面盾构刀盘主要由液压马达、减速器、转盘、伸缩缸、摆动缸、套筒、切割臂、盘形滚刀、压力传感器、位移传感器、角度传感器、比例阀、电液比例控制系统、电气控制系统及其他相关附件等组成。液压缸采用双耳轴液压缸，两端均通过销轴铰接，且两缸均带有位移检测仪器。其摆动缸一端与刀盘上的轴套铰接，另一端与套筒上的固定轴套铰接。伸缩缸一端耳轴伸出套筒与刀盘轴套铰接，另一端与切割臂连接。套筒底部通过双耳轴与刀盘铰接，内部通过导向套与切割臂配合，保证套筒与切割臂的同轴度。

As is shown in Fig. 8-29, special-section cutterhead is mainly composed of hydraulic motor, reducer, turnplate, stretch cylinder, swing cylinder, sleeve, cutter bar, disk cutter, pressure sensor, displacement sensor, angle sensor, proportional valve, electro-hydraulic proportional control system, electrical control system and other accessories. Both cylinders are equipped with displacement sensor and double trunnions. The rear-end of swing cylinder is hinged with the turnplate, and the fore-end is hinged with the hinge support on the sleeve; while the rear-end of stretch cylinder is hinged with the turnplate through the sleeve, and the fore-end is hinged with the cutter bar. Inside, the bottom of the sleeve is jointed with turnplate by a hinge pin, and it fits with the cutter bar through sleeve guide to ensure the coaxiality.

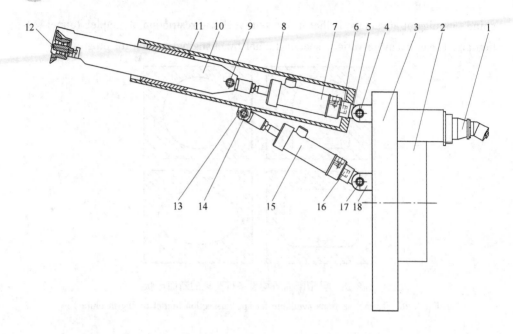

图 8-29 异型断面盾构刀盘结构示意图

(Fig. 8-29 Structure diagram of special-section cutterhead)

1—液压马达 (Hydraulic motor); 2—减速器 (Reduction gear); 3—转盘 (Turnplate);
4, 9, 14, 17—销轴 (Hinge pin); 5, 13, 18—耳环 (Trunnion); 6, 16—位移传感器 (Displacement sensor);
7—伸缩液压缸 (Telescopic cylinder); 8—套筒 (Sleeve); 10—切割臂 (Cutting arm);
11—导向轴套 (Guideway); 12—刀具 (Disc cutter); 15—摆动液压缸 (Swing cylinder)

8.10.3 几何关系推导 (Derivation of geometrical relationship)

如图 8-30 所示,以马蹄形隧洞断面为例研究异型断面盾构刀盘的控制过程,$O'B$ 为摇臂和切割臂在竖直面上的投影。设 P 为刀具切割轨迹的起始点,θ 是 $O'B$ 与 $O'P$ 之间的夹角,R 为马蹄形断面的圆拱半径,m 是马蹄形断面竖直线段长度,AD 为摆动臂的长度,BC 为伸缩臂的长度。则刀具点 B 的轨迹方程可描述如下:

As is shown in Fig. 8-30, we take horseshoe section tunnel for example to explore its control process. $O'B$ is the projection of one of the rockers in a vertical plane. Set P as the starting point of the cutter path; θ is the included angle between $O'B$ and $O'P$; R is the radius of the arc of the horseshoe section; m is the length of the vertical line; AD is the length of swinging arm; BC is the length of telescopic arm. Therefore, the locus equation of point B can be described as below:

8.10 异型断面盾构设备电液比例控制系统
(Electro-hydraulic proportional control system for special-section tunnel boring device)

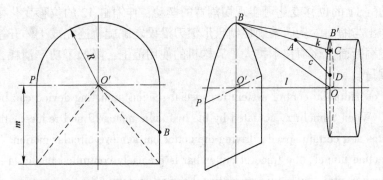

图 8-30 马蹄形隧洞断面参数及其与盾构机刀盘间的几何关系
(Fig. 8-30 Parameters of the horseshoe section and its geometrical relationship with the cutterhead)

$$O'B = \begin{cases} R, & \theta \in (0, \pi] \\ \dfrac{R}{-\cos\theta}, & \theta \in (\pi, \pi + \arctan\dfrac{m}{R}] \\ \dfrac{m}{-\sin\theta}, & \theta \in (\pi + \arctan\dfrac{m}{R}, 2\pi - \arctan\dfrac{m}{R}] \\ \dfrac{R}{\cos\theta}, & \theta \in (2\pi - \arctan\dfrac{m}{R}, 2\pi] \end{cases} \tag{8-1}$$

由上述几何关系可推导出摆动臂 AD、伸缩臂 BC 和转盘转角 θ 之间的相互作用关系，将其编译成相应的控制算法。因而盘形滚刀的切割轨迹可由 PLC 和工业控制计算机实现。

From the above-mentioned geometrical relationship, we can derive the functional relationship among the length of swinging arm AD, telescopic arm BC and the rotation angle of the turnplate θ, and then compile it into corresponding control algorithm. Therefore, the cutting trajectory of disc cutter can be achieved by PLC and industrial personal computer.

8.10.4 异型断面盾构设备电液比例系统 (Electro-hydraulic proportional system of special-section tunneling device)

图 8-31 为异型断面盾构设备部分液压控制系统原理图。盾构设备掘进工作时，通过液压马达 17 和减速器组件带动刀盘实现某一速度的旋转，从而带动切割臂实现圆周运动。为了获得各种不同的隧洞截面形状，刀臂轨迹通过一套比例闭环液压系统进行精确控制。按所需挖掘的断面形状，由计算机事先设定程序并计算出摆动缸 11 和伸缩缸 12 在刀盘各个转角位置的位移，通过位移传感器 15 和 16 反馈到比例阀上，实现流量、方向的控制，最终实现两个液压缸的位移控

制。摆动缸11的位移变化可实现切割臂的摆动，伸缩缸12的位移变化可实现切割臂的伸出与缩回。从而精确控制盘形滚刀沿设定螺旋轨迹完成土体开挖，获得所需的隧洞断面。该项技术扩大了盾构机的使用范围，降低了功率消耗，减轻了整机的重量。

Part of hydraulic control system for special-section tunneling device can be seen in Fig. 8-31. When tunneling, actuated by the hydraulic motor 17 and reducer, the cutter-head rotates at a certain speed, driving the cutter bar achieve circular motion. To obtain various section tunnel, the trace of cutter bar is precisely controlled though a set of closed-loop electro-hydraulic proportional control system. Set the program in advance

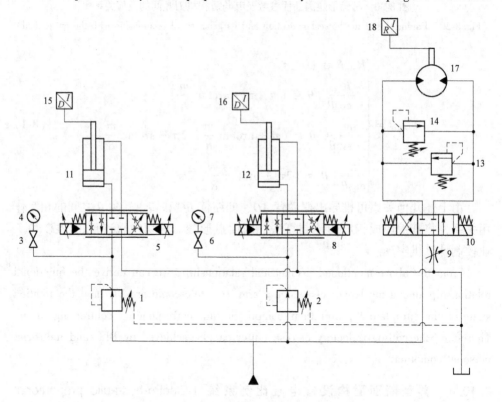

图 8-31 异型断面盾构设备液压系统原理图

(Fig. 8-31 Schematic diagram of hydraulic control system of special-section tunneling device)

1，2—减压阀（Pressure reducing valve）；3，6—截止阀（Shutoff valve）；
4，7—压力表（Pressure meter）；5，8—比例换向阀（Proportional directional valve）；
9—节流阀（Throttle valve）；10—换向阀（Directional valve）；11—摆动缸（Swing cylinder）；
12—伸缩缸（Telescopic cylinder）；13，14—叠加式溢流阀（Superposed relief valve）；
15，16—位移传感器（Displacement sensor）；17—液压马达（Hydraulic motor）；
18—转角传感器（Rotation angle sensor）

8.10 异型断面盾构设备电液比例控制系统
(Electro-hydraulic proportional control system for special-section tunnel boring device)

according to the required section and figure out the displacements of both swing and telescopic cylinders, 11 and 12, in each angle respectively, the data gained from the displacement sensors 15 and 16 is fed back to the proportional valve to regulate its flow rate and direction, and eventually achieve the displacement control of the two cylinders. Displacement variation of swing cylinder 11 can meet the requirement of swing of cutting arm, while displacement variation of telescopic cylinder 12 can achieve its extension and retraction process. Therefore, it can precisely control the disc cutter follow the designed spiral trajectory to excavate the tunnel, thus obtaining desirable tunnel section. This technology enlarges the range of application of tunnel boring machine and reduces its energy consumption and total weight.

附录　英语专业词汇

A		B	
安全阀	safety valve	比例容积控制	proportional volume control
安全回路	safety circuit	比例容积式调速回路	proportional volumetric speed regulating circuit
B		比例容积式调压回路	proportional volumetric pressure regulating circuit
摆动缸	oscillating cylinder		
薄壁孔口出流	discharge from thin-walled orifice	比例速度控制系统	proportional speed control system
饱和区	saturation zone	比例同步控制系统	proportional synchronization control system
保压回路	pressure retaining circuit		
背压	back pressure	比例位置控制系统	proportional position control system
比例插装阀	proportional cartridge valve		
比例差动控制回路	proportional differential control circuit	比例系数	proportionality coefficient
		比例压力控制系统	proportional pressure control system
比例电磁阀	proportional solenoid valve		
比例电磁铁	proportional solenoid	比例溢流阀	proportional overflow/relief valve
比例阀	proportional valve	比例溢流调压回路	proportional overflowing pressure regulating circuit
比例方向阀	proportional directional valve		
比例复合阀	proportional composite valve		
比例减压阀	proportional pressure reducing valve	比热容	specific heat capacity
		闭环	closed loop
比例减压回路	proportional pressure reducing circuit	闭环控制系统	closed-loop control system
		变桨距系统	variable propeller pitch system
比例节流阀	proportional throttle valve	变量泵	variable displacement pump
比例节流调速回路	proportional throttling speed regulating circuit	变量机构	variable mechanism
		变量马达	variable displacement motor
比例控制放大器	proportional control amplifier	变量柱塞泵	variable piston pump
比例力控制系统	proportional force control system	标准大气压	reference atmosphere
		步进电机	stepping motor
比例流量阀	proportional flow valve	**C**	
比例流量控制系统	proportional flow control system	参比弹簧	reference spring
		测速机	tachometer
比例流量压力控制系统	proportional flow-pressure control system	插装阀	cartridge valve
		差动放大器	differential amplifier
比例容积节流式流量控制回路	proportional volumetric throttling flow control circuit		

续表

C			D	
差动缸	differential cylinder		电流信号	current signal
差压式稳流量变量叶片泵	differential pressure steady flow variable vane pump		电容	capacitance
			电位器	potentiometer
			电压信号	voltage signal
颤振信号发生器	flutter signal generator		电液比例阀	electro-hydraulic proportional valve
车用变速器	automotive transmission			
齿轮	pinion		电液比例方向阀	electro-hydraulic proportional directional valve
齿轮泵	gear pump			
齿条	rack		电液比例方向控制回路	electro-hydraulic proportional direction control circuit
抽空	evacuation			
出口压力补偿器	outlet pressure compensator		电液比例控制技术	electro-hydraulic proportional control technology
传力弹簧	force-transferring spring			
磁滞特性	hysteresis characteristic		电液比例流量阀	electro-hydraulic proportional flow valve
D				
带钢对中装置	strip steel alignment device		电液比例速度控制回路	electro-hydraulic proportional speed control circuit
单出杆液压缸	single-rod piston cylinder		电液比例压力阀	electro-hydraulic proportional pressure valve
单通道力控制型比例放大器	single-channel force-controlled proportional amplifier		电液比例压力控制回路	electro-hydraulic proportional pressure control circuit
单相桥式整流电路	single-phase bridge rectifier circuit		电液数模转换器	electro-hydraulic digital-to-analog (D/A) converter
单向阀	check valve		电液伺服阀	electro-hydraulic servo valve
单作用缸	single-acting cylinder		电渣重熔炉	electroslag remelting furnace
弹簧预紧力	spring preload		叠加阀	stacked valve
弹性挡圈	circlip		定差溢流	fixed differential overflow
弹性模量	modulus of elasticity		定量泵	constant displacement pump
刀盘	cutterhead		动圈式矩马达	moving-coil torque motor
导向套	guide sleeve		动圈式力马达	moving-coil force motor
低通滤波电路	Low-pass Filter Circuit (LPF)		动铁式力马达	moving-iron force moto
			短管出流	discharge from shot-pipe outlet
			对称度	symmetry
电磁阀	solenoid valve		对中弹簧	centralizing spring
电磁吸力	electromagnetic attraction force		盾构机	tunnel boring machine
			多缸同步控制系统	multi-cylinder synchronization control system
电-机械变换器	electro-mechanical converter		多功能组合阀	multi-function combination valve
电极进给系统	electrode feeding system			

续表

D	
多级压力控制	multi-stage pressure control
多路换向阀控制回路	multi-way reversing valve control circuit

E	
轭铁	yoke iron
二极管	diode
二通比例调速阀	two-way proportional velocity regulating valve

F	
阀套	valve sleeve
阀体	valve body
阀芯	spool
阀座	valve seat
反馈信号	feedback signal
反相器	inverter
防尘圈	wiper scraper seal
放气阀	deflating valve
飞机拦阻系统	aircraft arresting system
非稳定流	unsteady flow
分辨率	resolution ratio
风力发电机	wind driven generator
缝隙流动	interstitial flow
幅值	amplitude
负载适应控制	load adaptive control
复位弹簧	reset spring

G	
干燥器	dryer
刚度	stiffness
钢管水压试验机	pipe hydrostatic tester
高频脉宽调制	high-frequency pulse width modulation

G	
格来圈	Glyd Ring
隔磁环	magnetism proof ring
功率	power
功率适应控制	power adaptive control
固体	solid
固有频率	natural frequency
管件	pipe fitting
管拧机浮动抱钳夹紧装置	floating clamping unit of pipe-wrenching machine
光编码器	optical encode
过载保护回路	overload protection circuit

H	
含湿量	moisture content
恒功率控制	constant power control
恒流量控制	constant flow control
恒压变量泵	constant pressure variable pump
恒压力控制	constant pressure control
后冷却器	after-cooler
滑阀	slide valve
缓冲装置	cushioning device
换向阀	reversing/directional valve
活塞	piston
活塞杆	rod
活塞行程	piston stroke

J	
机械杠杆式压力同步控制装置	mechanical lever pressure synchronous control device
机械效率	mechanical efficiency
积分环节	integral element
基准点	datum point
记忆元件	memory element
迹线	trace

续表

J			K	
加热器	heater		控制油道	control oil channel
减速回路	deceleration circuit		快速接头	quick coupling
减压阀	pressure reducing valve		L	
减压回路	pressure reducing circuit		拉力	tensile force
减压节流口	pressure reducing orifice		冷却器	cooler
减振油缸	damping cylinder		理论效率	theoretical efficiency
剪切力	shear force		理想流体伯努利方程	Bernoulli equation for ideal fluid
矫直机	straightening machine			
接箍	joint coupling		力矩马达	torque motor
节距角	pitch angle		两极同心式普通溢流阀	bipolar concentric relief valve
节流槽	metering groove			
节流阀	throttle valve		灵敏度	sensitivity
节流螺塞	orifice plug		零偏	null bias
节流器	throttler		零飘	null shift
截止区	cut-off zone		零位	null
解调器	demodulator		流量	flow
进口节流回路	meter-in circuit		流量表	flowmeter
进口压力补偿器	inlet pressure compensator		流量阀	flow valve
			流量适应控制	flow adaptive control
聚氨酯密封件	polyurethane sealing element		流体	fluid
			流体动力学	fluid dynamics
绝对大气压	absolute atmosphere		流体静力学	fluid statics
绝对湿度	absolute humidity		流体力学	fluid mechanics
K			流线	streamline
卡套式管接头	bite type fitting		流阻	flow resistance
开环	open loop		螺杆泵	screw pump
开环控制系统	open-loop control system		螺线管型电磁铁	solenoid electromagnet
抗污染能力	contamination resistance		滤波电路	filter circuit
可压缩性	compressibility		滤油器	oil filter
空气过滤器	air filter		M	
空气压缩机	air compressor			
空穴	cavitation		马赫数	Mach number
空载流量	no-load flow		脉冲电压	impulse voltage
孔口出流	discharge from orifice		密度	density

续表

M		**R**	
密封装置	sealing device	容积含湿量	volumetric moisture content
膜片	diaphragm	容积调速控制	volumetric speed control
N		容积效率	volumetric efficiency
内泄油道	internal leakage channel	**S**	
内置位置检测电子装置	built-in position detection electronic device	三通比例减压阀	three-way proportional pressure reducing valve
黏度	viscosity	射流管式电液伺服阀	electro-hydraulic servo valve of jet pipe type
P		射流元件	jet element
帕斯卡原理	Pascal principle	伸缩缸	telescoping cylinder
排量	displacement	声速	sound velocity
排气阀	exhaust valve	湿空气	humid air
排气孔	exhaust vent	手动电位器信号发生器	manual potentiometer signal generator
排气装置	air exhausting device	手动调节溢流阀	manual adjustment relief valve
旁路节流回路	bypass throttling circuit	手动调节装置	manual adjustment device
旁通调速阀	bypass speed regulating valve	数字缸	digital cylinder
配流盘	valve plate	双出杆液压缸	double-rod piston hydraulic cylinder
配油机构	oil distribution mechanism	双通路比例放大器	binary-channel proportional amplifier
平衡回路	balancing circuit	双作用缸	double-acting cylinder
Q		水平的位移-力特性	horizontal displacement-force characteristic
气动系统	pneumatic system	水平吸力特性	horizontal attraction characteristic
气马达	pneumatic motor	顺序动作回路	sequence circuit
气压	air pressure	斯特封	Step Seal
驱动室	driving chamber	四位四通比例方向阀	four-position four-way proportional directional valve
全波整流电压	full-wave commutating voltage	伺服变量	servo variable displacement
R		伺服电机	servo motor
热负荷阻抗特性	thermal load impedance characteristic	酸洗	pickling
容积	volume	梭阀	shuttle valve

续表

S		X	
锁紧回路	locking circuit	先导式减压阀	pilot operated pressure reducing valve
T		先导式溢流阀	pilot operated relief valve
调节螺塞	regulating plug screw	先导油道	pilot oil channel
调速阀	speed regulating valve	先导锥阀	pilot cone valve
调速回路	speed regulating circuit	先导锥阀芯	pilot cone valve core
调压回路	pressure regulating circuit	衔铁	armature
调压溢流阀	pressure adjustment relief valve	衔铁推杆	armature pusher
铁芯感性负载	inductive load of iron-core	限位环	stop ring
同步回路	synchronizing circuit	限压阀	pressure-limiting valve
凸轮机构	cam mechanism	线网电压	wire netting voltage
推杆	push rod	线性度	linearity
W		相对湿度	relative humidity
外围电路	peripheral circuit	相位	phase
微分调节器	differential regulator	相位裕量	phase margin
位移-力反馈式比例变量泵	displacement-force feedback proportional variable pump	消声器	muffler
位移直接反馈式比例变量泵	displacement direct feedback proportional variable pump	斜面吸力	bevel suction
		谐振	resonance
温度	temperature	泄压回路	pressure release circuit
温度膨胀性	thermal expansibility	信号-动作线图	signal-motion line graph
温漂	temperature drift	信号线	signal line
文氏电桥	Wien-bridge	行程程序	stroke program
稳定流	steady flow	行程控制型比例电磁铁	stroke-controlled proportional solenoid
无级变速（CVT）	Continuously Variable Transmission	虚地	virtual ground
无级调压	stepless pressure regulation	蓄能器	accumulator
物理量传感器	physical quantity sensor	**Y**	
X		压力	pressure
先导阀	pilot valve	压力表	pressure gauge
		压力表开关	pressure gauge switch
先导式比例方向阀	pilot proportional directional valve	压力补偿	pressure compensation
		压力补偿器	pressure compensator
先导式比例溢流阀	pilot operated proportional relief valve	压力负反馈系统	pressure negative feedback system

续表

Y		Y	
压力继电器	pressure relay/switch	原始信号	original signal
压力控制回路	pressure control circuit	圆管层流的沿程损失	frictional loss of laminar flow in round pipeline
压力适应控制	pressure adaptive control		
压力增益	pressure gain	圆管紊流的沿程损失	frictional loss of turbulent flow in round pipeline
压缩机	air compressor		
压缩空气	compressed air	圆形感应同步器	circular inductosyn
压铸机	die casting machine		
叶片泵	vane pump	运算放大器	operational amplifier
叶片马达	vane motor	Z	
液控单向阀	pilot operated check valve	增压缸	boost up cylinder
液体	liquid	增压回路	boost up circuit
液压泵	hydraulic pump	增压室	pressure-boosting chamber
液压冲击	hydraulic impact	增益	gain
液压阀	hydraulic valve	增益裕量	gain margin
液压辅件	hydraulic accessory	障碍信号	obstacle signal
液压缸	hydraulic cylinder	振荡器	oscillator
液压管件	hydraulic pipe fittings	蒸汽压	vapour pressure
液压管接头	hydraulic pipe joint	执行信号	executive signal
液压胶管总成	hydraulic hose assemblies	直动式比例方向阀	directly operated proportional directional valve
液压块	hydraulic manifold		
液压马达	hydraulic motor	直动式比例减压阀	directly operated proportional pressure reducing valve
液压密封件	hydraulic seal		
液压先导级	hydraulic pilot stage	直动式比例溢流阀	directly operated proportional relief valve
异型断面盾构设备	special-section tunnel boring device		
		直动式溢流阀	directly operated relief valve
		直流伺服电机	DC servo motor
溢流	overflow	制动回路	braking circuit
溢流阀	overflow/relief valve	质量	mass
油水分离器	oil-water separator	质量含湿量	mass moisture content
油温油位计	oil level indicator with thermometer	滞后	hysteresis
		滞后时间	retardation time
油雾分离器	oil-gas separator	滞环	hysteresis
油箱	oil tank	中位死区	median dead zone

续表

Z		Z	
轴承环	bearing collar	状态	state
主阀阀座	main valve seat	锥阀芯	cone valve core
主阀芯	main valve core	自由流动	free flow
主回油口	main oil return port	总效率	total efficiency
注塑机	injection molding machine	阻抗	impedance
柱塞泵	piston pump	阻尼孔	damping hole
柱塞缸	plunger-type cylinder	组合垫圈	bonded washer
转阀	rotary directional valve	组合阀	combination valve

参 考 文 献

[1] 宋锦春. 机械设计手册第 4 卷 [M]. 第 5 版. 北京：机械工业出版社，2010.
[2] 宋锦春，陈建文. 液压伺服与比例控制 [M]. 北京：高等教育出版社，2013.
[3] 成大先. 机械设计手册 [M]. 北京：化学工业出版社，2004.
[4] 雷天觉. 新编液压工程手册 [M]. 北京：北京理工大学出版社，1998.
[5] 路甬祥. 电液比例控制技术 [M]. 北京：机械工业出版社，1988.
[6] 许益民. 电液比例控制系统分析与设计 [M]. 北京：机械工业出版社，2005.
[7] 黎启柏. 电液比例控制与数字控制系统 [M]. 北京：机械工业出版社，1997.
[8] 宋锦春，苏东海，张志伟. 液压与气压传动 [M]. 北京：科学出版社，2006.
[9] 刘春荣，宋锦春，张志伟. 液压传动 [M]. 北京：冶金工业出版社，2005.
[10] 宋锦春. 液压技术实用手册 [M]. 北京：中国电力出版社，2011.
[11] 宋锦春. 液压工必备手册 [M]. 北京：中国机械出版社，2010.
[12] 关景泰. 机电液控制技术 [M]. 上海：同济大学出版社，2003.
[13] 董景新. 控制工程基础 [M]. 北京：清华大学出版社，1992.
[14] 张利平. 现代液压技术应用 220 例 [M]. 北京：化学工业出版社，2004.
[15] 张平格. 液压传动与控制 [M]. 北京：冶金工业出版社，2004.
[16] 李壮云，葛宜远. 液压元件与系统 [M]. 北京：机械工业出版社，2004.
[17] 杨逢瑜. 电液伺服与电液比例控制技术 [M]. 北京：清华大学出版社，2009.
[18] 王丹力. MATLAB 控制系统设计仿真应用 [M]. 北京：中国电力出社，2007.
[19] 薛定宇，陈阳泉. 基于 MATLAB/Simulink 的系统仿真技术与应用 [M]. 北京：清华大学出版社，2002.
[20] 王沫然. Simulink 建模及动态仿真 [M]. 北京：电子工业出版社，2002.
[21] 宋锦春，宋涛，王长周，等. 冷轧薄板酸洗生产线带钢对中装置 [J]. 重型机械，2006 (4).
[22] 马庆杰，宋锦春. 新型浮动式管拧机液压夹紧装置的设计 [J]. 机床与液压，2005 (1).
[23] 宋锦春，王艳，张志伟，等. 飞机拦阻系统电液比例控制研究 [J]. 航空学报，2005 (4).
[24] 陈建文，郝彦军，游显盛，等. 弧形连铸机液压伺服振动台的研究 [J]. 冶金设备，2006 (6).
[25] Song Jinchun, Wang Changzhou. Modeling and simulation of hydraulic control system for vehicle continuously variable transmission [C]. 2008 3rd IEEE Conference on Industrial Electronics and Applications, 2008.
[26] Song Jinchun, Wang Changzhou, Xu Dong. Dynamic simulation and control strategy of centrifugal flying shear [J]. Applied Mechanics and Materials, 2009.
[27] Song Jinchun, Chen Jianwen Zhang Zhiwei. IV Machine and Automation - The Constant Power Control of Hydraulic Grinding Saw [J]. Key Engineering Materials, 2006.
[28] Song Jinchun, Li Yiming. Study on the turret feeding servo system of precision CNC lathe [J]. Advanced Materials Research, 2010.

[29] 宋锦春,董嘉庆,张志伟,等.全液压稠油热力开采系统[J].东北大学学报(自然科学版),2006,27(2):221~223.

[30] 朱新才,周雄,周小鹏,等.液压传动与气压传动[M].北京:冶金工业出版社,2009.

[31] 杨征瑞,花克勤,徐轶.电液比例与伺服控制[M].北京:冶金工业出版社,2009.

[32] Song Jinchun. Simulation analysis of V-belt style continuously variable transmission with electro-hydraulic proportional control system [C]. ICMT2006, Science Press.

[33] 张志伟,毛福荣,宋锦春.电液伺服控制摩托车随机疲劳试验台的实验研究[J].东北大学学报(自然科学版),2006,27(8):903~906.

[34] S. H. Zhou, S. Yasunobu. A Cooperative Auto-driving System Based on Fuzzy Instruction [C], Proc. of the 7th International Symposium on Advanced Intelligent Systems (SCIS&ISIS2006), 2006:300~304.

[35] S. H. Zhou, S. Yasunobu. Design of a Cooperative Auto-driving System Based on Fuzzy Instruction [C], SICE Symposium on Systems and Information 2006 (SSI2006), 2006:331~336.

[36] 宋锦春,郝彦军,王长周,等.连铸冷床平移液压系统分析与改进[J].冶金设备,2006,(5):45~48.

[37] 宋锦春,王长周,宋涛,等.管拧机浮动夹紧装置液压减振系统分析[J].液压与气动,2006,(12):11~14.

[38] 宋锦春,郝彦军,王长周,等.钢坯连铸机液压伺服振动台的研究[J].冶金设备,2006,(5):32~35.

[39] 赵丽丽,宋锦春,柳洪义.电渣重熔熔速控制过程综合分析[J].冶金设备,2007,(5):20~24.

[40] 陈建文,宋锦春,李雪莲.油雾润滑中局部油雾损失实验研究[J].液压与气动,2007,(4):31~34.

[41] 宋涛,王长周,宋锦春.高压辊磨机液压系统动态分析[J].液压与气动,2007,(4):35~39.

[42] 郝彦军,裴嗣明,王长周,等.基于二次型理论的热轧带钢破鳞液压控制系统的改进[J].冶金自动化,2007,(6):26~30.

[43] Song Jinchun, Wang Changzhou. Modeling and Simulation of Hydraulic Control System for Vehicle Continuously Variable Transmission [C]. ICIEA2008.

[44] Chen Jianwen, Song Jinchun. Measurement of the Size of Oil Mist Droplets [C]. ICIEA2009.

[45] Song Jinchun, Wang Changzhou. Dynamic Simulation and Control Strategy of Centrifugal Flying Shear [J]. Applied Mechanics and Materials, 2009.

[46] Zhang Zhiwei, Song Jinchun. Study on Motorcycle Fatigue Test with Electro-hydraulic Servo System [C]. ICIEA2009.

[47] Chen Jianwen, Zhang Zhiwei, Song Jinchun. The application of optical rotation principle in density of oil mist measuring [J]. Applied Mechanics and Materials, 2009 (16~19).

[48] Song Jinchun, Wang Changzhou. Technical Principle and Design of Tubular Straightway In-Line Flow Meter [J]. SYSTEMS SCIENCE and APPLICATIONS.

[49] 张敬妹,周自强,王长周. 基于比例阀的气压伺服系统性能分析及仿真 [J]. 常熟理工学院学报, 2009, (2): 25~28.

[50] Zhang Zhiwei, Wu Zhongyou, Liu Fuchun, et al. Application of Symmetrical Proportional Direction Valve Controlled Unsymmetrical Cylinder in the Automatic Roller Gap Adjusting of a novel Straightener [C]. IECON2010.

[51] 王长周,宋锦春,张志伟,等. DSHplus 系统仿真软件在液压回路教学中的应用 [J]. 实验科学与技术, 2011, 11 (3): 58~60.

冶金工业出版社部分图书推荐

书　名	作　者	定价(元)
现代液压技术概论（本科教材）	宋锦春　编著	25.00
电液比例与伺服控制（本科教材）	杨征瑞　等编	36.00
液压与气压传动实验教程（本科教材）	韩学军　等编	25.00
液压传动与气压传动（本科教材）	朱新才　主编	39.00
机械振动学（第 2 版）	闻邦椿　主编	28.00
机电一体化技术基础与产品设计（第 2 版）(本科教材)	刘　杰　主编	46.00
现代机械设计方法（第 2 版）(本科教材)	臧　勇　主编	36.00
机械优化设计方法（第 4 版）	陈立周　主编	42.00
机械可靠性设计（本科教材）	孟宪铎　主编	25.00
机械故障诊断基础（本科教材）	廖伯瑜　主编	25.80
机器人技术基础（第 2 版）(本科教材)	宋伟刚　等编	35.00
机械电子工程实验教程（本科教材）	宋伟刚　主编	29.00
机械工程实验综合教程（本科教材）	常秀辉　主编	32.00
炼铁机械（第 2 版）(本科教材)	严允进　主编	38.00
炼钢机械（第 2 版）(本科教材)	罗振才　主编	32.00
轧钢机械（第 3 版）(本科教材)	邹家祥　主编	49.00
冶金设备（第 2 版）(本科教材)	朱　云　主编	56.00
冶金设备及自动化（本科教材）	王立萍　等编	29.00
环保机械设备设计（本科教材）	江　晶　编著	45.00
污水处理技术与设备（本科教材）	江　晶　编著	35.00
固体废物处理处置技术与设备（本科教材）	江　晶　编著	38.00
大气污染治理技术与设备（本科教材）	江　晶　编著	40.00
液压润滑系统的清洁度控制	胡邦喜　等著	16.00
液压可靠性最优化与智能故障诊断	湛从昌　等著	70.00
液压可靠性与故障诊断（第 2 版）	湛从昌　等著	49.00
液压可靠性设计基础与设计准则	湛从昌　等著	70.00
液压元件性能测试技术与试验方法	湛从昌　等著	30.00
冶金设备液压润滑实用技术	黄志坚　等著	68.00
液压元件安装调试与故障维修图解·案例	黄志坚　等著	56.00
液力偶合器使用与维护 500 问	刘应诚　编著	49.00
液力偶合器选型匹配 500 问	刘应诚　编著	49.00